Midnight Ride, Industrial Dawn

Johns Hopkins Studies in the History of Technology
Merritt Roe Smith, Series Editor

Midnight Ride, Industrial Dawn

Paul Revere and the Growth of American Enterprise

Robert Martello

The Johns Hopkins University Press
Baltimore

The Johns Hopkins University Press
2715 North Charles Street
Baltimore, Maryland 21218-4363
www.press.jhu.edu

Library of Congress Cataloging-in-Publication Data

Martello, Robert, 1968–
Midnight ride, industrial dawn : Paul Revere and the growth
of American enterprise / Robert Martello.
    p.   cm. — (Johns Hopkins studies in the history of
technology)
Includes bibliographical references and index.
ISBN-13: 978-0-8018-9757-3 (hardcover : alk. paper)
ISBN-10: 0-8018-9757-2 (hardcover. : alk. paper)
ISBN-13: 978-0-8018-9758-0 (pbk : alk. paper)
ISBN-10: 0-8018-9758-0 (pbk. : alk. paper)
1. Silversmiths—United States—History. 2. Metal-work—
United States—History. 3. Revere, Paul, 1735–1818. 4. Industrial
revolution—United States. 5. Industries—United States—
History. I. Title.
HD8039.S5262U665 2010
739.2'3092—dc22   2010006885

A catalog record for this book is available from the
British Library.

*Special discounts are available for bulk purchases of this book. For
more information, please contact Special Sales at 410-516-6936 or
specialsales@press.jhu.edu.*

The Johns Hopkins University Press uses environmentally
friendly book materials, including recycled text paper that is
composed of at least 30 percent post-consumer waste, whenever
possible. All of our book papers are acid-free, and our jackets
and covers are printed on paper with recycled content.

Contents

Midnight Ride, Industrial Dawn

Introduction

On the 18th of April in 1775, Paul Revere became a hero.

Shortly after 9:00 p.m., he responded to a summons from Patriot leader Joseph Warren. Warren's spy had just confirmed some gossip that had set the town on fire: a force of British regulars assembling on the Boston Commons would soon march to Lexington to capture Samuel Adams and John Hancock. Within the hour Revere was on the move. After asking several friends to notify waiting allies in Charlestown by lighting two signal lanterns in the steeple of the Old North Church, he met another two colleagues at the waterfront and retrieved his concealed rowboat. With muffled oars they rowed north across the silent Boston Harbor toward the Charlestown landing, cloaking themselves in the shadow of a British warship. Charlestown Patriots provided Revere with "a very good horse" and he raced toward Lexington, stopping to wake Patriots along the way. He soon ran into one of many British patrols, but maneuvered the first pursuer into a clay pond and left the second in his dust. By midnight he reached Hancock and Adams, who decided that the true purpose of the British must have been the capture of the cannon and ammunition at Concord. With two companions Revere rode toward Concord, but a second British patrol captured him, interrogated him at gunpoint, and took his horse. Thinking quickly, he bluffed them into believing that hundreds of minutemen would soon descend upon them and the British patrol fled the scene. Returning to Lexington on foot, he helped Hancock and Adams safely leave the town and then carried Hancock's chest of Patriot correspondence to a concealed hiding spot. As he struggled to lug the massive trunk into the woods he heard a single shot ring out across Lexington Green behind him, followed by mayhem. America's Revolutionary War had begun.

Henry Wadsworth Longfellow's inspirational poem "Paul Revere's Ride" ensured that the midnight rider is the Paul Revere that Americans have come to know and love. Americans idolize their heroes and use them as markers for larger trends, as symbols for sweeping cultural values that define and enrich national identity. The midnight rider might stand for resourcefulness, mobilization, and courage, but this is not the only Revere. Longfellow's simplified and occasionally inaccurate retelling of the Midnight Ride sidesteps several vital questions. Why did the Patriot leaders choose Revere for this mission, and what enabled him to do such a fine job? In a less positive vein, why was Revere carrying a trunk of papers at the end of the day while others took over the resistance movement's political and military leadership? The answers to these questions, which explain his successes and failures during and after the ride, lie in his artisan status. However spellbinding the Midnight Ride might be, it represents only a small piece of the larger story of Revere's life and times. One historian records the anecdote of a wealthy American woman returning to Boston after years lived in Paris. Upon discovering a statue of Paul Revere, she reportedly remarked, "I suppose it is proper to erect a monument to a silversmith, but why the horse?"[1] His manufacturing career began with acclaimed and sophisticated silverworking activities before and after the war, but silver was only the tip of the iceberg. An accurate accounting of Revere's lifelong contributions to his nation's welfare must start and end with his career as a craftsman, manufacturer, and entrepreneur.

On the 18th of April in 1800, twenty-five years after his midnight adventures, Paul Revere embarked upon his final "ride," in a manner of speaking, on behalf of his country. In the years following the Revolution he tried and failed to ascend into the gentry class by becoming a merchant or federal appointee, and then successfully redirected his efforts into manufacturing activities. Never content to rest on his laurels, Revere added new equipment, processes, and product lines to his business throughout his life. Beginning with a more mechanized approach to silverworking, he branched into iron casting, bronze bell and cannon casting, and the large-scale production of copper bolts and spikes for clients that included the young American government. By April 1800, at the age of 65, he set his sights upon the most advanced, elusive, and urgent technological limitation facing the young Navy Department: rolled copper sheathing. If he could master the complicated production process, he would enrich himself while helping his nation field a strong navy in a time of international turmoil. Success depended on his ability to apply all the resources he had amassed over his long career: technological tools, machines, and knowledge; a pool of skilled laborers as well as practical managerial experience; investment capital for the purchase of property and new equipment; and the ability to secure steady supplies of environmental resources such as raw materials, fuel, and waterpower. The story of his success, as well as the failures that surrounded it, provides the driving narrative of this book.

Revere's last ride received less attention than his earlier patriotic service, due to its complexity and lack of drama. In history, as in life, beginnings and endings often intertwine and it becomes difficult to point to one event, to one essential moment, that defines a person or an era. Did Revere's successful copper-rolling experiment take place on the unrecorded date when a first sheet of malleable copper emerged from his mill? Or from his meeting with Secretary of the Navy Benjamin Stoddert in 1800, when he first received the contract to roll copper? Should we celebrate the moment when he first hit upon the idea to roll copper, or when he painstakingly gathered sufficient experience from his many earlier endeavors to understand the concepts of annealing, work hardening, and malleability? Longfellow chose his subject wisely: his rousing poem succeeds because it describes an event of finite duration and obvious impact, whose Revolutionary context is immediately understood and appreciated by all American readers. The story of Revere's artisan and manufacturing exploits is not as simple or unequivocal as Longfellow's great work, and Revere's halting attempts at poetry (if one can call it that) are a far cry from "Paul Revere's Ride." Fortunately, the complexity and larger framework of this story only add to its value. We cannot separate Revere's lifetime of technological and managerial experiences from America's, and by dissecting the constantly evolving hopes and fears, the triumphs and setbacks, and the overlapping and occasionally contradictory identities of this highly intelligent entrepreneur, we learn about a different nation than the one inhabited by the elite statesmen in America's pantheon of founders.[2]

Paul Revere: Patriot, Artisan, Manager, and Recordkeeper

Throughout his life, Paul Revere benefited from a combination of external factors that fostered early America's voyage toward industrial capitalism as well as personal attributes that helped him exploit these ever-changing conditions. To put it more simply, he lived in exciting times and acted accordingly. Late eighteenth-century Americans undertook two simultaneous revolutions: a political struggle against British domination and the ongoing industrial revolution that transformed all aspects of its society and culture. Revere had the rare opportunity to play a pivotal role in both revolutions, and the synchronicity of these great social upheavals was no coincidence: freedom from British regulation inspired a number of manufacturing endeavors as well as a general spirit of enterprise and experimentation.[3] America's growing population, maturing market economy, precedent-setting government, tumultuous relationships with European powers, fluid and ill-defined social classes, and nascent manufacturing institutions produced a turbulent and uncertain climate. Revere's intelligence, drive, and creativity allowed him to identify and seize the opportunities presented by his surroundings: he adeptly learned new technologies, networked, and sought new business venues even at times when his operations flourished. Versatility, perhaps his most defining and prevalent trait, allowed him to circulate freely in the worlds of mechanic and merchant, politician and Patriot. This versatility and the consequent social mobility make Revere such a compelling historical subject: he walked in many worlds.

Revere's life's work helped America close the technological gap with Britain and moved the nation closer to the ideals of industrial capitalism. Similarly, his growing professionalism in both technical and managerial arenas became a microcosm of America's. His ability to epitomize larger national trends seems somewhat peculiar if one considers that he was a highly atypical American in almost every way. Revere crafted or manufactured metal goods and lived in a large city at a time when most of the country's population practiced agriculture in rural areas. He hired and managed a relatively large labor force when most workers were self-employed in small shops or farms. And yet, the minority segment of society that he did represent formed a dynamic and growing constituency that played an essential role in political, economic, social, and technological movements throughout America's history.

Revere offers one final, invaluable credential that makes him a wonderful historical subject: he was a meticulous and literate pack rat with accounting and organizational practices ahead of his time. Revere's recordkeeping methods help us re-create his footsteps. After finishing his breakfast he might head to one of his workshops to supervise his workers and interact with customers. On any given morning he could grab his notebook and jot down instructions for making silver paste or the recipe for mud for a bell mold, next to a doodle of a furnace design or an informal labor agreement specifying the salary and terms of service of one of his workers. When clients arrived he opened his "wastebook," a record book used to document orders taken, payments received, or goods dispensed. Every few months he tallied up his wastebook entries for each client in a ledger, recording all payments and the value of the goods received in order to compute his accounts receivable. On many days he traveled to the dockyard to take orders from the naval purchasing agent or discuss contracts already in progress, and on the way home he might visit the bank to make a deposit or withdrawal. After dinner or as time permitted he drafted a few letters in his letterbook, a bound volume that allowed him to polish up the writing on each piece of outgoing correspondence before he copied the final version onto a pristine sheet of paper to be mailed. These letters ranged the gamut: personal messages to family and friends, letters soliciting raw material shipments, requests for loans or payments of past debts, responses to customer complaints, or the exchange of technical advice with fellow practitioners. All of these activities generated a paper trail, and Revere's surviving correspondence and business records, along with those of his sons and other family members, now occupy several shelves at the Massachusetts Historical Society. Collectively titled the Revere Family Papers, these fifty-seven volumes and boxes of loose manuscripts provide an invaluable study of pre- and post-Revolutionary business and manufacturing methods. In spite of the great size of this collection, Revere could not and did not document everything. He never recorded information that he considered a trade secret, for example, nor would he write down anything that struck him as obvious or banal, never guessing the inestimable value of this information today. He also had no reason to commit to paper anything constituting a verbal agreement or "handshake deal," business methods that Revere learned in his youth and continued to use throughout his life.

Paul Revere's many surviving records provide the vital raw materials for a study of larger societal questions. He could not answer, and indeed, could not even frame such questions within his own lifetime because he was too focused on practical matters and too caught up in the turbulent flow of events to grasp the larger picture. Revere became one of the first Americans to shift from the role of a skilled artisan-laborer into the new one of a manufactory owner-manager, and many of his colleagues attempted to duplicate his transformation, with varying degrees of success. More than anything else, his example shines a light on America's late eighteenth- and early nineteenth-century transition from the world of crafts to industrial capitalism.

Craft, Industry, and the Proto-industry Transition

The term *industrial revolution* conveys an important message at the cost of oversimplification. Hindsight confirms that the long process of industrialization had certain revolutionary facets, as it upended the earlier craft-based production system and irrevocably altered nearly every aspect of human existence. But industrialization was in other respects non-revolutionary, as the emerging industrial system took shape in a gradual and inconsistent manner. For these reasons I portray industrialization as a *transition* with craft-based roots in the distant past, industrial implications that continue to unfold, and a middle "proto-industrial" state that mixes elements of both crafts and industry. Each of these stages involves a complex set of intertwined factors: *craft* and *industry* describe comprehensive systems of production that include labor, managerial, technological, economic, and cultural practices. A craft-centered manufacturing system generally denotes a pre-capitalistic mode of fabrication and management with features such as barter exchange; personal connections between owners, workers, and customers; apprenticeships; custom-made output in small shops; the predominant use of hand tools by skilled workers; and a dependence upon agricultural employment and spending cycles. In contrast, an industrial manufacturing system typically involves extensive use of machinery by wage laborers, cash transactions, large-scale factory production, high volumes of inexpensive standardized goods, division of labor, extensive market networks, and the establishment of separate classes of owners, managers, and workers.[4]

These generalized definitions do not adequately explain the confusing, changing, heterogeneous conditions that characterized the first fifty years of the new republic. We must not understate the longevity and importance of America's twilight interlude between crafts and industry: neither Revere nor many of his contemporaries comfortably fit either model, and all the forces of historical hindsight cannot decisively cleave one of these systems from another. Some colonial-era craft shops had already adopted many "industrial" characteristics, such as the use of wage labor, the division of labor, subcontracting, and the move toward standardized output. And early industrial endeavors retained many traditional practices, including the employment of skilled laborers, paternalistic managerial practices, and the continued reliance upon barter transactions. While craft practices gradually became more industrial, this evolution took place at different rates in different regions and trades, and the conversion often involved harsh upheavals or throwbacks to earlier ways. Indeed, elements of craft production still exist today.

Instead of attempting to demarcate beginning and endings for craft or industrial practices we must instead define the characteristics of a transitional, proto-industrial state that combines earlier and later methods in the same way that individuals such as Revere combined these practices in their workshops. America's industrialization depended upon the confluence of capitalized corporations and family-run small shops, hand tools and machines, workers who played multiple roles as the needs arose, and so on. Early manufacturing establishments defy attempts at generalization, and the capitalist investors, businessmen, mechanics, and wage laborers of the nineteenth century founded an array of shops and manufactories with dissimilar goals and methods.[5] As a result, America's industrial development cannot be easily encapsulated in a simple theory or defined by a small number of causes and effects. Economic historians in the 1970s created a proto-industrial theory that attempted to explain the development of European manufacturing and economic systems: this theory said that rural household manufacturing, often concentrated in workshops under the direction of merchants, paved the way for factory expansion by creating large quantities of investment capital and a non-landholding laborer class.[6] This theory came under attack because it fails to accurately explain why certain regions industrialized while others did not, and it also understates the importance of urban workshops and guilds.[7] Nearly all experts have abandoned this early theory, but proto-industry deserves to live again in a broader form that describes the fluid combination of old and new methods that fostered American industrial growth, with a particular emphasis upon how the tradeoffs between these methods shaped entrepreneurial decision making.

Paul Revere's story suggests a new version of proto-industry more applicable to America. Rather than showcasing the impact of the larger agricultural setting, we learn more about proto-industrial practices and attitudes by focusing upon the early manufacturing community and everyone affected by it: artisans and their apprentices, journeymen and other wage laborers, investors ranging from merchants to farsighted government officials, harvesters and processors of raw materials, shopkeepers and other retailers, the customers comprising the market for the growing torrent of goods, and, of course, entrepreneurs like Revere who attempted to weave all of these constituencies into harmonious and productive enterprises. We can best describe America's proto-industrial period by dividing these transitional changes into the four essential components of industrial success: capital, labor, technology, and environmental resources.[8] Contemporary manufacturers might not use these terms to describe their practices, but they would recognize the connections between these four factors of production, since those connections occupied much of their attention. For example, Revere did not think about environmental resources without considering practical issues such as the trade networks that brought him these materials or the technologies he used to manipulate them. His frustrating efforts to raise investment capital took place in the context of the many pressing needs for that capital, for machinery, raw materials, and the wages of his workforce. When Revere or his contemporaries attempted to change the way they managed one of these factors it tugged at all of the others, and the evolving combination of old and new strategies employed by increasing numbers of entrepreneurial manufacturers broadly defines the nation's proto-industrial pathway. The challenges posed by limited quantities of investment capital, labor, technology, and environmental resources determined whether Revere succeeded or failed, and shaped all his technical and managerial decision making. The combination of capital, labor, technology, and environmental practices at a given time becomes a useful benchmark of America's industrial evolution.

Paul Revere's lifetime of technical and managerial enterprise made him a midnight rider once again: emerging from the sunset of the craft system, his example and efforts heralded the dawn of industrialism. Over his long life, Paul Revere trained as an apprentice, achieved some fame as a skilled artisan, and began a manufacturing career that resulted in his ownership of a mill complex employing a dozen or more workers. He lived and excelled in the proto-industrial world, and altered his own status and methods while observing the changes occurring around him. His career trajectory raises several questions. Why did Revere choose to abandon his craft roots? At the heart of this question lies Revere's value system, his estimation of the costs and benefits of different occupations and the goals he wished to achieve by the end of his life. Many of his choices resulted from his assessment of Revolutionary-era social status, and the relatively small degree of social or political influence accorded to citizens seen as laborers, including skilled craftsmen. He strove to improve his position in the fluid social and economic hierarchy by leaving his craft roots behind and ascending into the gentry class, but political views also played a significant role in his career choices as he continually sought to further his country's development. One of the novel facets of post-Revolutionary culture was the remarkable overlap between private self-interest and public service. We must also ask about his methods: which craft elements did Revere retain, and which industrial elements did he adopt? Particularly at the end of his career, Revere had evolved a creative combination of old and new practices that enabled him to maintain a high degree of productivity and quality while avoiding excessively risky or controversial alterations to long-accepted work methods. A third question shifts from intentions to execution: how did Revere successfully switch from craft to industry when many others failed to make the transition? A broad look at Revere's entire career turns up certain patterns and practices that time and again allowed him to thrive at each individual endeavor. In particular, the startup process he followed each time he initiated a new product line or facility illustrates the combination of intentional preparation, composite experience, and innate talent that enabled him to expand his production. His experience and skills also emphasize the growing distinction between managerial and technical aptitude, both of which proved vital for success.

Paul Revere's manufacturing journey was at least as meaningful as his actions in 1775. His Revolutionary ride became the stuff of legend, but his metallurgical ride created a legacy not only for his family and business, but for his community and nation, basking in the first rays of the impending industrial dawn. We only begin to understand the real Revere by studying his work and his writings, encapsulating the context and goals of an entire life and career. Revere's accomplishments might have drawn more attention over the years if he fit the classical idealized mold of an inventor, but in lieu of that single label he took on a variety of identities: patriarch, manufacturer, entrepreneur, manager, borrower, networker, researcher, optimizer, disseminator. Versatile complexity makes him an authentic representative of a generation of aging

Revolutionaries who all attempted to understand and channel the vast societal changes they unleashed during the age of American proto-industry.

But we have gotten ahead of ourselves. Before he took on these identities or dreamed of the industrial changes he helped inaugurate, Revere was a silversmith, proudly carrying on his father's trade and upholding longstanding artisan traditions. In spite of all his later accomplishments, most twenty-first-century Americans will always picture Paul Revere as John Singleton Copley chose to capture him on canvas, the only colonial American artisan ever immortalized in the middle of the creative act. The year is 1768, and we find him at his workbench, pensively pondering a teapot in need of engraving.

CHAPTER ONE

Artisan, Silversmith, and Businessman (1754–1775)

The year is 1768 and the place is the living room of Paul Revere's comfortable house in the North End neighborhood of Boston. After weeks of anticipation, Paul excitedly gathered his family for the great unveiling. It all started a few months earlier when John Singleton Copley, a reputable local painter and professional acquaintance, made Revere an irresistible offer. Seeking the services of a masterful silversmith, Copley ordered several silver frames and gold cases from Revere and amassed a moderate debt. In accordance with the barter economy that governed many colonial transactions, Copley proposed to paint Revere's portrait if Revere cancelled his debt. Revere knew an incredible deal when he heard one: portraits adorned the houses of the well-to-do and typically depicted the calling and reputation of the upper-class subject. Merchants posed in front of dockside windows that showcased their sailing ships; generals wore gilded swords at the scenes of famous battles; and statesmen gestured at scrolls or held books. Silversmith subjects, however, were unheard of. Revere accepted the offer and posed for Copley in his silversmith shop. At last the artist completed his work.

Surviving records do not capture how Revere's family responded to this

Figure 1.1. John Singleton Copley (1738–1815), oil on canvas portrait, *Paul Revere*, 1768. Gift of Joseph W. Revere, William B. Revere, and Edward H. R. Revere. Photograph © 2010 Museum of Fine Arts, Boston (image number 30.781). Unique in early American portraiture, this image captures Paul Revere in the act of creation, highlighting and celebrating his artisan status.

painting. Fortunately, anecdotes recounted by Revere's modern-day descendents tell us that Revere's wife and children looked upon the painting . . . and hated what they saw.

In this portrait, which now hangs in a place of honor in Boston's Museum of Fine Arts, Revere sits at a polished wooden table, his white shirt in sharp contrast with the dark background. His right hand supports his head, which thoughtfully and directly meets the gaze of the viewer, while his left hand holds a polished silver teapot that lacks engraving but is otherwise complete. As with Copley's other surviving works, the execution is dazzling, ahead of its time. The tilt of Revere's head and penetrating clarity in his eyes shatters the boundary between subject and audience. His teapot is flawless and shining; his hands strong and graceful. This invaluable painting is one of the few—and certainly the finest—surviving illustration of colonial artisans at work, representing the pinnacle of Copley's talents. Any reception short of praise seems absurd. That is, absurd to us.

The family of a late eighteenth-century artisan would view the portrait differently. Its magnificent accuracy, its accurate and uncanny encapsulation of Revere's identity, led to its downfall in their estimation. Unlike the gentry and statesmen in Copley's other portraits, Revere is not garbed in elegant clothing. Instead, he wears the clean and functional dark leather vest of a workman, and his white shirt is unbuttoned. His sleeve has fallen slightly, revealing some of his forearm and wrist, and traces of dirt are visible under his fingernails. The presence of engraving tools on the tabletop leaves no doubt that this man works for a living, frozen by the artist in the middle of his work procedure. Revere's family loved its patriarch, a respected silversmith whose versatility, artistic skill, entrepreneurial insight, and network of patrons and colleagues facilitated his financial security and rising social status. His family knew he stood at the top of the craftsman hierarchy as a prince of skilled laborers, the biggest fish in the working-class pond. But we might forgive them their disappointment at seeing the family patriarch so truthfully reflected in his portrait. All of Revere's financial success, technical and entrepreneurial skill, and artistic reputation could not change his artisan status. Changes, for both Revere and the world he lived in, would have to wait a few years.[1]

The Revolutionary War divides Paul Revere's long silverworking career almost exactly in half: he first entered the field as an apprentice around 1748 and ceased working on silver items in the late 1790s. A detailed study of the nature of Revere's work before and after the Revolution offers a vivid illustration of the changes that took place in the American workplace following this incredible political, economic, and cultural upheaval. Revere's writings, activities, and body of work highlight his changing methods and goals, which had their origins in much earlier times, in European craft traditions that defined the education, practices, and social status of artisans. Early colonists carried their traditions to colonial America but many of the details failed to take root in the cultural climate of the New World. As a result, a different system evolved, and continued changing throughout the colonial period. Some of the elements of industrial capitalism already existed by the time Revere entered the silverworking field, and others soon developed. Revere was a trendsetter and entrepreneurial artisan, and his life offers an ideal window into this transitional world, beginning in the colonial days of his youth.

In order to understand the meaning behind Revere's pre-Revolutionary career, we must first unravel the threads that comprise his personal and professional identity, explaining how he categorized himself and which traditions, both explicit and tacit, shaped his skills, beliefs, and aspirations. In order to simplify the complexity of human identity, we can focus our analysis upon three major elements of Revere's background.

First, Revere was a Bostonian, a citizen of the British Empire raised in one of the three largest cities in colonial America. As such, he enjoyed constant exposure to people, goods, and ideas circulating around the Atlantic seaboard: he regularly encountered merchants, tradesmen, artisans, and working men; material goods produced by local manufacturers as well as overseas imports; and church doctrine, political polemics, scientific ideas, and a bewildering buffet of personal viewpoints. Early America resists generalization or simplification: the majority of farmers engaged in market activity and entrepreneurial pursuits, and urban workers combined old European conventions with newer ideas formed as responses to the colonial American setting. To appreciate the relevance of Revere's career one must understand this societal and cultural context—the rich and turbulent world of colonial Boston. Even as a child, Paul Revere became accustomed to a certain style of conducting business and dealing with his peers: he learned the similarities and differences between various social groups, where he fit into the picture, and where he wanted to fit in.

Second, Revere was an artisan. Artisans are the makers of goods, craftsmen who produce the many items that societies and individuals need on a daily basis. But in addition to their important practical role as society's producers, artisans are also the keepers of tradition, beneficiaries of a long and proud heritage stretching across hundreds of years and various European nations. Revere's identity as an artisan linked him with numerous colleagues who shared common experiences, such as an apprenticeship education, working-class lifestyles and work patterns, and a middle-class position in their social networks. Even as Revere and his colleagues perpetuated these attitudes and practices, various factors would soon overturn traditional artisan roles and expectations forever.

Third, Revere was a silversmith, proficient in creating and repairing silver and gold objects.[2] The most successful silversmiths occupied the top of the informal artisan hierarchy, earning prestige from the trade's relatively high capital costs and skill requirements. Revere's silverworking career illustrates the importance of his technical skill and judgment, best characterized by the incredible versatility that distinguished him from most of his peers and set the stage for a lifetime of experimentation and boundary crossing. But his silverworking endeavors also depended upon his business acumen, managerial proficiency, and ability to form and exploit social networks. Revere's prewar silver activities laid the foundation for a lifetime of creative entrepreneurial activity.

Copley layered some of these influences and perspectives throughout his masterful portrait. The man looking out from the canvas is many things, and his hands clue us into the two most important aspects of the image. One hand holds his head, framing a thoughtful, intelligent, confident, and focused subject. The other hand holds a beautiful object, a flawless, shining teapot about to receive an engraved touch that will make it truly unique. And Revere's hand is reflected in his work, reminding us that the hand of the maker is visible in all of his creations. Regardless of how his family might have reacted to this painting, Revere could not have been too surprised; after all, he posed for it. His work for Copley bestowed a rare honor upon him, the ability to be immortalized in canvas in spite of the fact that he claimed neither wealth nor social prestige. Even if "only" a silversmith, he knew he was a competent manager and skilled craftsman, and in eighteenth-century Boston this was certainly a good thing to be.

Growing Up in Colonial America

Before his Midnight Ride, before working with silver or other metals, and before beginning his apprenticeship, Paul Revere came of age in colonial Boston, a child of French and English descent. Born in December 1734 to silversmith Apollos Rivoire, a first-generation French immigrant, and Deborah Hitchborn, a daughter of a fairly well-off New England family, the relationships within and between the two sides of his family offered him an early understanding of social classes and privileges.

Revere's mother's branch of the family tree dated back to the beginning of colonial settlements. The Hitchborns migrated from England to America in 1641 and by the eighteenth century owned and lived on a small wharf in Boston's North End, too small to even warrant inclusion on contemporary maps. This wharf housed various Hitchborn-owned businesses, including boat repairs, cargo loading, liquor sales, and possibly shipbuilding. Wharf and shop ownership placed the Hitchborns in the upper tiers of colonial society but well below the larger merchants whose wealth and influence truly drove Boston's economy and culture. Deborah Hitchborn was born into this established New England family in 1704, but received few financial benefits from her well-off relations after she married. Following the standard practice of the times, Deborah's oldest brother Thomas inherited the bulk of the Hitchborn property and his ten children, Paul Revere's cousins, entered various professions. Oldest son Thomas eventually inherited the Hitchborn wharf and the lion's share of the family wealth, and probably employed several of his younger brothers, who received training in relevant fields such as shipbuilding and sailmaking. Benjamin, the second youngest Hitchborn son, received a gentleman's education at Harvard and became a prominent lawyer and member of the Boston upper class. Samuel, the youngest Hitchborn son, became a silversmith, possibly an apprentice to Paul Revere.[3] The Hitchborn family illustrates the male-centered nature of early businesses, the importance of kin networks, and the porosity of the home and work boundary. Living on their wharf and training sons to take on different branches of the family's endeavors, the Hitchborns worked hard to increase their influence and holdings. The Hitchborn family had a profound lifelong influence upon Revere, who had many children of his own and involved several sons in his own enterprises. Many of Revere's children and nearly all of his siblings received first names commemorating Hitchborn relations; his uncles and cousins purchased silver from him and loaned him money at critical points during his career; and Thomas's example constantly showcased the advantages of property and social standing that Revere relentlessly strove to attain.

Revere's father, Apollos Rivoire, was born in 1702 in the Bordeaux region of southwest France. The Huguenot (Calvinist Protestant) majority of this region had a long history of religious and political strife with the Catholic kings of France, causing hundreds of thousands of Huguenots to leave France beginning in the late seventeenth century. Only a few thousand, including Apollos Rivoire, immigrated to British North America. Rivoire traveled to Boston in late 1715 or early 1716 and became an apprentice to John Coney, the finest American silversmith of his day. Coney serves as almost a professional grandfather to Revere: even though the two never met, the techniques, mannerisms, and beliefs of Coney greatly influenced Apollos, who passed these lessons on to his son. Apollos established his own silversmith shop in the 1720s in Dock Square near the center of Boston and anglicized his name to Paul Rivoire, and eventually to Paul Revere, "merely on account the Bumpkins should pronounce it easier."[4] Apollos died young, in 1754, cutting short Revere's own apprenticeship and thrusting a family's worth of responsibility upon his shoulders at the age of 19. We will return to the details of this silverworking heritage—to the great reputation and artistic strengths of John Coney, to Apollos's curtailed but still promising career, and to the training that Apollos proudly bestowed upon his oldest son—in the silverworking section below. Despite his early death, Apollos Rivoire had a far greater impact on his son than the entire Hitchborn family. Paul Revere literally followed in his father's footsteps, serving as an apprentice under his watchful eye, mimicking his techniques, practicing with his tools, and patiently awaiting the day when he would work alongside his father as a fellow master craftsman.

While Revere's formal education began in a classroom and continued in his father's silver shop, his broadest learning experiences took place in countless conversations and interactions in many locations across Boston's North End, a marine-centered working- and middle-class neighborhood. Boston resembled an English market town, with its dark and narrow winding cobblestone streets, numerous shops, and large artisan population.[5] Revere frequently encountered his father's employees, clients, and colleagues; his family intermingled with practitioners of many other trades, such as shopkeepers, teachers, and members of the clergy; and he could not fail to notice the less fortunate individuals, ranging from the poorest laborers to those dependent on town support for housing, firewood, medical treatment, or other expenses. Through ongoing contact with his Hitchborn grandparents, uncles, aunts, and cousins, as well as less frequent encounters with merchants, lawyers, doctors, and government officials, he received a tantalizing introduction to a different Boston, a luxury-filled intellectual world promising greater economic and political power. Revere's personal and professional interactions constantly highlighted his city's close connection to England, the source of soldiers, government policies, artistic styles, and all sorts of imported goods. These projections of the far-off mother country had a more visible impact on Paul's life than the local farms and fisheries that provided his food.

By the middle of the eighteenth century, the British Empire stood on top of the world. Great Britain had gradually increased its military, economic, and industrial power over the course of several centuries, culminating in its victory in the Seven Years' War that added Canada to its extensive colonial posses-sions. A series of early eighteenth-century agricultural improvements often referred to as the "agricultural revolution," combined with an increasing num-ber of productive overseas colonies, brought great surpluses of food to England and thereby enabled more of the population to switch to non-food-producing pursuits such as manufacturing.[6] Britain's industrial revolution went hand in hand with the agricultural boom, and vast increases in textile production, iron manufacturing, and steam engine efficiency catapulted England, and eventu-ally the rest of the world, into the industrial age. Britain's power was most explicitly embodied in its large and well-equipped navy, feared throughout the world, as well as its enviable array of high-quality manufactured goods. The impacts of the industrial revolution reverberated throughout the British Empire, even in far-off America, and although these impacts became far more profound by the mid-nineteenth century, industry's shadow lay over the colo-nial period as well.

As colonial citizens of the empire, eighteenth-century Americans reaped the benefits of their imperial membership and appreciated them . . . some of the time. All Americans of this day and age were either recent immigrants or descendents of immigrants drawn to the possibility of enjoying the untamed continent's freedom and opportunity while benefiting from the economic and military security offered by a powerful mother country. Colonists naturally favored laws and policies that offered protection and support, and resisted anything that limited their actions.

The British Empire's predominantly mercantilist policy placed England at the hub of a powerful wheel of commerce and production, often to the detri-ment of its colonies. According to mercantilist theory, the home country is the proper site of manufacturing and the appropriate repository for the ac-cumulation of capital. Nations developed extensive trade networks to carry food, raw materials, and other resources from their colonies into the home country, where skilled and unskilled laborers converted them into finished goods for sale elsewhere. The American colonies found themselves placed in the position of raw material providers and markets for English-manufactured items. Some theorists portrayed this relationship as one of mutual support and dependence, recognizing the important role played by colonial raw materials and markets. But American discontent, peaking in the years leading to the Revolution, reflects the growing dissatisfaction along the fringes of the em-pire.[7]

Colonists benefited from Great Britain's mercantile policies whenever they purchased cheap and high-quality British goods, but chafed when they attempted to produce and sell their own goods. New England, and Boston in particular, experienced recurring currency, capital, and credit problems during the seventeenth and eighteenth centuries. Some New Englanders attempted to correct these limitations by importing forge, millstone, and spinning wheel equipment to augment their income and build manufacturing expertise. Fortunately for them, the thriving and loosely regulated nature of the colonial economy allowed these activities to prosper even if they defied the spirit of mercantilism.[8] As a silversmith who later entered other metalworking fields, Revere spent much of his career competing with inexpensive British goods while benefiting from precision equipment, materials, and even re-sellable items that he could import from abroad. The patriotic fervor of many Americans, artisans and others originated as a response to restrictive British policies, and ironically, these same Americans had an easier time implementing their patriotic agendas thanks to the economic and political rewards accrued from their association with the empire.

Although Revere identified himself as a craftsman and silversmith, he never referred to himself as what we in the twenty-first century term a *businessman* or *entrepreneur*. Colonists often neglected and downplayed the business and managerial aspects of most trades in spite of their obvious role in the success or failure of so many endeavors. A closer look at the world of eighteenth-century business clarifies the role Revere filled as an artisan and silversmith by highlighting the vital nontechnical aspects of his work. Most citizens in eighteenth-century America fell into four broad societal positions: farmers or other food producers; artisans and other producers of nonfood goods and services; merchants and shopkeepers who bought and sold goods; and workers who either hired out their labor to others or who worked against their will as slaves.[9] Individuals generally understood how their social position related to that of others, and the vibrant economy offered many opportunities for intensive interactions spanning different trades and social classes.

The growing eighteenth-century market economy fostered a dynamic form of capitalism that added fuel to America's spreading entrepreneurial flames. A capitalist approach to business includes calculation, risk taking, and profit maximization, traits closely related to the concept of entrepreneurship since entrepreneurs know how to identify opportunities and create new ventures, happily risking their capital and labor in order to accumulate wealth. With few exceptions, all society members dating back to the first colonists reflected entrepreneurial market and capitalist values. Many British emigrants had already lived in a market economy and undertook the risky voyage to America in search of cash crops that would provide opportunities to purchase more land, hire additional labor, and expand their holdings. Farmers and non-farmers exhibited many capitalist and entrepreneurial tendencies: they increased and diversified their production; engaged in complex exchanges of goods, services, cash, and debts; decreased their leisure time to allow more time for work; shifted production from consumable to marketable goods; and adopted accounting and recordkeeping practices that quantified the value of their property and time.[10]

Paul Revere epitomized the capitalist mindset, ahead of his time though not radically so. Revere's entrepreneurial inclinations visibly shaped his decision making throughout his entire life, and he adopted practices such as double entry accounting, wage labor, and diversification into new business endeavors before most of his contemporaries. However, as an artisan living in one of the three largest colonial urban centers, Revere already belonged to the most dynamic and forward-looking component of his society, distinct from the rural agrarian majority. The very existence of urban artisans testifies to the power of the market economy, because such workers can thrive only if others grow and sell surplus food and use the proceeds to purchase professionally made goods. Revere did not use terms such as *markets, capitalism,* or *entrepreneurial,* but he experienced their impact wherever he looked.

Paul Revere: Artisan

Revere's early education as a silverworking apprentice focused upon countless technical and artistic details inherent in the production of beautiful items. But in a larger sense he received an education in the artisan tradition, inherited from his father, who learned from John Coney, whose own educational lineage extended to England. By working at his father's side, witnessing the daily rhythm and routine of the shop, interacting with workers and clients, carousing with fellow artisans after hours, and seeing the many ways that his trade connected to others in the thriving town of Boston, Revere duplicated the experience of countless apprentices and internalized the practices and attitudes of his craft.

The term *artisan* (often used interchangeably with *craftsman* or, in later years, *mechanic*) typically describes all skilled technical craft practitioners in preindustrial and some postindustrial societies who make a living by producing the nonfood necessities and luxuries used in daily life. Vast social and economic differences separated artisans in different crafts, places, and times. Given the diversity of status, experience, and education that separated silversmiths, printers, shoemakers, blacksmiths, coopers, and other craftsmen, we cannot generalize too much about the viewpoint or goals of a typical artisan at any point: the typical artisan simply does not exist. And yet common traditions and attitudes separated all artisans from farmers, laborers, and members of the gentry. Artisans do more than merely manufacture, and it took more than productive output to make someone an artisan. "Artisan" not only defined what a person did or made, but rather, offered a way for craft practitioners to define themselves and be defined by others on an individual and societal level. "Artisan" is, more than anything else, an identity.[11]

Artisans' camaraderie and mutual heritage added meaning to their careers. Most artisans shared similar experiences, such as an apprenticeship education, functional dress, residence in workers' neighborhoods, membership in fraternal societies and social clubs, participation in public rituals and processions, and a unified political identity. Members of the same trade often clustered their shops and homes on a single street or in one section of the "artisan quarter" of town. This proximity enabled consumers to compare prices and services and encouraged the artisans to learn from each other's practices. Artisan neighborhoods also facilitated social and business interactions between neighbors and the continued development of a common identity. Technical and social networks grow in parallel, and people who exchange ideas also develop personal connections.[12]

In an urban center such as Boston, Revere realized, even as a child, that he had more in common with the children of other artisans than with the children of laborers or gentlemen. As he learned to work silver he also learned to perpetuate artisan traditions by training and providing for apprentices, interacting with customers and other silversmiths, and participating in the activities of his community. Even at the end of his life, when he no longer worked with silver or considered himself a skilled craft practitioner, his words and actions echoed the values of the artisan heritage.

Artisans have existed since antiquity, and some of the earliest written documents call attention to their skills and products. Homer's *Iliad*, for example, brims with descriptions of crafty or cunning Vulcan, the blacksmith of the gods, not to mention longer and surprisingly detailed descriptions of specific items, accompanied by tributes to their makers:

> Quickly Sarpedon swung his shield before him—
> balanced and handsome beaten bronze a bronzesmith
> hammered out with layer on layer of hide inside
> and stitched with golden rivets round the rim.[13]

Many characteristics of the English and American artisan traditions of Revere's time existed in ancient civilizations. For example, the Greek and Roman empires created a perpetual demand for skilled workers that led to an apprenticeship training system, specialized craft workshops in large urban centers, small operations that often employed fewer than a dozen workmen, and the gendering of craft labor that led to the exclusion of women and de-emphasis of their skills.[14]

Nearly all artisans in Europe, Britain, and America were men. Women had limited formal craft and guild opportunities in medieval times but found themselves increasingly excluded from the status of artisans, while their own contributions to the manufacturing or administrative aspects of running a shop became increasingly marginalized. Common doctrine in Europe and England modeled the ideal form of a well-ordered society upon the family, with an adult male at its head in a fatherly role. Societal norms expected women, apprentices, and employees to accept their subservient position in the larger hierarchy and show proper deference to the master of the house. But as Revere's example illustrates, societal attempts to marginalize or downplay the role of women does not change the reality of their contributions. The male-dominated demographics of the artisan field led to a highly masculine style of work and socialization, featuring hard drinking on and off the job, fraternal societies drawing their membership from one or more trades, and public brawls between artisans or apprentices from different professions.[15]

The history of artisans in Europe centers on the dominance and activities of urban guilds. Craftsmen organized guilds in the Middle Ages to protect the interests of all practitioners of a particular craft by defining, monitoring, and regulating acceptable standards and practices through collective action. Guilds amassed numerous regulatory instruments and powers: for example, they usually made guild membership a requirement for all artisans, and individuals who deviated from guild rules often faced eviction from the guild, the loss of legal status, and punitive boycotts. Despite the manufacturing focus of most guilds, the business and managerial aspects of crafts became at least as important as the technical aspects from an early point.[16]

Guilds regulated artisan training in order to create a closed labor market with a high demand for skilled labor and unchallenged authority for masters. All artisans first served as apprentices who received training from a master craftsman. "Graduated" apprentices became journeymen and earned wages for several years by working in the shops of master artisans until the journeyman spent enough time and earned enough money to set up his own shop and apply for master craftsman status. Guilds enforced policies to restrict the number of new apprentices, imposed fees upon the families of apprentices, prevented trainees from leaving their masters, maintained stable wages for journeymen, and slowed the rate at which journeymen could become masters. Apprenticeship practices also carried symbolic meanings: for example, apprenticeship promoted the elitism of skilled crafts by serving as a highly selective and public rite of passage. Guilds and craft shops were microcosms of society, illustrating the hierarchy and deference that connected families and craft shops with larger guilds and even governments. Artisan traditions and guild practices created a common ideology that highlighted the importance of maintaining one's personal honor by exhibiting honesty, independence, and a strong work ethic.[17]

Guilds also monitored and regulated the quality and consistency of craftwork, craft standards, and prices. Many guilds conducted inspections or required a conspicuous assessment of crafted items as a way of maintaining the virtuous reputation of all guild members. For example, England's powerful silversmiths' guild used its complete monopoly to define standards for "sterling" quality silver and required a guild assayer's mark on all finished items to certify compliance with these standards. These practices also helped to justify the high monopoly prices charged for guild-sanctioned products by guaranteeing quality and reliability. Many guilds fostered technological development by creating a stable work environment that encouraged technical experimentation and disseminated the ensuing innovations. Guilds preferred technology that enhanced skill and saved capital, valuable innovations from a master's perspective. Traditionally, artisans and guilds resisted riskier technologies that increased production, created new products, or explored new production techniques, preferring to rely upon their established practices and monopolistic control over all aspects of production.[18]

To be an artisan, therefore, implies many things: one is an individual practitioner of a trade; a member of a possibly organized body of practitioners of that trade; and a constituent of a productive segment of society whose members pride themselves on their training, independence, work ethic, and honor.

When colonists crossed the Atlantic to exploit the opportunities of a new world they carried their culture, a rich collection of familiar practices and attitudes, with them. But culture and society depend upon context, and change in new settings. The natural abundance of land and resources in America, as well as the opportunities and challenges posed by a sparse and distributed population, altered traditional political, economic, and social institutions and created entirely new ones. Artisan traditions evolved on America's shores while maintaining many connections to their origins.

Despite their low numbers, artisans had sweeping impacts upon American society. Of course, America's economy in colonial times, and indeed, its very identity, depended upon the farmers who constituted a vast majority of the population. Although American farmers produced many goods to satisfy their own needs or sell on the market, their demand for more specialized items often exceeded the output of local "jack of all trades" producers and overseas imports. Artisans met this need, and also became important links in growing trade networks. For example, artisans converted raw materials into more valuable and easily transported finished goods, repaired imported items, and produced or repaired tools to increase the productivity of other workers.[19] Merchants, planters, farmers, and shopkeepers all depended upon artisans.

Many craft practitioners may not have exclusively worked as artisans because they also dabbled in other fields. For example, most artisans living in small towns or rural areas doubled as farmers, owning a few acres of land to grow crops or keep livestock. Many others, including Paul Revere in the years after the Revolution, also ran their own retail shops and even referred to themselves as merchants. Southerners often trained slaves as artisans, to the great anger of white craftsmen unable to compete. Estimates of the number of artisans in colonial society vary widely, based on the definition of artisan, the time, and the location. By the time of the Revolution, artisans probably constituted between 10 and 20 percent of the population, and this percentage increased in urban centers to something between one-third and one-half of the population.[20]

Urban artisans, the closest American parallel to European and British artisans, usually centered their lives on their craft. Northern craftsmen often worked out of their own houses, which typically had a shop on the first floor and living quarters on the second. Most urban artisans operated on a modest scale: for example, a selection of ninety-six artisan shops in Boston employed a median of three workers and an average of thirteen, often including sons and apprentices in this number. Compared to rural practitioners, urban artisans had steadier employment, served a larger market, worked on a wider range of more specialized and elaborate tasks, and kept their shops open for more of the year. Urban craftsmen usually described themselves with their craft titles, such as "Benjamin Franklin, Printer."[21]

American master artisans had far more authority than their counterparts in England and Europe. Craft guilds never formed in America due to the small number and scattered distribution of artisans, and individual master craftsmen filled the void. The master owned and managed the shop, purchased tools and raw materials, managed apprentice and journeyman labor, designed products, determined what to produce, handled sales and advertising, and oversaw quality control. As head of the household, he also took on a paternalistic role, feeding, clothing, and supporting his family and apprentices. The master craftsman served as a teacher and role model to his charges, and accepted the full and solitary responsibility for planning and delivering an education in the technical, business, and cultural aspects of a craft.[22]

Urban artisans constituted a broad segment of American society that included struggling members of the working poor as well as economically prosperous and highly skilled or acclaimed individuals. In spite of the many differences between members of this group, nearly all artisans collectively distinguished themselves from both the merchants and gentry above and the "preindustrial pre-proletariat" laborers and sailors below.[23] These poorly defined social classes constrained and shaped the daily life and career expectations of all members of society.

Pre-Revolutionary America's class system combined freedom and hierarchical rigidity. Clever artisans and enterprising farmers had the opportunity to amass fortunes, and their non-privileged birthrights did not ruin the possibility of financial or entrepreneurial success. Foreign observers frequently described America as a middle-class world, meaning that poorer members might hope to enter the middle class while wealthier members had to keep the interests of less well-off groups in mind or risk reprisals. At the same time, the upper class remained a breed apart, gifted with exclusive privileges that success in one's trade could never bestow. Revere became aware of Boston's amorphous but still influential social classes at an early age, recognizing that his countrymen would never view or treat him as they did his wealthier "gentleman" cousins. He grew familiar with terminology such as the *better, middling,* and *poorer* (or *meaner/inferior*) sorts, and in particular, he knew that the title of gentleman carried with it a thrilling connotation of authority and advantage. He also encountered tangible manifestations of class differences all the time, for example, through sharp differences in clothing styles: most merchants dressed in velvet or imported wool clothes, silk stockings, linen, knee breeches, and waistcoats, while workers wore buckskin breeches, leather aprons, flannel, felt hats, wool stockings, and cowhide shoes.[24] While Revere accepted the existence of deference and privilege, he spent most of his life rebelling against the forces keeping him in a lower role, never ceasing his attempts to raise himself and his family to positions of societal influence.

Most people recognized the existence of social classes without ever firmly defining or clearly understanding them. Americans drew upon their British heritage, which inculcated deference and respect for someone's lineage, wealth, land ownership, and education. America's highest social classes consisted of both the largest property holders, such as landowners and merchants, as well as the most successful professionals, including doctors, lawyers, and government officials. These categories also varied significantly by region. Southern society enforced the most rigid class divisions, with a minority of aristocratic landowners occupying the top of the social hierarchy. Merchants represented the upper crust of northern urban society, and the fairly stable hierarchy of merchant elites possessed nearly as much political and economic power as their southern plantation-owning counterparts.[25]

All aspects of early America reflected the prevailing belief in a hierarchical world. Pre-Revolutionary society recognized one division above all others: the vast and almost unbreachable gulf between gentlemen and everyone else. The hazy concept of a gentleman referred to individuals or families who—in the eyes of society—possessed a combination of education, manners, wealth, virtue, and personal honor. Although most colonial gentlemen would merely have fallen into the upper middle class in Britain, they enjoyed many advantages in America.[26] Gentlemen had much freer access to education, credit, and influential administrators than others, and used these opportunities to enter prestigious professions befitting the leaders of society. They held a virtual monopoly on higher colonial political offices beginning in the mid-1700s, and the low turnover in these positions restricted members of the middle class to minor local positions. Gentlemen also amassed a large fraction of the total capital in America: in Boston, for example, the top 10 percent of taxpayers owned about two-thirds of the wealth in 1771, and this percentage only increased with the passage of time. The upper class often segregated itself socially and demanded respect and deference from what they termed the *middling* and *lower* segments of their communities. The life of a gentleman featured privileged private seating at all public events, membership in exclusive clubs, and special favors from local government officials.[27]

No factor divided gentlemen from all other members of society more than the unrecoverable stigma of performing manual labor, which implied the need to work for a living. If a gentleman ever performed nonintellectual labor he lost status in the eyes of other gentlemen, who found the thought of manual workers receiving gentry privileges both ludicrous and terrifying. In the words of political philosopher John Locke, "Trade is wholly inconsistent with a gentleman's calling."[28] This sentiment had a strong basis in prevailing British attitudes about public service and personal independence. Some contemporary political theorists believed that gentlemen and gentlemen alone should lead the society because of their freedom from occupational constraints and monetary worries.[29]

In light of colonial America's conception of physical labor, artisans could never achieve the full array of upper-class privileges based solely on their craft success. Several routes to social advancement did exist for a small number of well-off artisans, and unsurprisingly, none of them involved continued labor. Artisans most frequently improved their standing by beginning a successful merchant career, often using artisanal contacts with merchants and members of the gentry as well as capital amassed from a prosperous trade. Land speculation, a government appointment, membership in prestigious organizations, or marriage into a prominent family could also convert wealth into social power. Although most artisans never entered the upper classes, some managed to achieve large fortunes and at least adopt the trappings of the gentry.[30]

Artisans had a dual and almost contradictory role in society, as manager-entrepreneurs on one hand and as laborers on the other. They worked to distinguish themselves from the non-artisan laborers beneath them, while identifying with those laborers on issues such as dislike of aristocratic pretensions. Master artisans had to pass through apprentice and journeyman stages before earning the right to practice their trade, and as a result they greatly valued their autonomy and responsibilities. In his 1758 essay, "The Way to Wealth,"

Benjamin Franklin—an idol to many artisans—wrote, "He that hath a Trade, hath an Estate." Even though many members of society, particularly its elite members, treated all forms of physical labor with condescension, many artisans took pride in their social role: after all, their productivity provided necessary goods and services that directly contributed to their communities' well-being. Labor therefore had moral and social dimensions beyond the economic. Artisan organizations increasingly advocated for respectful treatment, including a voice in societal decision making and the prospect for security at the end of a lifetime of skilled labor. This pride played a major role in the Revolution and its aftermath.[31]

Master artisans, happily independent and in charge of their own craft shops, fell into different categories based upon the status of different craft trades and the reputation of individual artisans within each trade. For most of the eighteenth century, artisans primarily classified themselves according to their craft. These divisions resulted in an informal craft hierarchy that roughly correlated to the capital and skill requirements of each trade. At the bottom of the craft pyramid were the trades that involved the smallest startup costs, material costs, and skills, such as tailoring, shoemaking, and candle making. These crafts often featured smaller apprenticeship periods and lower-priced goods. Blacksmithing, carpentry, and other fields requiring greater skill, risk, and investment capital to purchase more expensive tools and raw materials filled the middle of the hierarchy. Since silversmiths typically produced highly visible objects for the most wealthy and powerful segments of society, earned high salaries, and required extensive skill and capital, the silverworking trade lay near the pinnacle of the pre-Revolutionary craft hierarchy along with printing. This craft hierarchy was poorly defined at best, and tradesmen such as millwrights, cabinetmakers, and clockmakers also carried considerable prestige commensurate with their high skill and capital requirements.[32]

Most social and economic variation took place between individuals in each trade. Artisans spanned a vast economic range, with some staying one step ahead of abject poverty and others achieving tremendous wealth and prosperity. Many artisans in colonial times could afford to set up their own shop after earning several years of income as journeymen, although limited quantities of investment capital posed a great challenge, especially during economic downturns. Artisans aspired through hard work and astute business instincts to become "respectable" or "reputable" tradesmen, which usually involved ownership of one's own shop and property, acknowledged craft skill, the ability to read and write (if not a greater education), material success, and participation in charities or service organizations. In contrast, less skilled and poorer "inferior mechanics" often performed simpler tasks or even worked for other artisans. The wealthiest and most skillful artisans supervised larger operations and dressed and lived in ways that set themselves apart. Although practitioners of the same craft often shared an esprit-de-corps in spite of their differences in status, by the end of the colonial period the division between the haves and have-nots began to widen. Wage labor grew increasingly common, and workers and craft masters formed different groups to protect their separate interests.[33]

Boston's 1789 parade, in honor of George Washington, clearly and publicly illustrated societal classifications and ranks. The parade organizers carefully arranged groups of marchers according to their profession, and the order of these professions says much about their prestige and precedence in the eyes of society. "Professionals," the educated group of gentlemen consisting of town officials, clergy members, doctors, lawyers, merchants, and ship owners, led the procession. The second group consisted of forty-six different artisan trades, and the final group consisted of sailors. Common laborers had no place in the parade at all. Interestingly, the numerous artisan groups within the second category of marchers appeared in alphabetical order to avoid giving any trade a distinction over the others, although the wealthiest practitioner of each craft prominently led his group.[34] This parade stands in for the overall social climate in colonial and Revolutionary America: everyone recognized the differences between individuals and professions, but calling attention to these differences was not always proper or even possible.

Educational pedigree held the same importance in Paul Revere's day as it does today, and society judged young artisans according to the reputation of the craft master who instructed them. Paul Revere fared very well in this regard because his pedigree led to his father's instructor, the esteemed John Coney. From the 1690s until his death in 1722, Coney's reputation among his peers had no equal. He mastered three artistic styles throughout his long career, and surviving pieces attest to his versatility and productivity. In particular, his skill at engraving distinguished him from his contemporaries and earned him a commission to engrave the paper money for Massachusetts in 1702. Other silversmiths purchased more ornamental hollowware items (large, hollow-centered cups, bowls, and teapots), the most lucrative of all silver commissions, from Coney than from any other practitioner. Governments, churches, and colleges repeatedly turned to him for their most elaborate orders, and his output even remained high during the economic instability between 1710 and 1720. A valuation of his estate placed him among the wealthiest 10 percent of all Bostonians.[35]

Surviving records do not describe John Coney's opinion of his 13-year-old, French-speaking apprentice. Coney assigned Apollos a nine-year apprenticeship term, longer than the average but consistent with Coney's reputation and demand for excellence. We can imagine the partnership as an amicable one, due to the fact that Apollos spent six years working for Coney without attempting to run away, a common recourse for disgruntled apprentices. Apollos's apprenticeship ended when John Coney died in 1722, three years before it had officially completed. The remaining three years of Apollos's indenture appeared as a commodity on the inventory of Coney's estate, and Apollos "graduated" by paying 40 pounds to Coney's widow. Details of his life become unfortunately sketchy at that point, but it is safe to assume that he worked for other silversmiths for one or more years until he saved enough money to establish and run his own shop.[36]

Coney bestowed two great gifts upon Apollos that paid dividends throughout his career. First, Coney's years of tutelage and command of the silverworking field rubbed off on Apollos, making him a skilled craftsman, fluent in all contemporary styles though never reaching Coney's level of mastery. Museum curator and historian Janine Skerry describes Apollos as a "capable, even talented, silversmith, who produced a variety of objects in an economical fashion."[37] His work illustrates the Bostonian set of stylistic and fabrication choices common at this time. The few surviving examples of Apollos's work, all museum-worthy, testify to the impressive engraving skill one would expect from a student of Coney. Because he was a known student of Coney, Apollos's reputation received a considerable boost even at the beginning of his career. Although Apollos could have benefited even more had Coney lived longer and helped him to get established, Coney's death created a void in the Boston silverworking community and patrons might have sought a student of Coney's to meet their continuing silverworking needs.

Apollos held a respected role in his local community, where he served as a financial supporter of Boston's New Brick Church, as well as among his fellow craftsmen. One tankard produced by Apollos indicates that he subcontracted some of his work to other silversmiths. The back of the tankard handle is stamped with the initials "WS," probably referring to the contemporary silver-smith William Simpkins. Apollos contracted with Simpkins and other silver-smiths to produce specialized components for some of his products, and prob-ably performed engraving and other tasks for them. Since each silversmith had a limited quantity of molds and other equipment, widespread "jobbing" of tasks enabled them to specialize while still handling all their clients' orders. Metalworking craftsmen in large urban centers often minimized competition by focusing their training and equipment purchases on a smaller number of operations. These tightly knit communities formed a collective versatility to compensate for their individual specialization: they sold their goods not only to the public, but also to each other. Subcontracting relationships extended far back into European craft history and demonstrated the communal nature of the silverworking field and the importance of reputation, because each sil-versmith only assigned work to competent colleagues and only accepted work from someone guaranteed to pay the agreed-upon wage.[38] Apollos would not subcontract if he could utilize his own equipment and have his apprentices perform some of the work as a part of their training. Following a common practice, he took on his oldest son as an apprentice, officially launching Paul Revere on his long career and ensuring both of their places in history.

Paul Revere became an apprentice to his father at age 13, joined approxi-mately five years later by his younger brother Thomas. Not a single surviving record describes Revere's apprenticeship or youthful experiences, but we can approximate these years from stories of "typical" apprenticeships throughout the colonies. Colonial American apprenticeship deviated from that in Great Britain by frequently requiring one or more years of "book learning." Revere received his classroom education at Boston's North Writing School before starting his formal apprentice training. Unlike the more prestigious Boston Latin School, whose curriculum was oriented toward upper-class children, the North Writing School focused on reading, writing, ciphering, and "man-ners."[39] Although Revere's schooling played a less visible role in his career than his training in his father's shop, his extensive lifelong correspondence, repeated reliance upon written information, and many pages of mathematical recordkeeping attest to the importance of this early knowledge.

Of all European artisan traditions, apprenticeship transferred to America with the fewest changes. Apprentices pledged a fixed period of labor and loy-alty, usually seven years in Britain, in exchange for a practical education in the "art and mystery" of a craft, a telling and often repeated phrase whose usage dated back centuries. "Art" in this sense referred to technical skill (and, in earlier times, to magical aid) while "mystery" suggested a combination of secret rites and essential knowledge that artisans must never share with outsiders. This fundamental relationship between knowledge and secrecy, drummed into the minds of all artisans throughout their apprenticeships, took on increased significance in the late eighteenth century.[40] Many apprentices and their fathers signed indenture papers, a binding contract with their new craft master that spelled out their rights and responsibilities. Close family members such as Revere and his father did not require written agreements, but surviving indentures offer insight into typical apprentice expectations. For example, Jacob TenEyck's seven-year apprenticeship to New York City goldsmith Charles LeRoux included the following stipulations:

> During all which Term the said Apprentice his said Master Charles LeRoux faithfully shall serve his Secretts keep his lawfull Commands gladly Every where Obey: he shall do no damage to his said Master nor see to be done by Others without letting or giving Notice to his said Master, he shall not waste his said Masters Goods nor lend them unlawfully to any, he shall not Commit Fornication nor Contract Matrimony within the said Term, at Cards Dice or any Other unlawfull Game he shall not play whereby his said Master may have damage, with his own Goods nor the Goods of others during the said Term without Lysence from his said master he shall neither buy not sell, he shall not absent himself day nor night from his Masters Service without his leave nor haunt Alehouses Taverns or Playhouses but in all things as a faithfull Apprentice he shall behave himself toward his Master and all his during the said Term, and the said Master during the said Term shall by the best Means or Method that he can Teach or Cause the said Apprentice to be taught the Art or Mystery of a Goldsmith. shall find or provide onto the said Apprentice sufficient Meat Drink and Washing in winter time fitting for an Apprentice and his said father to find him Apparell Lodging and washing in summer time and his said Master to suffer his said Apprentice to go to the winter Evening School at the Charge of his father. For the true performance of all and Every the said Covenants and Agreements Either of the said parties bind themselves unto the Other by these presents.[41]

Typical terms of indenture required apprentices to live with their master, serve him faithfully and obediently, keep his secrets, and protect his interests. More specifically, most contracts prohibited apprentices from engaging in "harmful" activities, such as marrying, fornicating, drinking, playing dice, leaving his master's service in the day or night, wasting or lending his master's goods, or disobeying his master in any way. Masters, in turn, had to provide food and drink, washing, accessibility to evening school, and an education in their craft. Of course, a world of difference lies between the theory and reality of laws that regulate human behavior, as attested by numerous complaints from participants on both sides of this relationship. Multiple reports of wild apprentices caused many states to forbid taverns to serve alcohol to apprentices, but few officials enforced these laws. On the other hand, master craftsmen might abuse their apprentices by overworking them, withholding craft secrets, or insisting that they perform household or farm chores instead of activities with educational value. Disgruntled apprentices often tried to escape, leading local governments to pass laws to protect the investments of masters. Runaway attempts even took place when the apprentice worked for members of his own family, as demonstrated when Benjamin Franklin fled his brother's printing shop in Boston at age 17 in order to try his luck in Philadelphia. But young Paul stayed in his father's shop until the end, and never saw reason to complain about his training.[42]

Although most aspects of apprenticeship drew upon Old World practices, some traditions did change in reaction to the social, political, and economic climate of the American colonies. American colonists considered servitude of all forms, including apprenticeships, demeaning.[43] In a land of scarce labor and cheap land, employment in the service of others evoked comparisons with slaves. Many Americans accepted lesser positions, such as apprenticeships, indentured servitude, or wage labor, as short-term expedients that hopefully allowed them to save money and start their own businesses or farms. The enormous labor shortage, prestige associated with independence and land ownership, and lack of guild regulation produced an apprentice deficit and higher demand for craft practitioners, which often compressed the seven-year apprenticeship into four or even three years. The number of craft apprentices remained small during the colonial years, giving some bargaining power and extralegal authority to those who knew they could run away and either find a new master or set up their own shop prior to their "graduation." Apprentices also used their bargaining power to end the British custom of families paying masters to train their sons, with occasional exceptions surfacing among the most elite colonial trades and most acclaimed masters. In addition, authorities often made compulsory apprentices out of orphans, representing colonial America's matter-of-fact attempt to address both labor and child welfare problems. With the spread of capitalist attitudes throughout the eighteenth century, apprenticeships and all other forms of written contracts became more formal and explicit, often delineated in monetary terms. In earlier times, these contracts defined a spirit of personal obligation that bound both parties, rather than specific promises of action and compensation. Apprenticeship changed to conform to the modus operandi of the land of opportunity, and masters and apprentices each tried to optimize their prospects. Contracts, monetary quantification, and legal enforcement of obligations marked the beginning of the shift from camaraderie and fraternal relations toward employer-employee expectations.[44]

While serving as an apprentice, Revere worked with other artisans in his father's shop, older and more experienced journeymen who received an hourly, daily, or monthly wage in exchange for their labor. Journeymen occupied the middle step of the craftsman ladder, possessing sufficient skills to begin practicing their craft but lacking the capital needed to set up their own shop. As their name implies, they often traveled in search of the highest wages, a practice that infuriated masters in need of steady labor, and they settled down as master artisans after amassing sufficient money. Journeymen lacked the status and independence of master craftsmen, but did play a role in shop operations: many craft shops allowed workers at different levels to collectively negotiate the division of labor or pace of work. Journeymen rarely received much credit for their work because the master craftsman typically absorbed the praise or blame for all his shop's output. Journeyman silversmiths, for example, frequently assisted master silversmiths with complex tasks and even produced their own silverware, but only the master artisan placed his maker's mark on the final products.[45] Skilled craft workers such as journeymen earned 20 to 100 percent more than unskilled workers in the 1770s, but all wage laborers experienced great vulnerability in early America, a time prior to long-term contracts and social security. In light of their lack of status, high wages, or job security, it is no surprise that wage laborers, including journeymen, remained relatively rare in America prior to the Revolution.[46]

Within the craft shop, masters, journeymen, and apprentices evolved work patterns that melded republican ideology with workplace realities. Many artisans read Benjamin Franklin's *Poor Richard's Almanack,* which sold approximately ten thousand copies a year until 1758, and agreed in principle with its stated values of frugality, hard work, and excellence. Artisans truly valued the personal independence they felt entitled to as skilled craftsmen, which translated into the ability to work for themselves, own their tools, set the pace of their work, and determine for themselves the best way to craft each item. In the time before widespread watch or clock ownership, many workers performed their jobs in a task-based manner, receiving a fixed salary in exchange for a specified quantity of work each week, but retaining the right to choose which days of each week and hours of each day to spend in the shop. These workers appreciated the option to add work breaks or vacations into the daily and weekly schedule, understanding that they had to work harder at other times to finish the task. Artisans filled their days and weekends with activities, including wagering on animal contests or races, billiards, dice, cards, heated political discourse, bouts of violence, and of course, rampant alcohol consumption.[47] Even when these pastimes took place in the shop, artisan employers trusted their skilled workers to exercise discretion and responsibility, and as a result the workplace became more personalized and lively, to put it mildly.

Young Paul Revere began his silversmith apprenticeship with routine duties such as cleaning the shop, collecting silver dust and filings for reuse, and tending the fire, and his responsibilities increased as he grew more skilled. Craft skills develop from the process of organized and open-ended repetition, which allows apprentices to observe their progress, learn from mistakes, and apply their existing knowledge in new directions.[48] But Revere's apprenticeship taught him more than just a set of silverworking techniques. From his lessons, his father's example, and the conversations of other apprentices, journeymen, master craftsmen, and customers, he learned what it truly meant to be an artisan. Because he was an independent master craftsman Revere's future largely lay in his own hands, and the practices he witnessed in his father's shop served as models for his own choices and values. For example, Revere always took on more than financial obligations toward his workers, and he felt compelled to ferociously defend his honor and independence to the end of his days. Many of the artisan traditions reproduced in colonial America had more in common with industrial practices than one might think, representing the early onset of certain proto-industrial practices. In particular, the use of wage laborers, the division of labor among different skilled and unskilled workers, and the subcontracting of labor among different shops based on their specialties, while not common, occurred with increasing frequency. Revere's later manufacturing career built heavily upon these earlier practices, and he benefited from many of the business and management customs that began in artisan shops.

As an apprentice in his father's shop, Revere focused upon a seemingly endless series of short-term goals that culminated in his attainment of master status. A tragedy soon replaced these goals with an even larger and more important series of responsibilities, which affected not only his own future, but that of his entire family. Apollos Rivoire died on July 22, 1754, and 19-year-old Paul Revere found himself the oldest male in the family, responsible for his own shop, apprentices, and dependents. It was time to become a master silversmith.

Paul Revere: Silversmith

More than a quarter of a century later, Revere wrote the only surviving description of his father's passing in a letter to his cousin in France: "My father was a Gold-smith. He died in the year 1754, he left no Estate but he left a good name and seven children, 3 sons & 4 daughters. I was the Eldest Son. I learned this trade of him and have carried on the business ever since."[49] Neither this letter nor any other surviving writings flesh out this meager description to show us the anguish and uncertainty that Revere undoubtedly felt when his father, his mentor, and his family's provider abruptly exited his life. Apollos Rivoire's death impacted the Revere household well beyond the distress caused by the loss of a beloved patriarch. Paul was technically still an apprentice, too young in the eyes of the law to inherit the family silver shop until he turned 21. In the absence of records before 1761 we can only speculate whether he ran the shop in his own name, in his mother's name (a common practice at the time, helping widows retain some control over a family business), or worked under a more experienced silversmith until he came of age. One way or another, he soon operated his father's shop and had to confront the realities of management. Even if he had fully mastered the technical aspects of silverwork by this time, which is by no means certain, he now had to learn the business aspects and keep his shop profitable during the economic downturn still gripping North America since the end of King George's War in 1748. Initially, the going was tough. Newly widowed Deborah Revere made her first rent payment in rum, cash, and a silver thimble, and Paul paid some of the next quarter's rent by making ten rings for his landlord. Revere's business had a slow start following his father's death, one of the slowest periods of business in the entire period covered in his shop records.[50]

Fortunately, Revere could draw upon several powerful assets. In addition to inheriting knowledge and training from his father, he also received a fully stocked silversmith's shop containing all the tools he needed for the next ten years at least, and some equipment, including his father's molds, which he used for his entire career. Unlike most artisans, who had to work for years to save enough money to set up their own shop, Revere avoided the journeyman stage entirely. He also benefited from his father's network of business connections with customers, merchants, and other silversmiths. In some cases these categories meld; for example, Boston merchant Benjamin Greene both supplied and bought from Revere and his father. Apollos recorded sales to neighbors, Hitchborn relatives, friends, and loyal clients, and Paul benefited from many of these relationships as he built his own base of friends and customers.[51] The importance of networks in early America cannot be exaggerated, and Revere's ability to forge connections and identify new communities that might yield new customers and allies played a major role in his rapid success.

Revere's early silverwork period portrays the rising fortunes of a young, ambitious, and talented artisan. In the late 1750s, when he first entered the field, he had to continue his technical and managerial education on the job while supporting a growing family in an uncertain economy. By the eve of the Revolution he enjoyed a solid reputation, large and steady sales, and even some extra time for other pursuits in his different organizations. Revere's success resulted from a combination of artistic talent, technical versatility, and good connections, and bore many similarities to the experiences of his brothers in trade, countless silverworkers who practiced their crafts since the beginnings of human civilization.

From the earliest times, the scarcity and beauty of gold and silver caused their value and prestige to soar above that of more utilitarian metals such as iron or copper. The scarcity of these metals, along with their resistance to chemical corrosion, led to the nickname "noble metals." Even though silverworking processes have at least as much technological complexity as that of most other crafts, the luxury status of silver placed silverware into the category of art, earning silversmiths a reputation for both technical and aesthetic skill. Returning once more to the *Iliad*, we see that the stature of silver is reflected repeatedly in lovingly descriptive passages ranging from individual titles such as "Apollo of the silver bow" and "silver-footed Thetis" to longer descriptions of items such as the mixing bowl of Achilles:

Achilles quickly set out prizes for the footrace.
A silver bowl, gorgeous, just six measures deep
but the finest mixing bowl in all the world.
Nothing could match its beauty—a masterpiece
that skilled Sidonian craftsmen wrought to perfection,
Phoenician traders shipped across the misty seas
and mooring in Thoas' roads, presented to the king.[52]

Silversmiths across far-flung times and places served as a bridge between the worlds of manufacturing, metallurgy, and art. Silversmiths traveled to America on some of the earliest colonization voyages to Jamestown in the 1600s, but truly cemented their role in American society in the more stable New England Puritan villages later that century. Throughout the later colonial period, only blacksmithing had more artisan practitioners than silversmithing.[53]

Changing silver styles respond to America's evolving societal values. New England silver during the early 1600s featured heavy and plain pieces, corresponding to the Puritan emphasis upon simplicity. Many of the pre-1640 Massachusetts Puritans belonged to the middle or upper middle classes and brought family silver with them, ensuring a steady demand for the services of silversmiths to repair or add to their collections. With the rise of merchant prosperity in the mid- to late 1600s foreign silver flowed into New England, Puritan restrictions upon conspicuous consumption relaxed, and silver became a status symbol representing the affluence and good taste of its owner. By the end of the seventeenth century, the increasingly cosmopolitan colonists embraced the vivid, three-dimensional complexity of the baroque style emanating from the court of Charles II, but in the early eighteenth century the Queen Anne (or early rococo) style achieved dominance, promoting a return to simplicity, elegance, and formality. The fully developed rococo of the mid-eighteenth century expanded upon the Queen Anne style with richer decorations usually favoring "natural" forms, such as shells and flowers. Finally, the neoclassical or federal style gained acceptance after the Revolution, and emphasized a return to purity, restraint, and geometric forms.[54] American society vacillated in its preferences from one generation to the next, often switching between plainer and more embellished fashions. The finest silversmiths demonstrated their skill and improved their reputation by mastering multiple artistic styles.

Colonial silversmiths served several important functions above and beyond their production of luxury items. The upper classes in urban centers and on southern plantations viewed silver plate as a luxury item and status symbol, but upper-class patronage alone cannot account for the large number of silversmiths, volume of sales, or existence of silversmiths in smaller towns and less settled areas. In addition to luxury silver items, colonial silversmiths also produced many pieces for middle- and even lower-income families as a form of wealth storage and security. American colonists regularly encountered English, Dutch, French, Mexican, Portuguese, Spanish, Arabic, Peruvian, and other silver and gold coins. This confusing array of currency was easily stolen and non-identifiable, major problems in the days before insured banks or a reliable national currency. Customers hired silversmiths to assess the value and purity of coins or other silver sources, and then melt and cast them into usable new forms. Silversmiths customized their work with makers' marks (the silversmith's name or initials, which he stamped onto all large silver items) and engravings (such as a family crest or the owner's initials), thereby adding to the value of silver. Silver plate frequently served as both a mortgageable item and as a form of payment, and colonial newspapers occasionally contained advertisements attempting to identify and reclaim lost silver. The maker's mark became even more important in America than in England: only a silversmith's judgment certified the quality and integrity of finished pieces in the absence of English regulatory mechanisms such as guilds and assaying offices. The maker's mark had more personal value as well: Paul Revere placed six stamps on one beautiful inscribed salver sold to William White in 1760, a clear indication of the pride he took in it.[55]

Most American silversmiths lived and worked in Boston, New York, and Philadelphia, the three largest cities. Urban centers offered access to wealthier and more numerous clients; a larger supply of silver coins and pieces; a network of fellow practitioners to lend labor, tools, and advice; and better information about changing artistic styles and preferences. These cities had more contact with London than with each other until the eighteenth century and quickly adopted and emulated the newest British fashions, although silversmiths in large cities often produced distinctive local interpretations of prevailing styles. Boston and New York led the colonial silverworking movement in the seventeenth century, and Philadelphia joined them in the eighteenth century. In spite of Philadelphia's leadership in most colonial crafts, many experts believe that Boston produced the finest silver objects, particularly during the late seventeenth and early eighteenth centuries. From these cities the craft spread to smaller urban centers throughout the colonies.[56] Throughout his career, Revere benefited more from the support network centered in well-populated Boston than he suffered from competition. He found himself particularly dependent on others at the start of his silversmithing.

Paul Revere turned 21 in 1756 and could finally run his own silver shop in his own name, formally joining Boston's silverworking community. But instead of launching his career, he enlisted in the Massachusetts militia in February of the same year as a second lieutenant in an artillery train, to fight in the French and Indian War. Although he might appear to have shirked his duty to run the shop and support his family, Revere probably had their interests in mind when he enlisted. As an artillery lieutenant he received 5 pounds, 6 shillings, and 8 pence per month, an ample cash wage roughly double the salary of a typical enlisted man. In six months he could earn 32 pounds, or two years' rent on his family's house, more than a novice silversmith in an uncertain market. Revere joined the only artillery unit in the Massachusetts force, an elite skilled group twice the size of an infantry company, allowing him to gain a prestigious rank at a young age. Impressive military titles certainly carried over to private life, as illustrated by numerous instances of Revere referring to himself or being referred to by his militia title. He remained in the military until November 1756, and although his regiment at Lake George never took part in combat he learned enough about military matters and artillery usage to lay the groundwork for military service during the Revolution and post-Revolutionary cannon-casting work. In addition, his experience serving under the condescending British officers probably altered his perceptions of status and colonial-metropolitan differences as it did for many of his contemporaries. This brief military stint delayed the start of Revere's formal silver career: although he began recording his silver shop's operations in September 1757, his records do not show a regular stream of output until 1762.[57]

In August 1757, less than a year after returning from his military service, Revere married Sarah Orne of Boston, who also moved into the cramped accommodations of his mother's house. Less than eight months later Sarah gave birth to their first child, a daughter named after Revere's mother Deborah. Sarah, or "Sary" as he recorded in the family Bible, gave birth to a new child every second year for the remainder of her life. With a family including a new wife depending on him, with his military adventures behind him, and as a master craftsman legally running a shop in his own name, he could finally devote all of his time and attention to his craft.

From the first day of apprentice training, the complex details of silver fabrication occupied the forefront of every silversmith's mind. Revere's surviving records do not contain an inventory of the equipment or raw materials present in his silver shop or of the fabrication procedures he followed at any point in time. He is not alone in this oversight: not a single American silversmith codified these technical details in any records, because this knowledge constituted part of the "art and mystery" of their trade, passed on to apprentices verbally and through observed practice but never in writing. The creative use of historical sources such as probate records or studies of finished silver products strongly suggests that silversmiths' equipment and procedures remained fairly constant throughout the colonial and early federal periods despite dramatic changes in artistic styles. These tools and the knowledge of their use defined the silversmith profession and circumscribed Revere's daily routine for many decades. To place ourselves in his shoes and experience the methods and challenges of silverworking as he did, we must re-create the workings of an eighteenth-century silver shop from surviving clues.

Upon first entering a silver shop, one would immediately observe the variety of tools required for common practices, including many versions of certain instruments. For example, an inventory of the estate of Boston goldsmith Richard Conyers performed by two other silversmiths in 1709 yielded eighty unique tools, including anvils, punches, compasses, vices, hammers, swages, stamps, punches, tongs, bellows, files, gravers, chisels, patterns, scales, and many others. John Coney had, at the time of his death in 1722, 116 hammers, 127 nests of crucibles, 80 anvils, and enormous numbers of other tools. Other typical equipment owned by seventeenth- and early eighteenth-century silversmiths included burnishing stones and pumice powder, furnaces, soldering equipment, wire-drawing benches, drop-press and screw-pressure dies, and molds. The primary tools of greatest importance to most silversmiths included hammers, punches, anvils, files, and shears, and each shop often owned dozens of each of these mainstays in different shapes and sizes. Apprentices learned to maintain their tools: for example, they had to keep hammers flawlessly polished and rust-free because they transferred any of their own imperfections to the metal. This vast selection of tools enabled skilled workers to produce silver objects of all shapes and sizes, but fortunately we can simplify the repertoire of silversmith's fabrication procedures into four basic categories: casting, hammering, seaming, and finishing.[58]

Casting. First, silversmiths learned to cast silver. Many customers offered silver coins, silver bars, or obsolete silver items to silversmiths for reuse in new purchases. A silversmith placed this silver in a special graphite crucible, along with any copper needed to strengthen the metal and raise it to the "sterling" standard.[59] He then placed this crucible in a small, charcoal-fueled or bellows-pumped furnace, or held it in the furnace with a skillet until the metal lique-fied. The craftsman poured the liquid metal into a mold, an iron-framed box of dense sand containing an indentation pressed or sculpted into the shape of the final object. The most common molds had simple rectangular indenta-tions that produced silver bars, but more complex molds formed silver into the shape of teapot handles, the legs of a water pitcher, and so on. Instead of sand, some molds consisted of valuable blocks of hard wood or iron with pat-terns cut into them. The silver cooled and solidified within the mold and the silversmith removed it when ready to shape it into the finished form. Silver-smiths rarely formed an entire silver object by casting, but might use molds to produce certain portions, such as handles or legs.[60]

Hammering. After casting the silver, silversmiths had to hammer and re-shape it. Labor costs accounted for most of the value of silver plate, and ham-mering represented the most time-consuming aspect of the silverworking pro-cess. Hammering encompassed a series of related processes, including forging, raising, hollowing, and creasing, that made use of numerous hammers and an-vils with different-sized and -shaped striking edges. Blows from a flat-headed hammer made a thick piece of silver wider and thinner, while curved hammers and anvils created rounded shapes, indentations, or other three-dimensional forms. Silversmiths used other tools to add depth to a flat item. For example, all cast spoons and other silver pieces began as flat, two-dimensional objects. A silversmith rounded and deepened the bowl of the spoon by placing it over the lower part of a die, a lead block containing a depression in the exact shape of the spoon. He then struck the spoon with the upper part of the die, an iron mallet also curved in the spoon's shape, and the spoon's bowl conformed to the shape of the die. Hammering took place at room temperature via a pro-cess called cold-working. When cool metal was hammered it became "work hardened," which means it responded to stress by growing hard and brittle. Fortunately, silversmiths could reverse these undesired effects via the anneal-ing process: they carefully heated metal at a specific temperature to partially soften it, and then plunged it into water or acid. A balanced application of annealing and hammering produced strong metal, neither soft nor brittle. Silversmiths and other craftsmen developed an instinctive understanding of metal properties and practices, developing and passing on these skills through observation, practice, and hands-on understanding.[61]

Seaming. In the seaming stage, silversmiths created closed hollowware forms such as cups or teapots by fastening different pieces together or connecting the edges of a single curved sheet to form a cylinder. Although silversmiths occasionally used rivets to attach pieces to each other, soldering remained the sealing technology of choice into the nineteenth century. Silversmiths made their own solder in a charcoal furnace from an amalgam of four parts silver and one part brass, combined with borax paste that served as a "flux" material to help the solder flow throughout the seam. Silversmiths applied the pasty solder mixture along the connecting seam between two silver sheets. Because solder melted at a lower temperature than the sheet silver, a precise application of heat from a furnace, blowpipe, or hot soldering iron melted the solder without affecting the silver object. The molten solder filled the space between the silver sheets and fused them together when it cooled. An experienced practitioner such as Revere then used sulfuric acid to remove all traces of the borax paste and produced a virtually invisible soldered seam. Cups required the least time and effort of all hollowware items: silversmiths used one seal to form a curved sheet of silver into a cylinder and a second seal to attach a silver disc to the bottom. A teapot, in comparison, had an unusually shaped body, legs, a handle, and a curved spout that all required careful attention.[62]

Finishing. Finally, a silversmith needed to finish his silver item by polishing, filing, and adding embellishments such as engraving or chasing. All silversmiths learned to even out the many irregularities marring the surface of unfinished silver items by hammering out indentations (called "planishing"), filing off burrs with small and unusually shaped files, and polishing the entire item with pumice stones or some other abrasive. Once the silver shone in a smooth and unblemished state, silversmiths had a choice of ornamentation techniques. Engraving, the most common of these techniques, involved cutting a design into the silver by gouging lines with either a pointed tool called a burin or a short *V*-shaped tool called a graver. Instead of engraving, silversmiths could add chasing to an item. This painstaking process used a steel punch and hammer or dull chisel to dent the outside of a silver sheet with repeated blows, forming a continuous indentation design without removing any metal. To add the far more difficult repoussé ornamentation, silversmiths hammered indentations into sheet silver from the inside with a curved rod called a snarling iron, producing a raised design on the outside of the object.

Figure 1.2. Wire-drawing machine, from *The Pirotechnia*, Vannoccio Biringuccio's comprehensive Italian metallurgy text first published in Venice in 1540 by Curtio Nauo & Fratelli and printed by Venturino Roffinello (reprinted in Cyril Stanley Smith and Martha Teach Gnudi, trans. and eds., *The Pirotechnia of Vannoccio Biringuccio: The Classic Sixteenth-Century Treatise on Metals and Metallurgy* [New York: Dover Publications, 1990], figure 72 on p. 379). The illustration indicates that, even in 1540, wire-drawing machinery automated many aspects of the wire production process. The different devices in this image allowed silverworkers to turn cranks and pull a tapered silver wire through a hole. Workers repeated the process with smaller and smaller holes until the wire achieved the desired diameter. Paul Revere probably owned similar equipment, and his familiarity with silver wire drawing helped him master more complex processes many years later.

Many silversmiths also made their own wire from thin silver sheets by using a wire-drawing bench, consisting of a huge crank attached to a rope on one end, and a huge iron dye containing different-sized holes secured to the other end. Silversmiths tapered a small bar of silver so that it barely fit through the largest hole in the dye, and secured it to the rope with a clamp. A strong apprentice turned the crank and pulled the silver through the hole, which compressed the silver into the diameter of the hole. This compression made the silver brittle, and the silversmith heated it after each "draw," enough to restore its ductility without making it too soft. The silversmith then tapered the end of the wire enough so it fit through the next smallest hole, and repeated the process. When the wire attained the desired diameter, the silversmith soldered it onto flatware as an extra embellishment.[63]

These descriptions only brush the surface of the intricacies of silversmithing, but reveal some of the skills and metallurgical knowledge required for competent practice. Silversmith tasks may have fallen into these four broad categories, but each piece of silver carried a set of attendant challenges and unusual constraints, not to mention customized respects that required cre-

Table 1.1.   Crafting a Silver Teapot

Process	Description
Melt silver	Melt a set of silver coins (provided by the customer) and bars (purchased from a merchant) in the furnace
Refine silver	Refine the silver to the proper standard by adding copper, and possibly add other substances (such as sulfur) to remove impurities
Cast an ingot	Pour the molten mixture into a mold to form a silver ingot
Form a sheet	Hammer the ingot into a large sheet of the desired thickness, length, and width, using the large forging anvil
Inscribe and cut pattern	Use a compass and ruler to inscribe the proper pattern onto the sheet of silver and cut it into the proper shape with shears
Anneal silver	Anneal the metal by holding it with tongs and heating it in a charcoal forge fanned by hand bellows
Form base	Shape the warm metal on the raising anvil using a series of hammers, forming it into the proper spherical shape for the base of a teapot
Anneal silver	Use repeated annealing to remove stress and brittleness
Form other teapot parts	Follow a similar process to form separate parts of the object: a teapot includes a cylindrical or spherical base and also a spout, cover, legs, and handles
Make solder	Prepare solder by creating a mixture of silver, brass, and borax paste
Solder pieces together	Apply solder to the seams between different pieces of silver; heat and melt the solder and remove traces of the borax paste with sulfuric acid
Cast other parts	Make small parts such as hinge plates, thumb pieces, or small handle tips by casting silver into a mold
File and polish	Remove hammer marks and imperfections with a file, planishing hammer, and pumice stone
Engrave	Add detailed engraving with a burin and graver
Add maker's mark	Strike a maker's mark onto the vessel

ative combinations of processes or new ones altogether. Table 1.1 illustrates one trajectory that a silversmith might take, applying tools and methods to a complex task.[64]

The elite status of silverworking resulted from both the amount of equipment required and the diverse range of skills and knowledge needed to make so many different objects. Revere's lifelong pride in technical work and desire to improve his ability to produce useful and beautiful items stems from the many skills he developed at this early time. Processes such as annealing and wire drawing not only served a variety of silverworking purposes but also prepared him for many of his later metalworking endeavors: artisan knowledge and experience surpassed scientific knowledge in silverworking and other metallurgical crafts until the late nineteenth century. All Americans at this time knew how to judge the quality of the items they purchased and even assess the quality of the materials composing them, and the entire consumer population placed a high value on artisan skills. Revere's skills quickly established his place in the Boston craft community, for he had many visible gifts, beginning with his versatility.[65]

Although the bulk of Revere's early silverworking activities involved routine work such as cleaning, dent removal, and small commissions, he provided a wide range of products and services more diverse than in any other period in his life. In response to the lack of guilds, the shortage of skilled labor, the highly unpredictable colonial economy, and the changing needs of a widespread population, most American artisans had to take a flexible approach toward their trade. Even so, it is hard to imagine a more versatile craftsman than Paul Revere in any place or time. As a young silversmith supporting a growing family and attempting to build a client base and reputation, Revere accepted virtually any opportunity for work. He produced more than ninety different kinds of objects, ranging from spoons, buttons, buckles, and teapots to candlesticks, thimbles, medical instruments, and children's whistles. He also fulfilled unusual requests, such as putting silver handles on seashells; making a dish out of an ostrich egg, a funnel, and a silver chain for a pet squirrel; cutting a branding iron for his cousin William Hitchborn; and mending glass objects. Revere even worked on small gold objects, such as buttons, rings, and bracelets. This diversity illustrates his combination of design skill, the ability to conceptualize and plan the form of the object, and craft skill, the ability to execute the design and actually create the final object. A broad range of products and services provided a steady source of income that helped him weather economic and political hardships. He worked on these unusual customized orders with far greater frequency in the early years of his career than in the later years, when his reputation guaranteed a steadier stream of more profitable work that allowed him to standardize his product line somewhat. In terms of total output, Revere recorded sales of at least 175 silver objects throughout his prewar career, second only to the 185 items made by silversmith Benjamin Burt in the same period. This output is particularly impressive considering that Burt's greater age gave him more time to establish himself than Revere, who also lost many weeks of work to his Patriot activities. Revere got his business off to an excellent and rapid start.[66]

Revere's output resulted from a well-managed group effort. As with all craftsmen, he received assistance from numerous apprentices throughout his silverworking career, although his surviving recordbooks rarely identify these helpers or their activities. If Revere followed standard practices, a shop of his size would have a reasonable number of assistant silversmiths and regular turnover among them. For example, Zachariah Brigden, a silversmith working at the same time as Revere, relied heavily on repair work and produced a much smaller amount of hollowware than Revere, but recorded transactions with fifteen journeymen and apprentices. Revere's records, in contrast, only mention four assistants explicitly, and we can only imagine the full size of his shop. His younger brother Thomas, who continued his training with Paul after their father's death, became Revere's first apprentice. Thomas seems to have graduated to the status of journeyman in 1761, when Revere began charging him for board and clothing, items provided free of charge to an apprentice living in the master's household. Another apprentice, the son of Josiah Collins of Newport, is identified in a letter from his father.[67] Two probable journeymen in his employ, Samuel Butts and Mathew Metcalf, appear in his account books when they paid him for board and shop supplies. In 1773–1774 Revere began receiving aid from his most important and valuable apprentice, his oldest son, Paul Jr. Paul Jr. never achieved the artistic or technical skill of his father, judging from his surviving silver pieces. But as a mature and trustworthy family member whose apprenticeship bracketed the Revolutionary War, he provided his father with services above and beyond those of ordinary apprentices, occasionally looking after the entire shop while his father devoted more of his time to Patriot activities. From this point forth, Revere's many endeavors became a family affair.[68] Revere certainly employed other apprentices and journeymen, but they do not appear in the surviving records.

Revere's versatility allowed him to apply silverworking skills to related activities in other fields when the opportunity arose, and colonial America's constant economic and political upheavals offered many opportunities. When silver scarcity became a concern during the postwar depression of the mid- to late 1760s, Revere branched into dentistry and false teeth construction. This field served as a second career for many American silversmiths because it used thin silver or gold wires to fasten the artificial teeth. Dr. John Baker, a "surgeon-dentist," taught Revere the basics of dentistry around 1767, and

Revere started advertising his services in 1768.[69] Revere also entered the field of copperplate engraving and printing prior to the Revolutionary War. Engravers often allied themselves with printing establishments and their work was always in great demand, particularly in the decade before independence. Around 1762 Revere made his first copperplate engravings—an outgrowth of his ability to embellish silverware—but quickly learned to produce a variety of images and texts for portraits, church hymnals, political cartoons, advertising cards,[70] bookplates, and many other media. In addition to mastering the technical intricacies of this new line of work, he also drew the illustrations that appeared in print. Several aspects of his engraving work typify his lifelong career trajectory and operating methods. Although he became quite proficient in reproducing images in a variety of formats, Revere was more of a borrower and adapter than a creative artist. He created most of his political cartoons by copying or modifying British illustrations, a plagiaristic action by today's standards but a common publication practice of Revere's time. He also demonstrated a remarkable ability to master the many technological aspects of this job. Many entries in his silver shop daybooks contain prices for engravings, including charges for "cutting a copper plate," "smoothing plate," "to the copper for a plate," "to preparing plate for engraving," and "cash paid for 4 letters." Although he never described his engraving in further detail, the evidence suggests that he purchased copper for plates, smoothed and cut it to the proper size, and engraved it with designs. He also carried charges for the printing of numerous copies of certain documents, explaining why he might have purchased letters of printing type. Similarly, when he produced "metal cut" illustrations and mastheads for mass-produced newspapers and almanacs, he almost certainly made his own "type metal," a compound of lead, antimony, tin, and other metals. He did not learn these skills during his apprenticeship because they do not fall within the purview of silverworking. They do relate to some of a silversmith's skills, such as cutting, polishing, and engraving, but require the silversmith to apply these techniques to a new metal with different properties and to new applications. Throughout his life Revere excelled at this type of technology transfer, and he also possessed the entrepreneurial acumen needed to identify each new opportunity and weigh the risks. Fortunately, he did not lose sight of the bread and butter of silverworking, the luxury items that commanded the largest profits and played the greatest role in establishing a silversmith's reputation. Revere may have loved trying new things, but he never lost sight of the old ones either.[71]

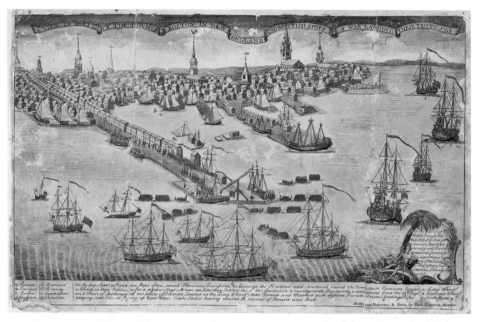

Figure 1.3. Paul Revere's "A View of Part of the Town of Boston in New-England and British Ships of War Landing Their Troops! 1768," colored engraving by Paul Revere, 1770, reproduced with permission of the Bridgeman Art Library. Paul Revere engraved this image, possibly from an original drawing by Christian Remick. This detailed view of Boston emphasizes the docks and North End neighborhood that circumscribed most of Revere's early activities, and highlights the British landing force that set the stage for increased Patriot resistance by marching with "Insolent Parade" into Boston.

In the famous Copley portrait described at the start of this chapter, Revere holds a teapot that reflects one of the most popular styles of the 1760s, described in one source as an "inverted pear-shaped body with stepped foot and a domed cover surmounted by a cast finial, with single or double scrolled wooden handle." Contemporaries and historians alike tend to judge silversmiths almost exclusively according to their ability to create elaborate luxury pieces, and according to this criterion Revere stood at the top of his profession. Luxury silver is doubly valuable to a historical study because it not only reflects the skill of the maker, but also the economic and artistic preferences of the community that purchased it. Mid-eighteenth-century America, and particularly its three largest cities, offered many opportunities to silversmiths: a general rise in the standard of living allowed many individuals to buy small quantities of silver, while the wealthiest classes sought to display their rapidly increasing affluence through major purchases.[72]

The strong economy in eighteenth-century America produced a steady rise in the amount of wealth at nearly all levels of society, as well as huge changes in the way people lived.[73] In 1700, "average" households owned two beds typically stuffed with wool or cotton fiber refuse, four brass pots, and a single table. At the start of the eighteenth century, people either shared or did without nonessential items such as plates, drinking vessels, and utensils. By 1750, chairs, chests with drawers, forks, knives, plates, and teapots became fairly common, and even luxuries such as clocks and mirrors grew in popularity. Greater overall wealth and an increased concern for status led to product differentiation: dining tables, tea tables, and sewing tables took the place of the earlier all-purpose table, and pots gave way to water vessels, teapots, and wine decanters. In addition to improving their material lives and standard of living, colonists hoped to raise their societal standing by assuming the trappings of gentility. Well-to-do colonists often displayed their status by conspicuously consuming luxury goods, and this practice grew more achievable and widespread as the market economy took root. Eighteenth-century Bostonians procured their material goods from the numerous merchant shipments from England as well as from domestic artisans, and former luxury items became increasingly commonplace. Silver in particular gained popularity as different classes tried to appear more genteel.[74]

Some of Revere's prewar success lay in his shrewd decision to produce a large quantity of lower-end silverware to appeal to customers from all economic strata. In spite of the high cost of silver, it leads the list of luxury items used by the general population: 20 percent of New England households owned some silver by the middle of the eighteenth century. Revere produced five small flatware items (spoons and other flat objects) for every piece of expensive hollowware. The emphasis upon both low- and high-end silver products increased Revere's ability to appeal to more clients by offering something for everyone, while also making better use of his labor pool: apprentices and hired workers had skills that better suited flatware production. The shift to flatware also revealed an early interest in producing standardized goods, something he avidly pursued after the war.[75]

Although Revere derived most of his income from the sales of smaller pieces of silver, large and costly items advanced his reputation and status in the community, showcasing his great skill at interpreting popular artistic styles. Expensive silver items, used for ceremonial occasions, gifts, or display purposes, were "bespoke," or custom-made to order. Customization ensured the buyer's satisfaction and prevented the silversmith from wasting time, materials, and effort on products that might not sell. At the beginning of Revere's career, the highly ornamental rococo style of decorative artwork satisfied a pervasive desire for novelty in the mid-eighteenth century, becoming somewhat popular in Boston but widespread in New York and Philadelphia. The rococo style is known for its curved shapes, use of sinuous and leaf-like heraldic engraving, and prominent display of coats of arms. Some modern experts consider Revere and Nathaniel Hurd the two best colonial practitioners of the rococo style, and one critic contends that Revere's rococo pieces stand well above other Boston rococo objects "in their exuberant interpretation and technical execution." Many other early Revere products display an above average to exceptional degree of workmanship, evidenced in terms of uniformity, artistic interpretation, and elegance. By the 1760s, he had matured as a silversmith and regularly turned out spectacular pieces of work to suit all tastes.[76]

Over half of Revere's total production of large silver items consisted of objects for drinking and dining, primarily on the tables of the wealthiest Bostonians. His output reflects changing social trends, such as the growing popularity of tankards and teapots for social drinking rituals. Throughout the eighteenth century, tea drinking grew in ceremonial importance, especially in the female-controlled domestic social world. Tea-drinking accessories, as well as tea itself, grew increasingly cheaper, eventually becoming available to all but the poorest members of society. Bostonians consumed more than 16,000 pounds of tea in 1759 alone, and a proper tea ceremony used teapots, creampots, sugar dishes, sugar tongs, tea tongs, and teaspoons. Revere made one or two teapots, which he often sold as one component of a larger and more profitable tea service, each year prior to 1767, but his production plunged when Boston boycotted tea in response to the duties imposed upon the colonies by the Townshend Revenue Act of 1767.[77] Many Bostonians who boycotted tea during this period turned to coffee, and correspondingly, Revere's coffeepot output increased from a total of two in the years before 1767 to six in the year 1769 alone. One of these, made in 1772, is the only known marked three-legged coffeepot made in America. Revere adapted the foreign form of the three-legged coffeepot to his own design by adding three large legs ornamented with shells and scrollwork in the rococo style to a basic coffeepot body. As much as any other single piece, this coffeepot exemplifies Revere's combination of artistic and technical creativity. Despite his increasing involvement in Patriot activities during this period, Revere did not let politics get in the way of a sale. Revere sold numerous pieces to his family physician, Tory Dr. Samuel Danforth, and one of the two teapots Revere sold in 1773 belonged to a forty-five-piece tea service, the single largest commission of his silverworking career, for Loyalist Dr. William Paine.[78]

Revere engraved nearly all his larger silver items, borrowing many creative flourishes from the work of other craftsmen, including his father, or from English sources. He perfected his engraving abilities throughout his career and mastered the art of combining different artistic forms into meticulous and balanced final products. Revere's outstanding embellishment work partially resulted from his craft lineage: John Coney, his father's mentor, also specialized in engraving. In fact, modern appraisers of silver often cite Paul Revere and John Coney as two of the finest colonial engravers. Engraved coats of arms personalized silverware and quickly became key components of many rococo items: the jurisdiction of the British College of Heralds did not cross the ocean, and colonial families could therefore design their own coat of arms to add to their social status, typically aided by advice books such as John Guillim's *A Display of Heraldry*. Revere added coats of arms to bookplates, salvers, teapots, and other large objects, and often engraved items that he did not make, reflecting the high esteem in which other silversmiths valued his abilities, including well-established ones such as John Coburn, John Symmes, and Nathaniel Hurd.[79]

A silversmith's reputation determined whether clients patronized his shop. This reputation depended upon a combination of the reputation of his mentor, his technical and artistic skill, the loyalty of his client base, and his perceived integrity. Revere's long record of silver output, ranging from small objects to high-prestige teapots and symbolic works such as the Liberty Bowl, provided him with an enduring reputation and a practice that remained profitable until his retirement.[80] His success depended on more than just technical skills, and some of these other abilities prepared him for his post-silverworking ventures.

Paul Revere: Networker and Businessman

Revere reaped great profits during his colonial silver years. He recorded income between 11 and 294 pounds in the years between 1761 and 1775, with an average yearly income throughout this period of 85 pounds.[81] In comparison, successful colonial journeymen might earn up to 45 pounds a year, full-time

Figure 1.4. Untitled and undated Paul Revere sketch of an engraving pattern from Paul Revere Cashbook for Goldsmith's Shop, volume 57, pages 41–42. From Revere Family Papers, 1746–1964, microfilm edition 15 reels (Boston Massachusetts Historical Society, 1979), reel 15. Paul Revere's engraving abilities earned him a number of lucrative subcontracting jobs and helped establish his reputation. This sketch of an ornate plate and its matching cover offers a detailed study of Revere's application of rococo ornamentation. Courtesy of the Massachusetts Historical Society.

manual laborers rarely earned more than 30 pounds a year, and the median free white income throughout the 1770s hovered around 12 pounds. Many of Revere's income fluctuations reflect external conditions, such as occasional recessions that slowed spending throughout the economy, as well as distractions minimizing his shop time. For example, his shop output and income took a dramatic plunge between 1767 and 1770, largely due to his political activities such as his many messenger rides on behalf of the Patriot cause, which could take him away from his shop for ten days or longer. This downturn did not have a major impact upon his confidence or his spending, as illustrated by his 1770 purchase of a house in the North Square area, about a block from his shop in Clark's Wharf, for 213 pounds. Business took a turn for the better in 1773–1774 due to prewar speculation and business from British officers.[82] His financial success and strong reputation among craftsmen and community members resulted from more than technical aptitude. Revere distinguished himself as both a spectacular networker and a shrewd businessman in a period often characterized by interpersonal relationships and reciprocal exchanges of favors.

In the midst of increasing financial security and prestige Revere lost his wife Sarah in May 1773, five months after she gave birth to little Isanna. Without a diary, without surviving letters, we can only imagine his thoughts at this trying time, as he faced the prospect of raising and supporting his seven children and elderly mother, alone. But Revere was no stranger to hardship or responsibility, and ever practical, he understood that life must go on. He returned to silverworking a week later, soon resumed his community activities, and even found love again. Within a few months he started courting 27-year-old Rachel Walker, whom he married in October 1773. This courtship process includes a love poem riddle that Paul wrote to Rachel, economically composed on the back of a bill. It reads:

> Take three fourths of a Paine that makes Traitors confess (answer: "Rac")
> With three parts of a place which the Wicked don't Bless ("Hel")
> Joyne four sevenths of an Exercise which shop-keepers use ("Walk")
> And what Bad men do, when they good actions refuse ("Err")
> These four added together with great care and Art
> Will point out the Fair One nearest my Heart.[83]

Putting aside Rachel's imagined response to pain, traitors, hell, and other romantic imagery, this poem is one of the two happiest items in all of Revere's surviving records, confirming a genuine affection for his lively young bride. His marriage to Rachel undoubtedly benefited his entire family. Revere had six surviving children from his first marriage, not counting the infant Isanna, who died four months after Sarah. Rachel immediately took up the slack on the home front, in addition to giving birth to eight more children over the next fourteen years. For the next four decades she stood by her husband through tea parties, midnight rides, court-martials, and countless other adventures.

Revere recorded a huge number of transactions with many customers throughout his silverworking career. Combining the customers recorded in his record books with the names of other known purchasers of his work yields a total of 757 known patrons. His business productivity owed much to his ability to meet and form relationships with numerous individuals. A number of neighbors, lifelong friends, and Hitchborn relatives appear in his sales records with great frequency, clearly indicating that they sent him as much of their business and support as possible.[84] Throughout his life Revere held important (often leadership) positions in many groups and associations, including the New Brick Church, activities in and around the North End, artisan groups, the St. Andrew's Lodge Freemason organization, the Massachusetts Charitable Mechanics Association, and various Patriot affiliations, including the Sons of Liberty, the North End Caucus, and the Long Room Club. Revere consistently converted approximately 20 percent of the members of each association into his customers. For example, of the 474 men commonly identified as prominent Revolutionaries and members of these Patriot groups, Revere enlisted 122 (22%) of them as clients. While no one group monopolized Revere's sales or allowed Revere to dominate their own patronage, he succeeded because he belonged to a large number of groups and drew some support from each one.[85]

The Freemasons organization contributed the largest number of Revere's customers, accounting for one-third of his total client list. Revere joined the St. Andrew's Freemason lodge in 1760 and won his first election as lodge master by 1770. Early colonial Masonic organizations catered to the highest echelons of society and emphasized the elite privileges their members expected. Beginning in London and spreading first to Philadelphia in the 1750s, a new branch of Masons called "the Ancients" opened their doors to artisans, lesser merchants and shopkeepers, and other members of the middle classes seeking to improve their economic and societal position—an exact match with Paul Revere's aspirations. Revere's Freemason career at St. Andrew's "Ancient" lodge brought him social entertainment, status in the eyes of the community, leadership responsibilities, business contacts, and lifelong friends. In particular, Revere became close friends with Masonic brother Dr. Joseph Warren, a doctor and extremely prominent Patriot who issued the marching orders for his Midnight Ride. More than a dozen of Revere's fellow Masons produced transactions listed in Revere's very first daybook, ordering Masonic medals, Masonic jewels, engraved notices of meetings or other business matters, personal items for lodge brothers and their families, and even items for neighboring lodges. Many of these friends loyally purchased goods and services from Revere for many years.[86]

Revere's good standing and contacts throughout the artisan community offered another important support structure. Artisans comprised slightly more than half of the purchasers of his silver items, accounting for the majority of his flatware sales and also a few expensive teapots and lodge paraphernalia. Artisans, and silversmiths in particular, also added stability to Revere's business by forming subcontracting partnerships with him, just as they had with his father. Some of his surviving silver pieces were stamped with his maker's mark and then overstruck by the mark of other masters, indicating that he sold silver parts or entire pieces to other silversmiths who took the final credit. Revere's pre-Revolutionary accounts record transactions with at least thirteen silversmiths as well as other craftsmen such as jewelers. Revere used these relationships to farm out work he preferred not to handle in his own shop: perhaps he lacked specialized equipment, felt others could do the work more efficiently, or faced time pressure for large orders. Other silversmiths often asked him to perform certain work for them in return. Revere's reputation and skill are illustrated by the fact that the work they sent him, such as engraving, chasing, and repoussé ornamentation, required advanced design and implementation skills. Silversmiths also exchanged goods for services in an extensive barter system. For example, between 1760 and 1765, Revere received important shop supplies, such as gold foil, saltpeter, files, borax, pumice stone, and binding wire, from John Welsh in exchange for engraving work and producing objects such as spoons, buckles, and spectacles.[87]

Revere maintained excellent ties with the moneyed elements of his community. Although he recorded many small sales to a wide range of clients, merchants and professionals purchased nearly all of his expensive hollowware and became valuable repeat customers. Only the wealthiest 5 percent of the population owned numerous pieces of silver, and these members of society also cared the most about their social standing.[88] Revere's records reveal a fairly consistent seasonal business pattern of peaks and lulls corresponding to spring and fall merchant shipments of new British goods, which completely dominated the colonial economy. New goods shipments provided Revere with raw materials and also increased the amount of specie in circulation, encouraging greater spending. Merchants also provided information about the latest British styles, which he quickly emulated, and close connections with merchants enabled him to import objects that he preferred not to make. For example, he made few silver forks or knives throughout his career, choosing instead to import them from Britain at relatively inexpensive prices. These merchant connections inspired him to serve as a retail seller of other goods in his shop, a practice he began before the war but greatly accelerated afterward.[89]

In addition to his networking ability, Revere possessed invaluable business and management skills. Despite the skilled labor scarcity, nearly half of artisan shops failed, usually due to some combination of the master's inadequate business acumen and generally unstable economic conditions that required a flexible and creative approach to weather the dry spells. This set of business skills included facility with procuring and managing lines of credit, extension of credit to customers, recordkeeping, and financial planning.[90]

Credit relationships allude to the importance and scarcity of capital in early America, and capital shortages generally produced the greatest limitations upon craft practitioners in Europe, England, and America. Silversmiths in particular needed to purchase and upgrade the many tools required for their practice, buy property, and purchase raw silver from merchants. Many artisans either used family connections and funds or received capitalization from merchant investors. Artisans, and especially those who made expensive products, had to offer credit to customers if they wished to make sales in cash-poor America, and long creditor lists show that compensation rarely took place at the moment of exchange. Many payments occurred "in kind" via a barter exchange of goods and services that did not involve cash or coinage, and payments ceased altogether during hard times. Artisans often had trouble repaying their own debts due to nonpayment by their customers and depended upon additional loans from local merchants or "moneyed" men, further extending personal networks of credit. In early colonial America, these personal exchange networks dominated all economic transactions and also solidified social relationships. Revere's records illustrate many cases of customers requiring extensions or special accommodations to enable them to pay for goods they previously received. Revere worked with customers to develop payment plans that might include a combination of paying via installments and compensation "in kind," such as bushels of corn or sugar, "old" silver, or something as creative as Copley's offer to paint Revere's portrait.[91]

Rigorous recordkeeping practices helped master artisans manage credit and debt, adjust to an increasingly capitalist society, and plan for the future. Throughout the colonial period, farmers and rural shopkeepers often tracked debts through oral agreements or informal account books. These books illustrate the network of exchanges that connected members of a community to each other, and since many of these exchanges involved two-way transfers of goods and services, the concept of debtors and creditors often did not apply. Most accounts lacked detail or oversight, using one page for each individual and occasional tallies of totals motivated by major life events such as a death or loss of a business partner. By the end of the eighteenth century, New England merchants, artisans, and eventually farmers led the way toward more rigid accounting methods, converting all debts and credits into monetary values in the ledgers. But in colonial times many artisans still kept records in the most rudimentary fashion, using long, single-entry columns of numbers to track stock and assets on hand without attempting to compute profits or net worth.[92]

Revere maintained records of all his silver shop transactions beginning in 1761 using the most advanced techniques of his day. Revere's earliest system consisted of two layers of recordkeeping: a wastebook and a ledger. The wastebook, also known as a daybook, allowed Revere to track shop activity as it happened: he recorded customer purchases and payments received in chronological order, with a "D" (for debit) next to purchases and a "C" (for credit, or "contra") next to payments received. Each entry contained the name of the customer and the value of the transaction, initially using the British system of pounds, shillings, and pence, with a switch to dollars in the 1790s. Revere subdivided many customer purchases based on the different components of the item: for example, he listed the value of silver on one line, a charge for his labor on the next (titled "To the making"), and charges for special work such as engraving on a third line. He periodically transferred wastebook entries into ledgers, separate books that collected all purchases and payments made by each customer. His ledgers used the "double entry" accounting system, listing all debits for one customer on the left side of the page and all credits on the right, thereby facilitating an easy comparison of each customer's indebtedness to Revere. This meticulous notation of all his activities shows that he affixed monetary values to all his items at the time of sale, which placed him slightly ahead of what soon became a widespread bookkeeping practice. At the same time, barter transactions, unexplained cash withdrawals, bizarre abbreviations, and incomplete entries create great confusion in these early ledgers. Revere's recordkeeping practices improved greatly after the war and he added new techniques, including receipt books, invoices, and bank accounts.

Double entry bookkeeping is a hallmark of capitalism. On the surface it is a practical system because it facilitates error detection: by tracking assets in one column and liabilities in the other, Revere could quickly compute sums to determine if customers paid their bills and whether the bookkeeper recorded entries properly. But double entry accounting also carried a subtle deeper meaning. Separating assets and liabilities into distinct categories brought the concept of profit into the limelight and caused owners to think of the success of their business in the more abstract world of numbers and net worth, instead of relying upon tangible physical assets. By adopting this system, even in a rudimentary form, Revere took a major step into proto-industrial capitalism, and his quantifiable mindset only grew stronger in the years that followed.[93]

Revere's recordkeeping techniques fail to shed light on an essential but opaque aspect of early American business, the unsung role of women in the workplace. Families served as the fundamental economic unit in American society, and all members of a family found ways to contribute to its economic and social success, for example, by endorsing each other's debts or helping to find jobs for relatives. Rural women typically farmed alongside the men while also raising children, cooking, cleaning, and producing household manufactures. In urban settings women took ownership of household upkeep, child care, and food preparation, but might also help in the shop. Revere's mother, first and second wives, and seven daughters who reached adulthood represented a potential labor force too promising to leave untapped. Although Revere's records and correspondence failed to discuss the contributions of women, one can assume that his family may have helped with bookkeeping—which would explain occasional changes in handwriting, waited on customers, or even offered technical aid.[94]

In a surprisingly short time Revere transformed himself from a novice master craftsman struggling to pay his rent into one of the most prolific, versatile, skilled, and reputable silversmiths in Boston. In addition to the numerous clients from wide-ranging walks of life who purchased his silver, Revere's own family bought some of his work, undoubtedly using the pieces as a conspicuous display of his skill and the family's material success. Artisan life clearly suited him, but Revere also encountered the limitations of his trade as well. In the hierarchical society of colonial America, artisans had a specific and circumscribed role to fill whether they wanted to or not.

Revere's entire career was in many ways a negotiation between his and society's view of his work. He developed an intimate awareness of class and hierarchy through close contact with his well-off Hitchborn relatives, service in the militia under the supervision of the British military hierarchy, ongoing interactions with wealthy clients, and membership in organizations that included individuals from different walks of life. He fully understood society's view of artisans and had frequent glimpses of the vistas that lay beyond his grasp. Even though his own position as a silversmith placed him comfortably above the median, this taste of success only made him wish for more, for the privileges and influence of the gentry. On the other hand, his devotion to his craft and pride in his skills reflects his love for the production process, a life-long fascination with experimentation and fabrication that must have begun in his apprenticeship. Science, technology, and the discovery and implementation of new processes intrigued him and occupied much of his attention, particularly after the Revolution. He had the ability to succeed as a first-rate craftsman, but could never accept craftwork as the endpoint of his personal or professional journey. The artisan in Copley's portrait is proud of his nearly finished masterpiece for the time being, but his next challenge is to help others appreciate it as well. Revere's penetrating gaze transfixes the viewer, not his product.

Revere's pre-Revolutionary silverworking experiences reflected the state of early American manufacturing, a blend of European tradition and New World modification that included hints of the onset of industrial capitalism. The pre-industrial, or "craft," method of production can be generalized as small-scale operations that depended on personal relationships between owners, workers, and customers to overcome capital, labor, technology, and resource limitations. Revere compensated for his early lack of access to capital by making the most of his connections, which enabled him to draw upon fellow silversmiths while steadily increasing his client base. He liberally combined old and new business and manufacturing methods throughout this period, dealing with craft apprentices and salaried employees, multiple silverworking styles, and a juxtaposition of affordable and luxury items. Above all else, Revere's early silver years demonstrate his impressive combination of technological and entrepreneurial acumen, which allowed him to identify and exploit new opportunities. This skill proved an essential survival characteristic in colonial Boston's turbulent society and economy, never more valuable than in the years to follow, when war brought the foundations of Revere's world crashing down.

CHAPTER TWO

Patriot, Soldier, and Handyman of the Revolution (1775–1783)

In November 1775, Paul Revere visited Philadelphia to learn how to produce gunpowder. He succeeded, but also received a more valuable education in a different subject.

Following his Midnight Ride, Revere kept busy outside Boston, biding his time in a boardinghouse in Watertown while British troops occupied his hometown. Separated from his silverworking shop, he repeatedly assisted the Patriot cause, but something was missing. Even after earning a position of trust in the organization, as illustrated by the importance of his midnight mission on the 18th of April, leadership positions remained outside his grasp, much to his chagrin. Most recently, the Continental army had denied his petitions for an officer's commission because he lacked specific military experience and the credentials of a gentleman. The opportunity to study and reproduce a gunpowder mill represented a new chance to prove his worth to the Patriot leaders by performing a service that few others could even attempt.

The Continental army perpetually faced shortages of weapons, ammunition, and gunpowder. With the British navy intercepting merchant ships and the former colonies struggling against a century of manufacturing dependence, the Patriots took desperate measures to keep their army in the field. The Massachusetts Provincial Congress dispatched Revere to a Philadelphia gunpowder mill in hopes that he could establish a similar facility in Massachusetts. Revere carried an official letter from influential congressional delegates Robert Morris and John Dickinson to mill owner Oswell Eve, assuring him that "A Powder Mill in New England cannot in the least degree affect your Manufacture nor be of any disadvantage to you. Therefore these Gentn & myself hope You will Chearfully & from Public Spirited Motives give Mr. Revere such information as will inable him to Conduct the bussiness on his return home . . . P.S. Mr. Revere will desire to see the Construction of your Mill & I hope you will gratify him in that point."[1]

At last, Revere could exercise authority and contribute to the military effort by providing a unique service, backed by the full authority of the Continental Congress. Unfortunately, Oswell Eve chose not to cooperate. Eve grudgingly took him on a quick walking tour of the factory but refused to give him a diagram of its layout unless Revere paid a tremendous bribe. We can imagine Revere's frustration and perhaps even his shock as he wondered why his vital mission and wonderful letter received such a harsh reception. Patriotism meant nothing to Eve, who only cared about maintaining his advantageous monopoly. Secrecy was the basis of his power.

But power flows in many directions. Revere's brief factory tour still provided him with a basic understanding of the shop's layout, and he shrewdly observed enough details of the gunpowder production process to begin his designs. Patriot leader Samuel Adams brought his own networking skills to bear, obtaining a set of plans for a different gunpowder mill from a "New York gentleman."[2] Between his personal observations and these borrowed plans, Revere helped design a new mill soon constructed in Stoughton, Massachusetts. In this case, true power did not lie in secrecy, but rather in the allied network of innovators, politicians, and influential people who worked together to pool their expertise and solve common problems. This lesson served Revere well.

Revere's period of patriotic service, including his famous Midnight Ride and other resistance activities in the years before the Revolution, his brief time in the Massachusetts militia, and the numerous technological endeavors he undertook on behalf of the Revolutionary leadership, served as the true turning point of his career. These activities increasingly pulled Revere away from his primary vocation and forced him to reevaluate his career goals, particularly in light of his frequent interactions with political leaders. The Revolution also provided several clues about the structure of American society, clues that eventually helped him identify his assets and liabilities. Revolutionary America's social, economic, and political upheaval gave rise to an overlap between the older system of patronage-driven hierarchy and the emerging democratic meritocracy. Revere's diverse wartime experiences, including his successful technical endeavors, his failed military career, and his multifaceted patriotic efforts, taught him the importance of operating in both of these worlds: his networking abilities paid respect to the existing hierarchy and opened doors, while his practical technical and managerial abilities allowed him to solve problems and make the most of his opportunities. As a successful artisan, Revere often took on an intermediary social and political role that, if he could play his cards correctly, might lead to grander things.

Patriot Resistance and the Role of Artisans

Throughout his colonial silverworking career, Revere participated in an increasingly important series of Patriot activities that brought him in contact with Revolutionaries who ranged from the poorest laborers to some of Boston's most elite lawyers and merchants. Revere's metallurgical and patriotic activities influenced each other. Although the Revolution frequently disrupted his silver operations by taking him away from the shop or disturbing the local economy, at the same time he steadily increased his social responsibilities, reputation, and network of potential patrons by growing closer to fellow Patriots. Silverworking influenced his patriotic exploits in return, because his career perfectly situated him for the role he played in America's Revolution. Although Revere achieved an unusual degree of prominence in the Revolutionary movement, his experiences and goals paralleled those of many artisans across the colonies who parlayed the conflict with England into a valuable opportunity for social advancement and public participation.

At the close of the French and Indian War in 1763, Revere had no inkling of the impending tide of resistance and revolution that would soon overtake his nation and his own life. America's pride in Britain's victory and its hopes for a peaceful, prosperous future gradually receded in light of a series of ominous developments. Both England and America suffered from depressed economic conditions when the end of wartime spending produced cash and credit shortages. Many merchants, artisans, and farmers fell into debt or even bankruptcy during this period, mirroring the staggering sums owed by the British government to the many creditors who had helped finance the war.[3] Britain's postwar regulation of its American colonies included new policies prohibiting colonial westward expansion into Indian-controlled lands, as well as increased efforts to raise colonial taxes that could help pay the ongoing costs of supporting a North American army. Parliament passed the Sugar and Stamp Acts in the mid-1760s in order to raise revenue: the Sugar Act taxed colonial imports and the Stamp Act required the purchase of tax stamps on most colonial printed items. Instead of raising revenue, these acts raised hell.

Even though the American colonies had a long and scandalous history of dodging customs duties, the shift in the 1760s from evasion to outright and widespread opposition represented a bold change that frightened many observers. Hindsight casts the colonial resistance movement in a deceptive light: far from being an irresistible force, colonial attempts at resistance to parliamentary acts began in a disorganized manner and faced colossal impediments. The divided American colonies lacked a common cause or political identity at first, and the majority of colonists looked to town authorities or occasionally to colonial capitals for leadership. While some colonists, particularly merchants or larger landowners, began interacting with counterparts in other colonies by the late eighteenth century, intercolonial trade remained minimal and most economies operated on local scales.[4] But in spite of prevailing isolationist tendencies, the colonists shared their British heritage and culture, which included common legal systems, traditions, and the widespread belief in America's untapped economic potential. In addition, the timing of the Anglo-American consumer revolution around the mid-eighteenth century helped align the experiences and expectations of many would-be Patriots. Most families at this time, regardless of their social or economic status, began acquiring inexpensive, high-quality British manufactured goods that made their lives more comfortable. American imports of British merchandise increased each year, and some colonists—such as Revere and other artisans who viewed cheap foreign goods as threatening competition—began connecting imports to American dependence and British oppression. Consumers in many colonies developed a strong distaste for further parliamentary attempts to exacerbate the disparity between the mother country and its periphery. Sugar and Stamp Act taxation fell immediately into this category and gave many colonists another reason to identify with one another and unite against England.[5]

During the early years of colonial resistance to British taxation policies,

Americans forged connections that began to cross geographical, social, and economic lines. Colonists during the resistance period saw foreign taxes as dangerous to liberty. The Stamp Act proved particularly incendiary because it attempted to pose an "internal" tax, or a tax on individual purchases for the purpose of extracting revenue from the colonies, rather than a less controversial "external" tax levying duties on trade for the benefit of all constituents of the empire. By targeting document users such as lawyers, merchants, and newspaper publishers, the Stamp Act inadvertently antagonized a literate and highly influential segment of colonial society, uniting certain master craftsmen, professionals, and merchants. This broad array of actors produced an equally varied set of responses, including ideological opposition, mob intimidation, and some of the earliest organized boycott activities in world history. Colonial rhetoric questioned Parliament's right to levy any tax upon British citizens against their will, and in May 1765 the Virginia House of Burgesses became the first official group to publicly protest the Stamp Act while more broadly attacking Parliament's right to tax the colonies. Before the end of the year, legislatures in eight other colonies had passed similar resolutions.[6]

In Boston, following news of Virginia's action, a small group called the Loyal Nine initiated more dramatic resistance methods. The Loyal Nine primarily consisted of middle-class working men, including lesser merchants, distillers, braziers, and a printer. Using their connections to laborers' organizations, they guided the activities of Boston's North End and South End workingmen mobs. The mob produced rapid and dramatic results: it nearly leveled the professional and personal residences of designated stamp collector Andrew Oliver on August 14, secured his promised resignation the next day, and ravaged the house of Lieutenant Governor Thomas Hutchinson eleven days later. Throughout this ordeal the sheriff and all other authorities proved powerless in the face of the multitude, an instructive development for would-be Patriots in other colonies who also started using force or the threat of force to persuade stamp collectors to resign. By December 1766, the Loyal Nine rechristened themselves the Sons of Liberty. This larger organization soon had branches in many other colonies, and generally consisted of respectable men in good standing in their communities. The Sons of Liberty gained influence with shocking rapidity: according to Thomas Hutchinson in 1766, "the authority of every colony is in the hands of the sons of liberty."[7] Paul Revere counted himself a member of this organization, and greatly benefited from the connections he formed with other members.

The Stamp Act struggle initiated a significant change in the membership and operations of the colonial resistance movement. Throughout most of the colonial period, artisans did not take on social leadership roles or collectively attempt to change policies. For the first time during the resistance period, agency started shifting from gentlemen to the prominent, influential, and growing middle class, including the large number of master artisans who joined Revolutionary organizations such as the Sons of Liberty.[8] Evolving relationships between social classes led to interesting politics, to say the least. Artisans grew more forceful in the pursuit of common goals and interests, beginning with their support for an egalitarian meritocracy. Proud of the societal value of their skills, artisans felt entitled to community respect and a role in public policymaking. Artisans in the 1760s and 1770s, in their efforts to join and lead resistance activities, asserted their right to advocate for themselves and hold public office in spite of, and even because of, the fact that they worked for a living. Gentleman officeholders found these suggestions threatening and resented the implication that members of the lowlier trades might make diplomatic or legal decisions. Some artisans, in turn, resented merchants for their role in importing British goods, accusing these merchants of placing profit over patriotism. In spite of these ideological differences, resistance groups exercised incredible restraint, avoided destructive dissent between their constituents, and minimized their use of violent public actions. Organizations such as committees of correspondence and the ever-growing Sons of Liberty gradually took on more leadership functions in a strategic manner that involved members from different societal groups.[9]

The Sons of Liberty, and Patriots in general, won an important victory when Parliament repealed the Sugar and Stamp Acts in 1766.[10] Following the resolution of this crisis, the remainder of the 1760s and early 1770s offered periods of cooling patriotic fervor punctuated by new conflicts that helped solidify the resistance movement. The British Parliament continued inciting the colonists, starting with the Townshend Revenue Acts of 1767, which raised new duties on colonial imports, and culminating with the Tea Act of 1773, which began the countdown to war. The situation grew uglier as British troops lodged themselves in various towns, particularly Boston, which had the dubious honor of hosting two regiments as early as 1768. Colonial resentment arose over what appeared to be a foreign occupying force in their towns, and injury added to insult when these soldiers competed with local laborers for hard-to-find hired work. Physical confrontations intensified during most of this period, peaking in 1770 when a small group of British soldiers triggered the "Boston Massacre" by firing into a huge mob of hostile colonists, killing five. Ideological resistance also increased at this time, with colonial assemblies and individuals such as John Dickinson, author of the highly influential "Letters from a Farmer in Pennsylvania," protesting Parliament's right to legislate without colonial consent.

The American resistance movement became the first to employ the voluntary consumer boycott strategy, using the market to punish Britain's economy. Patriotic non-importation associations encouraged, and in many cases forced, colonial merchants to boycott British goods, producing impressive reductions in import levels while also greatly increasing the quantity of home manufactures. The production of native goods such as homespun clothing became imbued with moral values such as industry, frugality, and independence.[11] Artisans, already enjoying greater political influence through their membership in resistance organizations, played a crucial role in supporting and enforcing the non-importation agreement. By the later days of the resistance, artisans effectively took action against rogue or Loyalist merchants who attempted to escape the non-importation movement and increased their political influence in certain cities and towns. Artisans and other pro-manufacturing members of the resistance movement took this opportunity to expand the scale of their goals: instead of simply wishing to overturn unfriendly legislation and restore the pre-1765 status quo, the non-importation movement now attempted to foster an economic climate more favorable to colonial interests, such as manufacturing. Non-importation opposed British taxation while improving artisans' ability to compete.[12]

Colonial resistance measures did not produce a dramatic shift in British policy but did impact the mother country, and by 1770 the new ministry repealed the Townshend duties for economic reasons, retaining only a token tax on tea. Tea became anything other than token in 1773, when Parliament passed a new Tea Act that granted the East India Company a favorable tea monopoly in the colonies. Not surprisingly, this infringement on colonial autonomy provoked a new round of protests that culminated on December 16, 1773, when angry Bostonians—including many artisans and Sons of Liberty—dumped ninety thousand pounds of tea into the harbor to protest Britain's attempts to raise taxes again. The Boston Tea Party marked a point of no return in the relationship between colonies and mother country. The next few years witnessed passage of the Coercive Acts (or, as they were more colorfully known in the colonies, the Intolerable Acts) intended to punish Boston by closing its port, revising the Massachusetts charter to minimize the role of popularly elected officials, shifting certain judicial hearings from Massachusetts to Britain, and making it easier to quarter troops in the colonies.

Boston became the primary stage of colonial resistance rhetoric and activity in these turbulent times, hosting the earliest mob protests, the Boston Massacre, the Boston Tea Party, two regiments of British troops, and eventually the start of the Revolutionary War. Britain's repressive actions had a particularly incendiary impact in Boston because of its long tradition of public participation in government. Any adult male with an estate valued at more than 20 pounds could vote at the Town Meeting, a local government institution loved by the people. Town Meetings often operated by seeking consensus rather than giving rise to conflicts, although much of this consensus came about through negotiated "behind the scenes" arrangements between groups such as the North End Caucus, which discussed and resolved pressing issues prior to the actual Town Meeting. This strong tradition of communal decision making partially explains why Boston artisans deferred to merchants far longer than the artisan committees in Philadelphia and New York. Boston artisans might not sway Town Meetings or hold prestigious offices in the same manner as merchants or professionals, but their voices and votes carried critical weight in important community matters.[13]

Boston's collective outrage and opposition to the "Intolerable Acts" did not come as a surprise. In keeping with Boston's proud tradition of resistance, indignant armed mobs quickly overthrew all royal authority not backed by British troops. But the sympathetic and supportive reaction of other colonies did astonish many observers. Rather than making an example of Boston and dividing the colonies, Britain's final move in this escalating series of hostilities ultimately united and emboldened the Patriot resistance. Patriot leaders in many colonies formed Committees of Correspondence to share information, leaders arranged for the meeting of a Continental Congress in September 1774 to coordinate resistance activities, and many colonies sent aid to Boston to make up for its lost ocean trade.[14] The Continental Congress called for enforcement of non-importation, non-consumption, and non-exportation actions in all colonies in order to economically injure the mother country, and sent a petition to the king, which was stubbornly ignored. England responded by ordering General Thomas Gage to suppress the rebellion with force, culminating in the British march to Lexington and Concord on the 18th of April in 1775 and the Midnight Ride of Paul Revere. However, Revere's service to the Patriot cause had long roots extending back to the earliest resistance activities.

Paul Revere took to the resistance movement with a vengeance, and Revolutionary roles soon dominated all his other activities. His ferocious opposition to British subjugation erupt from the pages of a letter he wrote to his cousin John Rivoire in 1782:

> [The British] were not contented to have all the benefit of our trade, in short to have all our earnings, but they wanted to make us hewers of wood & drawers of water—their Parliaments have declared "that they have a right to tax us & Legislate for us in all cases what ever"—now certainly if they have a right to take one shilling from us with out our consent, they have a right to all we possess; for it is the birth right of an Englishman not to be taxed without the consent of himself, or Representative . . . America took every method in her power by petitioning +c to remain subject to Brittain, but Brittain (I mean the Brittish King & ministers) did not want colonies of <u>free men</u> they wanted colonies of <u>Slaves,</u> . . . I do asure you the name of an Englishman is as odious to an American as that of a Turk or a Savage. You may depend that the Americans will never submit to be under the Brittons again.[15]

Along with many of his fellow artisans, Revere identified with the Patriot cause from the beginning, joining the Sons of Liberty in 1765 and actively serving the cause of the resistance in various ways. For example, on March 5, 1771, the one-year anniversary of the Boston Massacre, Revere placed large illustrations in the brightly lit windows of his house, such as the ghost of Christopher Seider (an 11-year-old boy killed shortly before the massacre) as well as a depiction of the massacre itself, under the title "Foul Play." This creative and moving memorial attracted thousands of spectators. He also took less creative actions, as when Boston merchant John Rowe recorded in his diary on May 27, 1773 that "Two commissioners were very much abused yesterday when they came out from the Publick dinner at Concert Hall . . . Paul Revere and several others were the principal Actors."[16]

Mystery and innuendo shroud Revere's involvement in Sons of Liberty activities, thanks to the organization's concern for secrecy and anonymity. We can assume that Revere participated in many of the group's activities, such as organized political protests, marches, and discussions over the years. He left a few records of participation in several Patriot actions, such as his name appearing on petitions and attendance lists for dinners or public meetings and his membership in a twenty-five-man volunteer watch over the *Dartmouth,* the ship carrying the Boston Tea Party tea. Various parties attested to Revere's participation and possible leadership in the tea party over the years, reinforced by the text of a popular rallying song at the time:

> Rally Mohawks! bring out your axes,
> And tell King George we'll pay no taxes
> On his foreign tea;
> His threats are vain, and vain to think
> To force our girls and wives to drink
> His vile Bohea!
> Then rally, boys, and hasten on
> To meet our chiefs at the Green Dragon.
> Our Warren's there, and bold Revere,
> With hands to do, and words to cheer,
> For liberty and laws;
> Our country's "braves" and firm defenders
> Shall ne'er be left by true North Enders
> Fighting freedom's cause![17]

All available evidence indicates that Revere both actively and passionately promoted the resistance movement from start to finish.

Revere's societal and intellectual credentials barred him from the highest ranks of Patriot leadership and from groups such as Boston's Committee of Correspondence, but his hard work and networking skills paid dividends and his influence grew. He joined the prominent North Caucus Club no later than 1772 and immediately served on subcommittees assigned a variety of tasks such as communicating and coordinating actions with the South End Caucus, and "to correspond with any Committee chosen in any part of the town, on this occasion; and call this body together at any time they think necessary."[18] The North Caucus Club interacted with similar clubs in other districts of Boston and sponsored policy positions and political candidates before Town Meeting votes, having a huge impact upon the outcome. One roster list for the North Caucus contained sixty members, including a number of prominent or rising merchants, doctors, lawyers, and successful artisans. Revere also received an invitation to join the Long Room Club, a secret seventeen-man society that led many resistance activities. Interestingly, Revere represented the only identifiable "mechanic" in this group of primarily lawyers, doctors, and ministers.[19] Revere also joined and supervised an information-gathering organization consisting largely of artisans, with the intent of observing and reporting on the activities of British soldiers. He later described this association to the Reverend Jeremy Belknap, founder of the Massachusetts Historical Society:

> In the Fall of 1774 and Winter of 1775 I was one of upwards of thirty, cheifly mechanics, who formed our selves in to a Committee for the purpose of watching the Movements of the British Soldiers, and gaining every intelegence of the movements of the Tories. We held our meetings at the Green-Dragon Tavern. We were so carefull that our meetings should be kept Secret; that every time we met, every person swore upon the Bible, that they would not discover any of our transactions, but to Messrs. HANCOCK, ADAMS, Doctors WARREN, CHURCH, and one or two more . . . In the Winter, towards the Spring, we frequently took Turns, two and two, to watch the Soldiers, by patroling the Streets all night.[20]

This narrative again illustrates Revere's penchant for leadership positions among members of the artisan class, as well as his networking and organizational abilities. His responsibilities likely included coordinating different artisan lookouts, establishing chains of communication and contingency plans in case of emergency, interpreting individual observations to discern big-picture conclusions, and conveying important information to Joseph Warren and other Patriot leaders, all of whom happened to be gentlemen.[21]

Revere laid the groundwork for his most celebrated patriotic service between 1773 and 1775, when he made many courier trips on behalf of the Patriot cause, at least five of which took him to New York or Philadelphia to communicate with sympathetic groups in those towns. Revere began his service as a courier in December 1773, carrying news of the Boston Tea Party to New York City and returning to Boston ten days later with news of New York's support. In the years that followed he made further trips but expanded his role, acting almost as an ambassador and interpreter of Boston's Patriot movement rather than as a mere courier. On these visits he carried information that helped Boston, New York, and Philadelphia coordinate their actions, advise one another, and transmit the most pressing news, with the appropriate spin, of course. Revere's loyal and intelligent service brought him into closer contact with Boston's Patriot leadership as well as several members of the Continental Congress meeting in Philadelphia, foreshadowing his upcoming role in the most important ride of them all. By the eve of the Revolution, Revere's

Patriot activities had received notice as far away as London, with his name appearing in angry newspaper articles in 1774 and 1775 that identified him as an "Ambassador from the Committee of Correspondence of Boston to the Congress of Philadelphia."[22] This was of course an overstatement, because he lacked decision-making authority. But in the days of horse and ocean transportation, when a message could take days or weeks to reach its recipient, any organization that hoped to keep up with rapidly unfolding events needed reliable liaisons able to implement open-ended instructions and exercise judgment. Paul Revere perfectly suited this role.

Prior to the Revolutionary War, Revere's resistance activities proved a mixed blessing to his silversmith career. His account books show major gaps that correspond to his courier rides and other periods of intensive resistance activities, but long-term benefits balanced these temporary deficits. Patriotic activities did not yield tangible economic rewards, but indirectly aided him by expanding his network of associates, improving his reputation in the eyes of his fellow rebels, exposing him to new fields and ideas, and increasing his organizational experience. Fellow affiliates of Patriot groups such as the Sons of Liberty and the North End Caucus repeatedly purchased silver from him, as did the members of other groups such as the Loyal Nine. Apart from his famous Liberty Bowl, none of Revere's silver pieces had any particular symbolic or patriotic importance. He used his craft to more directly serve the resistance by engraving and publishing various Patriot propaganda items. Revere had a good deal of experience in the engraving field prior to the resistance movement, as he embellished many of his silver items and also produced copperplate engravings of book plates, trade or advertising cards, psalm tunes, illustrations, and other printed items. Beginning in 1765, Revere turned these talents to the Patriot cause by printing a number of political images intended to rally support against the British. Following the common practice of his day, Revere often borrowed and adapted images from other sources, in one case causing Boston painter and engraver Henry Pelham to accuse him of dishonorably mimicking and distributing Pelham's original drawing before Pelham had a chance to do so. Many of Revere's illustrations, and primarily the Boston Massacre engraving, achieved widespread notice at the time, often accompanying political broadsides or newspaper text describing recent events in a manner favorable to the Patriots.[23]

Revere undertook his Revolutionary efforts, ranging from dramatic resistance activities to politically motivated engravings, to serve what he saw as the

Figure 2.1. "The Bloody Massacre perpetrated in King Street Boston on March 5th 1770 by a party of the 29th Rgt." Engraving with watercolor by Paul Revere, 1770. Library of Congress Prints and Photographs Division, LC-DIG-ppmsca-01657. Paul Revere produced this engraving shortly after the Boston Massacre took place on March 5, 1770, closely borrowing from an original drawing by Henry Pelham. This image served as highly effective Patriot propaganda, inaccurately depicting the British as an orderly line of soldiers commanded to fire into a helpless crowd.

Figure 2.2. "A View of the Obelisk erected under Liberty-Tree in Boston on the Rejoicings for the Repeal of the Stamp Act 1666." Engraving (etching) with watercolor by Paul Revere, 1766. Library of Congress Prints and Photographs Division, LC-DIG-ppmsca-05479 (the Library's impression is a restrike dating from 1849 or later). Paul Revere personally engraved this representation of a Sons of Liberty obelisk, erected on the Boston Common in celebration of the Stamp Act's repeal. After a fire accidentally destroyed the obelisk, this engraving served as the only memory of the obelisk's many patriotic messages.

public good. He also achieved this goal by holding positions on government committees. Many elected government jobs, such as constables or assessors, drew artisans and other members of the middle classes into public service even though they carried large workloads and minimal salaries, while more influential positions, such as selectmen and town clerks, remained within the upper class. In their public service, as in so many other ways, successful artisans served as societal intermediaries, trusted with the implementation of pragmatic policies while finding it difficult to ascend to the pinnacles of political influence. Revere's resume of minor civil service positions supports this general trend. The Boston Town Meeting tapped Revere for important committee service on multiple occasions, including membership in a group that collected and distributed relief payments to help Bostonians impacted by the Port Act; membership in the Committee of 63, a group appointed to enforce local compliance with the Continental Congress's non-importation, non-consumption, and non-exportation decrees; service on the committee that proposed members for the Ways and Means Committee; and others. His committee service increased after the war when it took the place of his Revolutionary activities, and even though he personally did not benefit from government positions or ascend to higher offices he took public service quite seriously. Revere's patriotic duty to his country did not end when Britain exited the battlefield.[24]

Revere's patriotic and government service mirrored the actions of many other successful artisans in large colonial towns. Capitalizing on their position as well-off workingmen with upper-class patronage and rising prospects, these artisans stayed active in town politics and played a key role in organizing independence gatherings and boycotts. These political services increased the unity, organization, and public prestige of the artisan class, leading toward a collective artisan identity and numerous prewar artisan associations. While not a Revolutionary leader, Revere proved his value to the cause by serving as the primary message courier for the Committees of Safety and Correspondence; prominently participating in groups such as the Sons of Liberty, North End Caucus, and Long Room Club; receiving mention in London newspapers; and engraving propaganda.[25] Because Boston relied upon overlapping committees instead of a single centralized command structure, Paul Revere's ability to share information and perspectives between different groups became essential. But all his complex and valuable services, so visible at the time, have faded in the eyes of history due to the vast symbolic importance of his Midnight Ride.

"Listen my children and you shall hear . . ."

On April 17, 1775, Boston became a hotbed of intrigue. Two days earlier General Gage had received orders to confiscate any military supplies held by the Patriots and to arrest anyone suspected of treason. Far from being a secret, Gage's preparations to carry out these instructions drew the attention of the Boston resistance, including Revere's watchful intelligence network, who "expected something serious was to be transacted."[26] Paul Revere rode to Lexington the day before his now-famous Midnight Ride to warn John Hancock and

Samuel Adams that the British might soon attempt to capture them, and either he or a different courier informed the Concord militia that their military stores could be in jeopardy. On his return trip, Revere and his Charlestown contacts initiated an additional measure to ensure that he could get a message out of Boston if trouble arose. British troops could easily lock down the town by blocking both land and sea egress: a guardpost at "Boston Neck" barred anyone from traveling through the one road leading out of Boston, and the warship *Somerset* watched for any boats attempting to row across the Charles River. Revere had a plan to circumvent these impediments if he needed to escape from Boston, but just in case—"for we were aprehensive it would be dificult to Cross the Charles River, or git over Boston neck"—he told a group of Patriots in Charlestown to watch for signal lanterns hung on the Old North Church: one lantern meant the British troops would march through Boston Neck, and two lanterns meant they planned to cross the Charles River.

General Gage tipped his hand on the evening of April 18, when a large force of soldiers marched to the Boston Common while others prepared a number of boats, signaling a Charles River embarkation. At 10:00 p.m., Patriot leader Joseph Warren summoned Paul Revere, his Masonic brother and friend, and asked him to slip past the British and make sure Hancock and Adams left Lexington before the soldiers arrived. And thus it began.

This book opened with an overview of the actual Midnight Ride, starting with Revere's successful escape from Boston by rowboat into Charleston, where he mounted "a very good horse" around 11:00 p.m. and headed to Lexington by way of Medford. In spite of the need to outmaneuver a small British patrol and alert important minutemen at many houses along the way, he made excellent time and reached Lexington around midnight to warn Hancock and Adams of the approaching soldiers, completing the first phase of his mission. Soon after that he met up with William Dawes, a tanner whom Joseph Warren had also dispatched on a midnight ride via the slightly longer "land" route out of Boston. The two men headed to Concord to complete their second task, the protection of the military supplies there. Joined by Dr. Samuel Prescott, a third Patriot rider familiar with both the roads and the residents of the Lexington–Concord area, they started on the road to Concord only to be stopped by a large British patrol that captured Revere, while Dawes and Prescott escaped. Of the three only Dr. Prescott arrived in Concord to warn the militia.

Alone and in enemy hands, Revere demonstrated why he held the trust of the Patriot leaders. His own words best describe the interrogation he faced:

one of them, who appeared to have the command, examined me, where I came from, and what my Name Was? I told him. He asked me if I was an express? I answered in the afirmative. He demanded what time I left Boston? I told him; and aded, that their troops had catched aground in passing the River, and that There would be five hundred Americans there in a short time, for I had alarmed the Country all the way up. He imediately rode towards those who stoppd us, when all five of them came down upon a full gallop; one of them, whom I afterwards found to be Major Mitchel, of the 5th Regiment, Clapped his pistol to my head, called me by name, and told me he was going to ask me some questions, and if I did not give him true answers, he would blow my brains out. He then asked me similar questions to those above.

Threatened at gunpoint, Revere calmly bluffed the soldiers into releasing him. He offered detailed knowledge of the soldiers' march from Boston and predicted that British reinforcements would be slow to arrive, while "five hundred Americans" would momentarily surround them all. In reality, Revere and the British patrol stood dangerously close to Hancock and Adams while far fewer than five hundred Patriots had gathered nearby. The agitated soldiers fell for his ruse, released Revere near the center of Lexington after taking his horse, and, in their own words, "galloped for their lives."[27] As a final coda to his long night's work, Revere helped John Lowell, Hancock's clerk, carry a heavy trunk of important Patriot papers out of the tavern into the woods just moments in advance of the soldiers' arrival. Revere lugged the chest right through the hastily formed lines of the Lexington militia, barely making it into the woods a mere hundred or so yards from the action when he heard the firing of a single pistol, then two more guns, and finally the roar of the many muskets that had just begun the Revolutionary War.

Revere had no way of knowing on the 19th of April in 1775 that he had already completed the activity that would immortalize him and outshine the rest of his life's work. In the years that followed, countless observers, Patriots, and historians reinterpreted the Midnight Ride and ensuing militia actions in Lexington and Concord in different ways. For example, Revere's compatriots did not approve of his narrative of the ride because of the way he planned certain measures in advance: the Patriot leaders preferred an image of American innocence, in which the British soldiers' secretive and provocative actions were only overcome via swift and virtuous American responses. Patriots also kept the names of individual actors secret, which explains why Revere requested in vain that his name not be included in printed versions of his 1798 retelling of the Midnight Ride. Revere remained silent on the matter through the end of his life, but myths of the founders and heroic inflations of the Revolutionary struggle had already taken hold of Revere's story by the turn of the nineteenth century, and even though he did not become famous outside New England until Longfellow wrote his poem, he had a reputation as a local hero as early as 1795.[28]

But back in 1775, Revere had pressing matters on his mind. The seven hundred British regulars had begun surrounding the seventy or so militia upon the Lexington Green when, in spite of orders on both sides to hold all fire, shooting quickly escalated into a confusing melee that killed eight militia and wounded nine. The British regulars marched on to Concord, where they encountered hundreds of American militia who grew ever more numerous as new companies joined their ranks from distant towns. Recognizing that time was not on their side, the regulars wisely decided to return to Boston before matters grew worse. Even though the British soldiers received support from a relief force sent from Boston, thousands of Patriots arrived throughout the next day and turned the British march to Boston into an exhausting and deadly rout.[29] By April 20, more than twenty thousand militia had surrounded the town of Boston, trapping General Gage, his army, and many Loyalists and other citizens on the inside, and Paul Revere on the outside.

Following his Midnight Ride, Revere searched for the best way to aid the Patriot cause and support his family. He took a forced vacation from silver-working for the next five years: he had no way of returning to his shop while the British occupied Boston, and besides, his skills were needed elsewhere. Revere ambitiously undertook an array of new activities that included militia leadership, engraving for the Commonwealth of Massachusetts, designing a gunpowder factory, and cannon casting. Revere moved in many directions at once, often switching between two or more tasks while planning his next opportunity, but for the sake of clarity we will untangle the threads of his Revolutionary activities and study them one at a time.

After the Ride: Martial Longings and the Pursuit of Honor

Revere began a period of exile from British-occupied Boston the morning after his Midnight Ride, lasting until the British soldiers finally left the city in

March 1776. The rest of his family joined him in a Watertown boarding house by mid-May, with the exception of 15-year-old Paul Jr., whom Revere asked to remain behind in Boston to look after their property. Revere's instructions to his teenage son blended affection and pragmatism: "My Son, It is now in your power to be serviceable to me, your Mother, and yourself. I beg you will keep yourself at home or where your Mother sends you. Don't you come away till I send you word. When you bring anything to the ferry tell them it is mine & mark it with my name. Your loving father, P.R."[30] This was not the last time Revere, and the entire family, leaned upon one of its younger members. Paul Jr. took on weighty responsibilities from an early age and soon stood at his father's side as co-owner, and eventually the solitary owner, of the silver shop.

Beginning on April 20, 1775, Revere earned much-needed money by continuing to serve as a courier for the Committee of Safety under the direct command of Joseph Warren. Warren exited the town of Boston on April 19 and immediately sprung into his leadership role by coordinating the actions of the many independent militia companies encamped around Boston, directing ongoing Patriot propaganda activities, and staying in touch with the Continental Congress and other groups. Revere's services as a resourceful and experienced courier offered great benefits to the overworked Warren.[31]

Revere lobbied for an artillery officer appointment shortly after the start of the siege of Boston, again showcasing his high aspirations. He certainly could have enlisted without delay as a soldier, but officers carried more authority, particularly the artillery officers whose specialized knowledge and control over the military's most advanced technology added extra prestige. Becoming an officer was easier said than done, as the military recruiting system did not follow merit-based guidelines. Officers overwhelmingly came from the upper classes of society: for example, 84 percent of New Jersey Continental officers originated in the richest one-third of the population and none came from the poorest one-third. Across America, nearly all officers were highly respected members of their community, if not leaders or the sons of leaders. George Washington later made this hiring tendency even more explicit when he told officer recruiters to "Take none but gentlemen." Revere possibly felt his many patriotic services and success as a master artisan earned him a leadership position, or might have counted upon his French and Indian War artillery experience to prove his military potential. But in spite of America's shortage of skilled artillery experts, Revere did not receive an officer's commission and moved on to other endeavors.[32]

Revere tried a different approach a year later. In March 1776, shortly after George Washington placed a number of artillery pieces on Dorchester Heights within firing range of Boston, the British troops damaged any military supplies that they could not carry and sailed off, leaving the town in the hands of the Patriots once more. The main theater of the war soon shifted to New York, but the Patriots understood they could not leave a town of Boston's importance undefended. Paul Revere received an appointment on April 10, 1776, to join the Massachusetts militia's first regiment with the officer's rank of major. Surviving records do not indicate whether or how Revere might have lobbied to receive this appointment, but he did have some useful local connections and his prior militia experience related to this appointment. A month later he transferred to the artillery regiment, a move that certainly reflected his interests and technical skills. He must have served well in this position because he received a promotion to the next highest rank of lieutenant colonel on November 27, 1776, a bit more than seven months after first joining the militia.[33]

Revere exercised real authority during his early service; following his promotion to lieutenant colonel he usually stood third in command behind Colonel Thomas Crafts and Major General William Heath. On several occasions he acted as the commander of Castle William, and on at least one occasion in March 1778 Major General Heath put him in charge of Hull, Long, and Governor's Islands as well. He also presided over many court-martial proceedings and issued a number of written orders to the officers and soldiers under his command, typically dealing with mundane issues such as maintenance, the granting of leaves of absence, punishments against deserters, and procurement of supplies.[34]

Revere's memoranda books reveal some of the effort he put into his command. In using and maintaining a variety of armaments, gunpowder, and ammunition that might have originated in France, Britain, or inexperienced colonial foundries, he had to improvise solutions to unpredictable problems and pass on these insights to his men. Revere's books reveal a painstaking attention to technical details, including directions for making gunpowder, signal rockets, fireworks, smoke bombs, fuses, and charcoal. He also recorded his ideas on the optimal use of many of these items, such as the mechanics of firing behind parapets and the use of delayed detonation shells. And he drew upon his manufacturing skills on at least one occasion when he engraved, and almost certainly constructed in their entirety, a pair of calipers for his own use. These calipers enabled him to measure the caliber of a cannon bore as well as the diameter of iron cannonballs, to ensure a tight fit that would not jam under fire. Without any significant military education, Revere had the sensibilities and methodical approach of a military engineer, and acquired an impressive quantity of technical knowledge in the process.[35]

Repetitive and quotidian militia responsibilities do not offer inspiration, and the situation grew more depressing when the primary theater of war moved to the New York area in the summer of 1776, leaving the real action to the Continental army. Revere revealed his disappointment in an April 5, 1777 letter to his friend John Lamb, commenting, "I did expect before this to have been in the Continental Army, but do assure you, I have never been taken notice off, by those whom I thought my friends, am obliged to be contented in this State's service." Shortly after that, he and Colonel Crafts, both chafing at their militia service, asked Samuel Adams to help the officers in their militia regiment receive rank privileges equal to officers in the Continental army. Samuel and John Adams discussed the request with their colleagues in the Second Continental Congress and had no choice but to reject it: state militia officers could not be considered equivalent to officers in the American army. At the same time Revere undertook several minor positions in Boston's restored town government, including repeated terms as fire ward; service on the Committee of Correspondence, Safety, and Inspection; and membership in various official and unofficial groups attempting to locate and punish Loyalists.[36] None of these positions proved particularly satisfying or noteworthy, accounting for Revere's continuing restlessness and search for new work.

Revere finally had an opportunity to lead soldiers on an expedition in 1777, when his regiment marched to Newport, Rhode Island, but they failed to see any action. They returned in 1778, hoping this time to engage and rout the British forces. Revere, enjoying the high point of his military career, demonstrated his great patriotism and familial affection in a letter to Rachel in August 1778: "It is very irksome to be separated from her whom I so tenderly love, and from my little Lambs, but were I at home I should want to be here. It seems as if half Boston was here. I hope the affair will soon be settled, I think it will not be long first. I trust that Allwise being who has protected me will still protect me, and send me safely to the Arms of her whom it is my greatest happiness to call my own."[37] But the 1778 expedition again ended when the American forces returned to Boston without engaging the enemy, due to British reinforcements, naval support, and American militia unreliability. Monotonous daily routines and aborted excursions only seemed to confirm the pointlessness of militia service. Little did Revere know, life would soon be far worse.

The following year Revere received the opportunity to command the artillery train in a major assault against a British fort located in Penobscot Bay, Maine. In late July 1779 Massachusetts launched a huge force of armed privateers, warships, marines, Indian allies, militia, and artillery, easily the largest American naval expedition of the Revolutionary War. The expedition misfired almost from the beginning. After some of the ground forces landed and secured their positions in opposition to the British fort, the commanders of the American land and sea forces began arguing over the chain of command, nervously avoided any major actions, and continually fortified their positions in lieu of attacking. After squandering the element of surprise, the long-delayed American assault finally took place on August 13, but soon after the attack began a British naval squadron consisting of seven warships appeared in the bay. The Americans hastily called off their attack and commenced a disorganized retreat that rapidly degenerated into an all-out rout, culminating in the loss of all American vessels and most of the equipment, followed by a humiliating march home. The Penobscot expedition was a fiasco from start to finish.[38]

The Massachusetts council relieved Revere of his militia command shortly after he returned to Boston and the board of inquiry censured him for contributing to the disaster, even placing him briefly under house arrest. Revere later contended that these attacks and charges against him originated from several other officers who had personal grudges against him. The hint of a scandal devastated Revere's aspirations: not only could he never achieve a higher military rank, but a dishonorable discharge would cast doubt upon his honor and prevent him from ever entering the ranks of the gentry. He tirelessly petitioned the Massachusetts legislature to convene a court-martial to rule on the charges against him and mounted a vigorous defense in several hearings. But to his horror the ambiguous first verdict condemned or exonerated other officers without mentioning Revere at all, and a more specific second ruling dismissed many of the charges against Revere but still found him guilty of two charges. Revere continued pressing the General Court for a formal court-martial to reassess these charges, and vindication finally took place in 1782, when the court formally rebuked all the charges against him, stating that Revere should be treated "with equal Honor as the other Officers in the same Expedition."[39] By this point the war and his military career had long since ended. Revere obviously considered his military reputation essential, as illustrated by the ferocity with which he attempted to defend his record as well as the fact that he sporadically referred to himself using his military title for the remainder of his life.

Of all Revere's experiences during the war, the premature termination of his military career became his greatest defeat, effectively closing off the one opportunity he had to advance himself through non-technological means. Unlike his aborted military career, his technological work took place in an environment that used a more transparent merit-based system to allocate responsibility. Revere continued offering his technical expertise to the Patriot cause and became quite successful in three productive endeavors, even though the artisan who wanted to be a gentleman found them far less glamorous than military command.

Mechanic for the Revolution: Engraving, Mill Design, and Cannon Casting

The Patriot establishment, if not Revere himself, deemed manufacturing and managerial experience nearly as important as military leadership because of America's scarcity of technical versatility. During the war, America suffered from a serious shortage of many manufactured goods previously imported from Great Britain, and prices skyrocketed. American manufacturing took on heightened practical and ideological significance when armament procurement became essential to the war effort. Individuals and local governments constructed iron, steel, and gunpowder plants to meet this new demand, but America's manufacturing efforts faced many obstacles: minimal supplies of raw materials and investment capital, chronic labor shortages exacerbated by military recruitment, and virtually nonexistent production and distribution networks. America's Patriots never reached self-sufficiency, always relying on imports of critical materials such as guns and powder in spite of determined efforts in each colony.[40]

Almost immediately upon resettling in Watertown, Revere worked as an engraver and printer for the Massachusetts Provincial Congress. His prewar printing and engraving work had already given him all the skills needed to cut copper plates to the correct size, smooth their surface, engrave desired text and images, and ink and print copies off them through the use of a press. Revere connected this new work to his prior printing by engraving several currency notes on the back of his earlier copperplates, including his famous

"Boston Massacre" plate. Even though he recycled these plates as a matter of efficiency—high-quality copper was hard to find and the existing plate saved the effort of making a new one—Revere hopefully appreciated the symbolism of this gesture, as the same technological artifact contained Patriot propaganda on one side and Revolutionary financial instruments on the other.

For his earliest work he used large plates to print notes for Massachusetts's bills of credit. In May 1775 the Provincial Congress decided to borrow 100,000 pounds by printing and selling a large number of colony notes to individuals, who could redeem them in June 1777 with 6 percent interest added for each year the bearer held the note. Loan and currency notes typically featured an elaborate title, one or more images such as a seal, standard text explaining the terms of the note, and blanks to be filled in with specifics such as the due date, note number, loan amount, and name of the recipient. These early notes had relatively high face values of 4 pounds or more apiece, and a second printing commissioned later in May featured several "soldier's notes" holding values between 4 and 20 shillings. The importance of these notes became evident on June 3, when the Provincial Congress asked Revere to print the soldier's notes overnight if at all possible, making two officers available to help him night and day until they could be distributed to the militia. The soldier's notes served an interesting role: in exchange for their services, each soldier chose whether to receive 20 shillings of salary immediately, or 40 shillings of currency notes payable in one year. The creation of these notes kept the Patriots afloat: they depended upon loans and paper currency for the payment of all of the expenses of waging war and running a government. Following the initial 100,000 pounds of loan notes and soldier's notes totaling 30,000 pounds, Massachusetts placed orders to print additional bills of credit to serve as currency, totaling 100,000 pounds (released in August 1775) and 75,000 pounds (released in December 1775).[41]

Revere's relationship with the Massachusetts interim government for these engraving contracts foreshadowed the rewards and headaches that accompanied his career-long transactions with government officials. He immediately discovered that government engraving provided steady and lucrative commissions: Massachusetts paid him more than 240 pounds in 1775 and more than 81 pounds in January 1776 for printing 100,000 extra bills of credit. His engraving work continued at least to the end of 1776, when the government asked him to replace the word *Colony* with *State* on new notes. But even this profitable employment could not compete with his desire to serve in the army, and Revere focused exclusively upon military matters from 1776 until November 1778, when he began a final engraving project that earned him more than 700 pounds. This early government work also carried occasional downsides. Revere received one embarrassing rebuke when the Provincial Congress refused to pay his bill for 64 pounds and instead paid what they considered the fair price of 50 pounds for his services. He also had to face the security concerns of the supervisory committee, which asked Revere to guard his press at all times, and which promptly relieved him of all plates as soon as he printed each set of notes. These interactions reflected reasonable concerns of frugality and security, but did infringe on Revere's methods of running a business.[42]

Revere put aside his engraving and printing work several times when other pressing concerns arose. The Continental Congress asked him to travel to Philadelphia in November 1775 to learn how to manufacture gunpowder, as mentioned at the start of this chapter. The Patriot leaders believed that creating additional gunpowder mills in New England represented an absolute necessity for their cause. In December 1774, well before the shot heard round the world, the Massachusetts Provincial Congress appointed a committee to look into the state of manufactures, and their report indicated "That gunpowder is also an article of such importance, that every man among us who loves his country, must wish the establishment of manufactories for that purpose; and as there are the ruins of several powder mills, and sundry persons among us who are acquainted with that business, we do heartily recommend its encouragement by repairing one or more of said mills, or erecting others, and renewing said business as soon as possible."[43] In the first months of hostilities, Patriot forces in different colonies scraped together all the gunpowder they could find, much of which came from captured British stockpiles or stores remaining from the Seven Years' War. The Continental army almost completely exhausted this initial supply, estimated at eighty thousand pounds of powder, within a few months. The shortage reached crisis proportions in early engagements such as the Battle of Bunker Hill, which the Patriots might have won if they had had access to more gunpowder. As George Washington wrote on Christmas Day of 1775, "Our want of powder is inconceivable. A daily waste and no supply administers a gloomy prospect." The Second Continental Congress discussed this emergency, but decided that individual colonies should take the lead in manufacturing large quantities of gunpowder. Some colonial governments took the hint, offering financial support and bounties to manufacturers willing to learn the process.[44]

In November 1775 the Massachusetts Provisional Congress contacted Revere and asked him, while in Philadelphia on his next courier run, to study a gunpowder mill and learn how to reproduce and run one in Massachusetts. The specific mill, run by Oswell Eve, was probably the first operational gunpowder manufactory in Pennsylvania, certainly one of the first in America.[45] His specific instructions outlined several of the established steps in any successful technology transfer: "Obtain an Exact plan of the best Constructed powder mill, the Quantity of powder that may be made in One day in said mill, the Expence of the powder mill, & Whether a person Can be Obtained who is well skilled in manufacturing powder, and the Expense of said man p^r ann°."[46] The Provincial Congress had to quickly meet the gunpowder shortfall while also setting up a sustainable operation that would not unduly strain the government's meager finances. The emphasis upon hiring someone "well skilled in manufacturing powder" also reflects the preeminent method of technology transfer at the time, that is, the emigration of skilled labor. Paul Revere's involvement, on the other hand, signaled a different strategy in which an interested individual visited the state-of-the-art facility and learned its operations well enough to reproduce them. Successful technology transfer depended upon two factors: a quick technical learner and an environment fostering the free exchange of knowledge. In this case the Provincial Congress had to settle for one out of two.

Revere had practical, social, and technical qualifications for this mission. His impending visit to Philadelphia on a courier mission may have proved the overriding factor due to the sheer convenience of the trip. But the Provincial Congress could not entrust this job to anyone who just happened to be passing through Philadelphia; it needed someone trustworthy, capable, and technically proficient enough to do the job right. Gunpowder manufacture carried notorious risks, and the government had to ensure that this powder mill followed the proper guidelines in order to prevent injuries or deaths. The qualities that made Revere a successful courier and silversmith overlapped nicely in this case because he could use discretion, act with some diplomacy, and relate his existing technical knowledge to this new task.

Of these skills, Revere's technical acuity made by far the largest contribution to his success. In spite of a letter of access intended to gain him passage through the plant, Oswell Eve only allowed him a quick walking tour of the building due to his suspicion and competitiveness. But once Samuel Adams passed along some gunpowder mill plans, Revere's experience and intuitive grasp of different production processes enabled him to use this limited information to help design a powder mill in Canton. As a silversmith he understood heat treatments, the operation and combination of certain chemicals, and other processes that, while not directly tied to gunpowder manufacture, related to the general activities and pace of a shop floor. He also worked on two engraving projects before the war that applied more closely to the task at hand. He engraved the first, *Plan of a Hemp Mill*, in 1765 on behalf of the Massachusetts House of Representatives to accompany a published "Treatise of Hemp-Husbandry." This engraving depicts a scaled blueprint of a manufactory, including front, side, and overhead views of water-powered machinery. At a minimum, the engraving illustrates Revere's familiarity with mill layouts and the use of blueprints. He also engraved a plate in 1774 titled *The Method of Refining Salt-Petre* for the August issue of the *Royal American Magazine*. This illustration portrays two men employing sieves, skimmers, ladles, vats, and a hearth to boil down and filter a nitrate-rich material, remove impurities, and produce a more concentrated form of nitre that eventually yielded saltpetre (potassium nitrate), the critical component of gunpowder. While Revere certainly did not claim expertise in this area, the fact that he produced this engraving less than two years prior to his mission illustrates some familiarity with the principles at work. He returned to Massachusetts and produced his designs, culminating in a successful gunpowder mill in Stoughton that began operating by May 1776.[47]

In addition to his engraving and gunpowder-factory-designing services, Revere soon served his Commonwealth by working in a cannon forge. Americans' almost total lack of experience in cannon casting in 1775 posed a huge problem for the rebels, who captured some cannon but needed to learn to cast their own as soon as possible. The Massachusetts Board of War decided to set up and operate a cannon-casting furnace at Titicut (Bridgewater, Connecticut) in December 1776, and asked newly arrived French engineer Louis de Maresquelle, also known as Lewis Ansart, to serve as state superintendent of furnaces with aid from Colonel Hugh Orr. The Board of War immediately asked them to seek Revere's advice concerning the best manner of brass cannon casting, in one case even using the phrase "under the Direction of Col°. Revere."[48] This request seems peculiar, considering that Revere had not previously cast any large copper objects, although he could draw upon his artillery company experience and his knowledge of silver casting. In contrast, Maresquelle came from a proud tradition of French cannon casting and could be considered one of the finest metalworkers on the American continent. Revere's silverworking experience may or may not have aided the experimentation process, but he certainly helped cast and test four cannon by March, and he undoubtedly learned much about working with metals other than silver on a large scale. For example, he had the firsthand opportunity to witness Maresquelle's innovative technique of casting the cannon as one solid piece and then boring a chamber in the center, a method that eventually became the industry standard. Revere successfully duplicated the cannon-casting process in his own forge seventeen years later.

In addition to the specific skills Revere contributed to the Patriot cause, his repeated technical services illustrated his great ability to learn new procedures. Others clearly recognized his technological competence, since Revere's name always seemed to come up whenever the Massachusetts Patriots quickly needed to learn or implement a cutting-edge manufacturing process.

Following Gage's exit from Boston, New England's role in the Revolutionary War became more marginal. The main theater of war first shifted to New York and the Atlantic states and then to the South, where Washington had his final victory over Cornwallis at the Battle of Yorktown in 1781 thanks in no small part to French assistance. Hostilities more or less ceased after that point, although the war technically continued until the Treaty of Paris laid out the terms of peace between America and Britain in 1783.

The Revolutionary War produced economic impacts on a scale impossible to imagine at the time. Lasting for eight years, the war put at least one hundred thousand men—close to 10 percent of the male population—into military service, and destroyed or dislocated large amounts of property. The war overturned all preexisting market patterns: armies and navies effectively acted as a protective tariff by temporarily removing most British competition, state and federal governments purchased enormous quantities of goods, and ongoing shortages of weapons or other products focused attention upon America's manufacturing shortfalls. The money supply also changed as a result of the war: state and federal governments issued more than $400 million of paper money to pay expenses, total quantities of gold and silver coin also increased due to purchases by foreign troops, and changing local conditions led to variable price increases and a pronounced inflationary spiral. These impacts touched all producers and consumers in America, providing a massive boost to the advance of capitalism until the war ended. In addition, as many as eighty thousand Loyalists, who generally occupied the highest rungs of the social and economic ladder, left America during the Revolution. The removal of a large segment of well-to-do citizens left a void in the economy and shattered many existing social networks, leaving openings for some of the newly rising entrepreneurs who improved their prospects during the war.[49] Most observers saw postwar America as a brave new world, even if few of them could explain the full depth of these changes.

The war also produced vast ideological impacts that, while less tangible than the economic changes, proved no less important. Following years of resistance and Revolutionary rhetoric, Americans had no shortage of ills they could blame upon Britain's overly restrictive policies, including unjust taxation, harsh legal traditions, discriminatory trade policies, suppressive actions against manufacturers, and many others. Victorious Revolutionaries equated their political victory with a newfound economic and social freedom. Americans often cited their new country's numerous real or imagined gifts, such as its seemingly boundless natural resources or the perceived virtue of its people, as heralds of impending national greatness. Even though many Loyalists still lived in America, the war gave rise to an American identity, with nationalist sentiments beginning to appear in art and literature as citizens reflected upon the triumphant shared endeavor they had so recently undertaken. The excited writers of the postwar republic produced an outpouring of optimistic predictions, and many Americans responded to this rhetoric by directing their energies into new political and entrepreneurial ventures.[50]

And what of Revere? His selection as a midnight rider and his success at that task resulted from the combination of his personal connections, overall competence, and position in society, factors that also explain his success or failure in many other endeavors during and after the Revolutionary years. Personal connections proved invaluable to Revere throughout his career, and Joseph Warren's position as one of Boston's highest-ranking Patriot leaders may have had the greatest influence upon his selection as a midnight rider. Revere's reputation improved his interactions with peers and clients, but patrons or friends in high places created the most dramatic opportunities. Warren knew Revere through numerous social and patriotic activities and even considered him a friend, someone he could summon and trust in a moment of need. While powerful connections opened doors, Revere had to depend on his own skills to succeed. A responsible professional with good judgment and a sharp intellect, he overcame all challenges in the rush of events on the 18th of

April, beginning with a flawless escape from Boston and ending with a shrewd and courageous bluff delivered at gunpoint. And finally, his success also drew upon his place in society and his ability to connect to fellow Patriots from the top to the bottom of the hierarchy. While Revere could not lay claim to being a "leader" in any official sense, the effectiveness of his midnight ride shows that he functioned well as a de facto field commander: he knew how working men operated and they listened to him. He also appealed to gentleman leaders such as Hancock and Adams who appreciated the judgment and responsibility demonstrated in his successful craft practices, shop management, and community service. Trusted by those above him and respected by fellow artisans and the working strata of society, Revere represented the perfect intermediary who strengthened the Patriot network by carrying messages, exercising limited authority, and rallying support. The networking abilities that served him so well in his search for customers came into play during the Midnight Ride, as he knew which people to wake and which houses to visit, lighting sparks in just the right way to ignite a firestorm of popular resistance.

At the end of the war he had mixed feelings about his experiences. The Midnight Ride placed him briefly in the center of events, but soon became eclipsed by a frustrating and eventually disgraceful militia career. Looking back, he could juxtapose military failures with his admittedly less glamorous technical work, accomplished by using his ingenuity and practical skills in ways that others, particularly the leaders of society, could not. Tension between his hopes for a high-profile societal leadership position and the reality of a pragmatic manufacturing career played out in the ensuing years. Revere's attempts to reconcile productive merit-based technical work and patronage-influenced political, military, and social hierarchies foreshadowed larger questions that soon concerned artisans across the nation. As social and economic intermediaries, artisans were among the first to chart pathways through untested waters.

Revere faced difficult and intriguing choices in 1783. He could return to his trade, but so much had changed since he had last practiced his silverworking that everything from his town's political structure to its physical landscape looked different. Such society-wide upheavals did not occur every day, and a sufficiently entrepreneurial approach might yield exciting rewards. Perhaps it was time to aspire to something more.

CHAPTER THREE

Mercantile Ambitions and a New Look at Silver (1783–1789)

In October 1781 Revere penned a letter to his cousin John Rivoire in France. As their correspondence had only recently begun, he brought his cousin up to date on the details of his life, in the process offering what might be the most revealing self-analysis contained in the entire Revere Family Papers. He began:

> My father was a Gold-smith. He died in the year 1754, he left no Estate but he left a good name and seven children, 3 sons & 4 daughters. I was the Eldest Son. I learned the trade of him and have carried on the business ever since till the year 1775 when the American Revolution began. From that time till May 1780 I have been in the Goverments service as the Lieu't Col of an Artillery Regt, the time for which that was raised, then expired, I thought it best to go to my business again, which I now carry on, besides which, I trade some to Holland.[1]

For a brief statement, this passage reveals much about Revere's priorities. Following a description of his father and his silverworking origins, we now find him at a crossroads, as he had just returned to his old profession after a five-year absence during the Revolutionary War. Much had changed in these five years, for his country as well as for himself. This letter made no mention of the Midnight Ride that ensured his everlasting fame, or the many technical endeavors he undertook, or the promise of life apart from British rule. Instead, Revere distilled his entire Revolutionary service into an allusion to his rank as an artillery officer. Unsurprisingly, the letter glossed over the details of his less than glamorous discharge and court-martial still in progress. And at the end, he "thought it best" to return to silverworking when other opportunities ended—hardly an enthusiastic return to an interrupted profession.

The letter continued:

> I did intend to have gone wholly into trade, but the principal part of my Interest, I lent to Government, which I have not been able to draw out, so must content my self, till I can do better. I am in middling circumstances and very well off for a Tradesman. I am forty seven years old. I have a wife + 8 children alive, my Eldest daughter is married, my oldest son since has learned my trade. Since we left the army [he] is now in business for him self. I have one brother + two sisters alive.[2]

The description ended on a positive note. His large family, his oldest daughter's marriage, and his son's success all pleased Revere and secured his identity as a traditional patriarch; he was especially happy that his son could continue his father's legacy independently. However, the happy tone of this passage is undermined at several points. Two uses of the word *alive,* even intended as positive statements, allude to the constant presence of death in the early American household. By 1781, Revere and three of his siblings were the only survivors of his father's twelve children and he had already lost four of his own twelve offspring: could he see the future he would sadly learn that only five of his sixteen "little Lambs" would outlive him. He also, perhaps unintentionally, glossed over the passing of his first wife by referring to Rachel as simply "a wife"—in this case, his second.

Prior to the discussion of his family, Revere offered a telling reflection upon his own place in postcolonial American society. His silverworking activities, apparently, were not his occupation of choice, because he really wanted to go "wholly" into trade. Because he had tied up many of his funds in government bonds and other securities he had to return to his earlier craftwork, at least until he could "do better"—clearly not a ringing endorsement of his artisan heritage. Revere further qualified his success as a silversmith by claiming he was well off "for a Tradesman," revealing the awareness that, in comparison with other professionals, he had not even begun to approach the summit of the social pyramid. He described his circumstances as "middling," an accurate though unflattering assessment. Revere's tepid assessment of his artisan status reminds us of his underlying ambition: he would never accept a skilled laborer's fixed status in the middle of the social hierarchy. In the new America, he could be so much more.

America's post-Revolutionary period featured virtually unrivaled social, economic, and political changes, and Revere longed to take advantage of these newfound opportunities. The new economic landscape created a tremendous upheaval: abundant quantities of dubious new currency, speculation in land and imported goods, and swings in the credit market enabled vast fortunes to be made or lost. The removal of the political restraints that the Patriots' propaganda had lambasted for so many years produced a new feeling of liberation that inspired new aspirations and endeavors. And the alteration of traditional privileges and institutions yielded a complex, society-wide reconfiguration of individual status, property distribution, and financial opportunities, one that many ambitious former colonists like Revere sought to exploit. Class divisions and societal attitudes toward elitism and labor changed after the Revolution. Social mobility for many white, male, and usually "middling" or better workers increased for a period following the Revolution in part because of new job opportunities, more specialization in many fields, growing geographical mobility, urbanization, and industrialization. Social mobility had its limits, of course, and anyone attempting to cross into the uppermost tiers of society encountered all sorts of impediments. While certain individuals had more geographic and economic mobility after the Revolution, this mobility had a darker side, evident whenever workers lost their jobs due to economic downturns or changes in their trade. Society, and especially urban society, became more stratified at the end of the eighteenth century, with ever-widening divisions between the visible super-wealthy minority and a growing population living at the poverty level. In keeping with Revolutionary rhetoric, artisans, businessmen, and "common" people gradually began to distrust and even condemn some of the trappings and attitudes of the gentry, while paradoxically going to great length to duplicate the upper class's material comforts. Samuel Adams lamented the erosion of social deference as well as the excessive consumption of luxuries in 1785 when he wrote, "You would be surprizd to see the Equipage, the Furniture and expensive Living of too many, the Pride and Vanity of Dress which pervades thro every Class, confounding every Distinction between the Poor and the Rich." As a result of these new attitudes, success through ambition, talent, and hard work became the birthright of a new generation of Americans. Hereditary social classes started giving way to a class system theoretically based on economic divisions, justified in the sense that society's leaders should be the most gifted and successful, with wealth serving as the clearest gauge of one's merit and status.[3]

To say that Paul Revere subscribed to these new sentiments would be an understatement, since nearly every aspect of his life embodied the rising expectations of the most dynamic members of society. He believed in the need for hierarchical leadership but wanted to select leaders according to their integrity and effectiveness, as opposed to their heritage. As his early story has illustrated, Revere's identity as a master craftsman placed him in the company of ambitious and independent men hoping to advance their social and economic position while serving the needs of their country. He joined other groups such as St. Andrew's Masonic Lodge and the militia in the continued move toward combining service and status.[4] By 1781, he longed for even larger changes.

After drying the ink on his cousin's letter, Revere returned to his daily shop routine, perhaps without consciously realizing how much his attitude about his role as a tradesman had changed since his pre-Revolutionary days. Up to this point silverworking provided a badge that connoted skilled status and qualified him for leadership positions among working men. Yet at the culmination of the Revolution this badge started to lose its luster. Since the field of silverworking had not changed in this short time, the differences lay in Revere himself. Ever since he put aside his tools to begin the Midnight Ride he had held lesser leadership positions, led his lodge, carried and interpreted intelligence reports, commanded men in battle, and trafficked with the leaders of society. He had gone as far as any practicing artisan could hope—farther, in fact, than almost all of them. If being an artisan no longer helped him rise, perhaps he needed to become something different.

Quest for Gentility: The Would-be Merchant

Reacting to the possibility of postwar social and economic advancement, Revere attempted to establish himself as a merchant and become a gentleman, a social rank that still held a huge appeal. To succeed as a merchant he needed to greatly expand the scale of his business and gain a broader awareness of consumer trends, while shifting his personal role from production to sales. He soon learned that mercantile activities invited stress and risk under the best of conditions, let alone the perilous postwar economy.

The growth of a colonial consumer society coincided with the appearance of layers of middlemen eager to satisfy the full range of postwar material needs. Buying and selling became a way of life for many individuals who took on different roles. Merchants sat atop the heap, as international wholesalers possessing enough capital to speculate in other commodities such as land, and they might sell to customers and other traders as well. Shopkeepers occupied the next level of the trade hierarchy, serving as general retailers who bought from merchants and sold to local residents. Subcategories of shopkeepers, such as grocers, tobacconists, and ironmongers, answered more specialized demands, while traveling salesmen, often known as hucksters and peddlers, carried their goods between cities and rural areas. As consumption increased, shopping districts with growing expanses of store windows and new retail shops arose in larger towns and cities. Many artisans doubled as specialized shopkeepers in the daily process of selling their wares, accruing various goods as barter payments for their own products and reselling them to other customers.[5]

Recognizing the potential for profit, Revere hoped to make retailing his primary occupation. He had some experience with overseas orders: as a silversmith he bought raw materials and other goods from overseas suppliers and sold occasional wares to Boston acquaintances who had moved abroad. On several occasions these dealings defied British law when he imported tools that Britain considered contraband. For example, in 1783 and 1784 Revere imported files, scissors, iron chapes, melting pots, borax, pumice stones, copper scales and weights, iron binding wire, saws, hand vices, hand lathes, shears, awls, and many other types of equipment. He always purchased more tools than he personally used because he sold or bartered many of them to other metalworkers.[6] So it might have seemed a small leap for Revere to deemphasize his work on silver items and thereby spend more time importing and reselling a wider selection of goods.

Revere entered the retail profession on a small scale in the early 1780s, using profits from his silver shop to set up a "hardware store" that sold local and imported goods. In many cases master artisans continued their successful craft shops even after diversifying into other fields such as property speculation or trade. Successful artisans often took on more of a managerial role to minimize the taint of manual labor, but many could not bring themselves to abandon a profitable high-profile craft shop. A steady income from silver sales proved extremely valuable to the many silversmiths—not just Revere—who aspired to lofty merchant status. Between 1783 and 1789, he withdrew hundreds of pounds of cash from the silversmith shop as he experimented with new commodities, and one shop inventory in 1785 reveals 12 pages of in-shop goods with a total value of more than 1,186 pounds. He almost exclusively imported his merchandise from England, including practical items such as wallpaper, pewter tableware, iron kitchen and fireplace implements, looking glasses, paper and related writing items, as well as luxuries such as nutmeg and other spices, raisins, rice, and silk handkerchiefs.[7] An advertisement placed in the December 10, 1785 issue of the *Massachusetts Centinel* portrayed his growing retail business:

> Imported and to be Sold, by Paul Revere, Directly opposite Liberty-Pole, A General assortment of Hardware, consisting of Pewter, Brass, Copper, Ironmongery, Cutlery, Jappaned and Plated Wares, Among which are a few neat Brass Sconces, of one, two, and three branches, 3-pint plated Coffee-Urns, and Goblets, Very neat japanned Tea-Trays, in sets Brass Candlesticks, Looking-Glasses, Patent-Jacks, Carving Knives, &c. &c. &c. All which will be sold at the lowest advance for cash. The Gold and Silver-Smith's business carried on in all its Branches.[8]

Revere's choice of words reveals the continued prominence of the Liberty Pole, a revolutionary icon, as well as the juxtaposition of his new retail trade with his silver shop's ongoing presence. In addition, Revere's comment regarding the "lowest advance for cash" hints at the nationwide shortages of specie that soon played a major role in his financial well-being.

A combination of personal and nationwide shortcomings prevented Revere from making his living as a merchant. Investment capital shortages became a major impediment to mercantile success. He did not have large quantities of specie or other liquid forms of wealth, in part because some of his funds remained tied up in irretrievable government securities for years, so he attempted to compensate by aggressively seeking credit from overseas merchants. Revere had ties with several British merchants and hoped they might loan him some of the value of his purchases until he had time to sell those goods for a profit. Unfortunately his contacts chose not to extend sufficient credit, probably because he lacked collateral and had a workingman's reputation that did not inspire trust in his ability to repay debts. At one point when an overseas purchaser failed to accept Revere's securities for a loan, he sent 340

Spanish dollars to cover the debt and quickly asserted the importance of his honor: "I should think very little of my self if I attempted to deceive any man in matters of this kind."[9] Honorable though this was, cash payments were not a sustainable business practice.

Revere also tried using his silver shop's steady income to guarantee a loan, as illustrated in an earlier letter to the same London purchaser: "You may depend that I shall pay the strictest attention to making you remittance and flatter my self, I shall be able to give you full satisfaction, as my dependance for a living will chiefly depend on the Goldsmith business, which will be carried on by my son, under my inspection."[10] Revere's earlier letter to his cousin mentioned his son's being in "business for him self," and this letter qualifies that description by illustrating the continuing professional relationship between father and son. While Paul Jr. certainly supervised shop operations in his father's absence, Revere also kept a close watch on the operation. More important, silverworking income served as his "dependance for a living" even while he functioned as a merchant, because he still had a large stake in the proceeds of the shop. This is the first indication of Revere's facility with diversification: sons, beginning with Paul Jr., took over some of his established operations while he applied his energy to the newest one, in this case mercantile dealings. Revere had already taken the first step toward a new paradigm of business operation, serving as a manager and owner rather than the chief skilled laborer. Despite steady income from a silverworking shop, Revere experienced repeated lifelong difficulties in securing lines of credit.

In addition to insufficient investment capital, inexperience also hindered Revere's ability to succeed as a merchant. Some of his letters indicate a lack of specific instructions for his London agents: "I could wish that the goods may be fashionable, tho' at the same time I should not prefer the extremes of fashion, as a medium will best answer here . . . Where I have not mentioned the number and peices of articles but the amounts of what I wish to have purchased, I must beg your kind advice; as your aquaintance with the business of this Town will enable you to give better directions than I am able to."[11] One can imagine the difficulty a merchant would have in filling this indecisive request, and as a result Revere did not help his merchandise stand apart from that of the many other local retailers. In his defense, many retailers depended on the purchasing advice of others, since the slow rate of information and merchandise flow across the Atlantic produced huge lags between orders and shipments, in comparison to more rapidly changing fashions and prices. This time lag hurt Revere on many occasions, demonstrated in numerous letters complaining of shipments arriving too late for the fall market, or seasonal fall goods arriving in spring, resulting in "a prodigeous damage to me in my business." The best merchants sent instructions that gave some freedom to trusted local representatives who could then exercise judgment within larger parameters. But many merchants, including Revere, had trouble finding a capable and honest foreign merchant to work for them. In this case family and personal contacts proved vital, and outside his silver shop and local community, Revere stood alone.[12]

The final, and perhaps most important, reason for Revere's mercantile difficulties relates to the nation's oppressive economic climate. The mid-1780s were a terrible time to request a loan or attempt to sell merchandise. The war had interrupted a period of colonial economic growth, and disruptions to commerce forced America back into a less efficient mode of limited trade, inconsistent local markets, and self-sufficient local production. The large military purchases made by the American, British, and French armies partially offset these depressed conditions, but the cessation of military spending at the end of the war deprived local producers of a major source of income, paving the way for the financial doldrums of the 1780s. Currency mismanagement added to the crisis, as the Continental Congress and state governments issued massive quantities of paper money, largely in the form of bills of credit printed by Revere and many others, in order to pay their creditors and soldiers. Governments eventually issued more than $400 million of colonial currency by 1781, and inflation had reduced $167 of these paper dollars to the value of $1 of specie.[13]

After the end of the war, British and other European merchants flocked back to American ports in ships laden with trade goods in order to attempt to recapture their lucrative colonial markets. Americans welcomed this opportunity to correct wartime deficits and raise their standard of living with prestigious high-quality conveniences and luxuries. American merchants and shopkeepers quickly exhausted their supply of specie and continued making large purchases on credit, borrowing from overseas merchants and extending loans to their own customers. More than 75 percent of American purchases relied upon credit, and a "network of debt" blanketed the country.[14] The flood of imports led to a speculative bubble that soon burst. One trader observed in 1783 that New York had imported enough goods to last seven years. Merchants faced with overstocks had to lower prices and eventually call in the loans they had extended to the many retailers who had purchased them on credit, who then called in the loans made to their customers. By 1784, the glut of goods, scarcity of specie, and unredeemable loans took their toll, producing a contraction of credit, bankruptcies, forced auctions, and a depressed economy, described vividly in one eyewitness account:

> The disagreeable state of our commerce has been the effect of extravagant and injudicious importation. During the war, our ports were in a great measure blocked up. Imported articles were scarce and dear; and we felt the disadvantages of a stagnation in business. Extremes frequently introduce one another.
>
> When hostilities ceased, the floodgates of commerce were opened; and an inundation of foreign manufactures overflowed the United States: we seemed to have forgot, that to pay was as necessary in trade as to purchase; and we observed no proportion between our imports, and our produce and other natural means of remittance.
>
> What was the consequence? Those who made any payments made them chiefly in specie; and in that way diminished our circulation. Others made no remittances at all, and thereby injured our credit. This account of what happened between the European merchants and our importers, corresponds exactly with what happened between our importers and the retailers spread over the different parts of the United States.[15]

The depression hit New England the hardest because the region's entire economy depended upon merchants who suffered grievously from the slump. The influx of cheap foreign manufactured goods also drove many artisans to ruin, and by the late 1780s one-third of Philadelphia workingmen needed help from poor relief to support families, a thousand homes lacked occupants, and poorhouses were inundated with requests for aid.[16] Paul Revere therefore received a double dose of economic injury: the glut of goods and collapse of credit crushed his mercantile ambitions, and his artisan shop also felt the pinch when customers cut back their purchases.

Revere's setback had implications for both his bank balance and his societal identity. Incomplete surviving records do not reveal Revere's profits or losses from these merchant activities, though it seems clear that he failed to turn a profit in the early years of the 1780s. The fact that he continued importing and reselling occasional shipments of goods until his retirement shows that he could secure some income through retail sales, particularly when the economic crisis abated. Revere also entertained hopes of joining the prestigious merchant elite and possibly parlaying that into a political career. In the 1780s he started signing his name "Paul Revere, Esquire," a gentleman's title, and referred to himself as a merchant instead of a goldsmith on all official documents.[17] Because America's social classes operated in a flexible manner, no one had the authority or inclination to prevent him from giving himself the title of Esquire or highlighting his merchant identity on legal documents. Regardless of how he chose to sign his name, circumstances informed all observant parties that Revere, in spite of all his hopes and dreams, remained an artisan manufacturer.

As his new official signature suggested, Revere attempted to distance himself from members of the artisan trades although the attempt amounted to little. After the war, artisans in New England formed organizations to promote their interests, though they remained courteous to the merchants who continued to dominate the intertwined political and economic spheres. Boston artisans formed the Association of Tradesmen and Manufacturers in 1785 to successfully lobby for a tariff to protect them against cheap foreign imports.[18] Revere avoided participating in this group while pursuing his merchant aspirations, and also refrained from involving himself in other artisan activities in spite of his earlier leadership role among Boston's tradesmen. He ended his vacation from manufacturing the decade after his merchant career had flopped by returning to join and lead mechanics' organizations. Regardless of his organizational memberships, Revere's politics never wavered: he always agreed with Federalist artisans who supported the integration of domestic manufacturing and commerce.[19]

Paul Revere's political activities reflected his shifting prioritization of mercantile and manufacturing goals. Tellingly, he produced the most tangible political impacts whenever he embraced his artisan roots. The most striking example of his influence among artisans, and indeed, among even wider circles of Boston society, took place during the constitutional ratification process. In response to the many weaknesses of the Articles of Confederation, the delegates of the Philadelphia Constitutional Convention submitted the new Constitution in September 1787 to the Articles of Confederation Congress, which sent it to the thirteen states. The Constitution would take effect, and subsequently dissolve the Articles of Confederation Congress, if nine of the states voted to ratify it. Heated public debates immediately took place in Massachusetts and other states between pro-ratification Federalists and opposition Anti-Federalists. An Anti-Federalist coalition of mechanics led by Benjamin

Austin Jr., a member of a mercantile family who based his artisan standing on the fact that he owned a ropewalk, began exerting its influence in late 1787 to urge rejection of the Constitution. Anti-Federalist activities prompted a larger coalition of Federalist mechanics to support the Constitution, and Revere became one of the three artisan leaders of this action along with baker John Lucas and publisher Benjamin Russell. These three artisans wrote resolutions urging support and ratification of the Constitution, which a group of Boston's "Tradesmen and Mechanicks" unanimously adopted on January 7, 1788, in a mass meeting at the Green Dragon Tavern, their old alehouse from Revolutionary days. The annals of the Massachusetts Charitable Mechanics Association recorded the outcome of this meeting in a poetic form that, while almost certainly fictional, reflects the emotions of the day:

> When the question of the adoption of the Federal Constitution was agitated, there was found to be great opposition to it in many of the States, and the course of Massachusetts was watched with anxiety. It was a time for prompt and decisive action, for her example might turn the scale. The mechanics of Boston sent a large committee of their number, headed by Paul Revere, to urge its adoption upon the convention then in session.
>
> The president, Samuel Adams, asked Mr. Revere, "How many mechanics were present when these resolutions were adopted?"
>
> "More than the Green Dragon could hold," was the reply.
>
> "And where were the rest, Mr. Revere?"
>
> "In the streets, sir."
>
> "And how many were there in the streets?"
>
> "More, sir, than there are stars in the sky."
>
> Their mission was successful.[20]

Whether or not the intervention of Paul Revere and the mechanics had a major impact, Samuel Adams and John Hancock did shift their stance from an Anti-Federalist position to one of qualified support of the new Constitution, and the Massachusetts Constitutional Convention eventually ratified it on February 6 by the slim vote of 187 to 168. The Federalists celebrated this victory several days later with a massive parade that symbolically featured all three branches of the balanced economy: large numbers of foresters, cattle, sowers, reapers, and other farming trades symbolized agriculture; a huge contingent of 250 merchants marched with the model ship *Federal Constitution* to signify commerce; and a column of artisans from forty crafts represented manufacturing. Paul Revere and the other artisan leaders enjoyed a position of honor in a sleigh pulled by four horses. All members of America's society needed to establish their places in the new political, economic, and social order, and Revere and the artisans had to appreciate the results of this early negotiation.[21]

Influence among artisans did not fully compensate for his failure to ascend into the merchant class, but Revere did not lose hope. In the years after ratification he tried his best to secure a federal appointment to a position of influence in spite of a well-meant warning from his friend Congressman Fisher Ames: "The number of expectants, however, will be considerable, and many have merit and powerful patronage." In 1791 he attempted to become the first director of the National Mint, a position of great prestige. Revere did not have the slightest chance of receiving this appointment due to his lack of gentleman status, intellectual credentials, and influential government contacts. Fisher Ames again tried to let him down easy, writing, "However your own ingenuity might qualify you for it, the circumstances will not much encourage the hope of an appointment."[22] Revere tried again, lobbying for an appointment as a federal customs inspector, and again failed to make any progress. In an exact parallel to his earlier attempts to receive an officer's commission in the Continental army, when he failed at the national level he turned to state and local endeavors with far more success. He eventually served the public by holding positions as county coroner, president of the board of health, and a member of the Boston Library Society, Boston Humane Society, and Massachusetts Charitable Fire Society. He continued his leadership in his Masonic lodge, including a three-year term as the grand master of the Massachusetts Grand Lodge, the highest Masonic office in the state.[23]

In light of his inability to enter the merchant class, Revere redirected his efforts into other ventures, a process that eventually ended where it began, in the operations of his silver shop. But Revere had dramatically changed his methods and goals since his prewar silverworking, and new technological adventures lay just around the corner.

Return to Silver: Products, Methods, and the Shift toward Standardization

In spite of his intentions, Revere's involvement in his silver shop expanded, eventually becoming a lengthy continuation of his prewar career. His second silverworking period lasted approximately from 1779 to as late as 1800, occupying him from age 44 into his early sixties. While his first (prewar) silverworking period clearly began with his entry into the business and ended when he devoted his full attention to Patriot activities, this second phase is more nebulous. After producing a small number of silver items in 1779 and 1780, Revere resumed full management of his silver shop in 1781 while dealing with fallout from his inglorious militia exit. Silverworking competed with many other endeavors: he initially divided his efforts between the silver shop and mercantile activities, and added iron and bronze casting to his repertoire in 1788. In spite of these diversions he continued to work personally on some silver objects at least until 1795 and supervised or helped produce many others. Revere's postwar silverworking practices represented his first and most significant step away from artisan traditions into the manufacturing paradigms of the future, and his shifting practices therefore offer a valuable window into the evolution of a proto-industrial mode of production. More specifically, he decided to increase his sales volume by developing more standardized products, adopting new business practices, changing his labor policies, and bringing in new equipment.

Several underlying factors put an end to the economic weaknesses of the mid-1780s and set the stage for eventual growth. By the late 1780s, the market economy had expanded to the point where most people turned the goods they produced into commodities for exchange, producing sweeping changes across most aspects of American life. Farms produced surpluses such as agricultural produce and household manufactures for sale, and used the income to buy manufactures. The expansion of America's market economy during the 1780s and 1790s corresponded with a much greater demand for goods; the extension of transportation networks; increasing numbers of corporations, banks, and factories; larger quantities of money in circulation, typically in the form of bank notes; and bigger and more numerous businesses in all fields. These remarkable changes affected every American citizen.[24]

Ambitious, forward-looking craft masters such as Revere realized they could benefit from America's rising consumerism by emphasizing production quantity, replacing personal trade and barter networks with monetary transactions, and introducing machinery and wage labor to cut production costs. All types of manufacturing expanded in the 1780s and the booming market economy paved the way for industrialization thanks to the division of labor and mechanization, two technological advances that used managerial supervision and discipline to ensure coordination and consistency among the labor force. Mechanization usually preceded the division of labor in labor-scarce America, and machinery increased manufacturing precision while also boosting productivity through the application of alternative power sources such as water or steam. The division of production into smaller subtasks enabled manufacturers to decrease their costs and increase the quantity and consistency of their output, which fostered standardization. As illustrated by Revere's example, some artisans already had extensive experience with these concepts before the Revolution, and further expanded their production in the late 1780s.[25]

Revere greatly expanded his shop's output from 1,145 prewar items made from 1761 through 1775, to 4,210 objects finished between 1779 and 1797 (see Table 3.1).[26] While he made more items in every category of silverware, by far the highest increases and largest overall production of silver pieces occurred in the category of flatware, referring to simple, flat items such as spoons or buckles. Revere made fourteen different types of spoons alone, including standard varieties for teaspoons, tablespoons, and salt spoons as well as rarer versions of marrow, capers, ragout, and dessert spoons. He also entered the field of harness fittings, which primarily consisted of buckles and other flat pieces. Simple, inexpensive flat items could be manufactured quickly and consistently by less experienced workers, and their production benefited from new equipment. Therefore they represented the perfect candidates for standardization.[27]

Revere did not neglect elaborate silver items after the war even though his emphasis shifted to cheaper ones. He produced thirty different categories of hollowware (large and usually cylindrical objects such as teapots, cups, and bowls) items after the war and thirty-four before it, a minor decrease. Similarly, the variety of unusual custom-made hollowware forms decreased as Revere focused his production on a subset of his preferred items. Revere's versatility peaked in the early years of his career, probably because he most desperately needed new work and most eagerly hoped to augment his reputation. In later years, after he had become more established, he less frequently agreed to make atypical objects requiring a substantial time investment, indicating the start of his shift toward the ideals of mass production and standardization.[28] Even though Revere's variety of complex silver products declined, his overall output rose, as illustrated in Table 3.1.

Revere's formidable networking skills contributed to his increased sales and production. His postwar customers included carryover clients from before the war, colleagues from his many organizations and affiliations, a moderate

Table 3.1.   Revere's Silver Production before and after the War

Category	1761–1775		1779–1797	
	Number of Items	Total Output (percent)	Number of Items	Total Output (percent)
Flatware	410	35.8	2069	49.1
Harness Fittings	0	0.0	1044	24.8
Tea and Coffee Wares	61	5.3	198	4.7
Table Wares	129	11.3	177	4.2
Personal Items	449	39.2	623	14.8
Miscellaneous	96	8.4	99	2.4
Total	1,145		4,210	

number of merchants, and a relatively large number of craftsmen who either bought his silver for their own uses or purchased items from Revere in order to resell them to other customers. He also forged new business relationships with people in different trades who might diversify and increase his sales. Beginning in 1787, he formed profitable business relationships as a supplier to four saddle and harness makers. By 1793, he switched to new alliances with a hat maker for whom Revere printed thousands of customized "maker's labels," and with a buckle maker in Newburyport who made buckles for Revere to sell in exchange for Revere's help in acquiring a rolling mill of his own. By broadening his product line, Revere was able to establish relationships that offered him security against the fluctuating post-Revolutionary economy. His postwar records show fewer income lows and fewer gaps in sales than his earlier silverworking, though this partly reflects his improved recordkeeping.[29]

Revere adopted several new administrative practices that aided his shift toward standardized mass output. The clearest example of how he embraced and implemented the ideal of standardization appears in a new advertising technique. In addition to continuing to use print ads, beginning in 1783 Revere produced more than 117 pounds' worth of representative items such as buckles, buttons, and spoons for display in his shop window. This modern technique represented an impressive investment of material and labor in the eighteenth century, underlying Revere's views of his business: hopefully, visible examples of some of his pre-selected styles and most common objects might help create a demand for these items, orienting customers toward the patterns his shop could most efficiently produce. At the same time, Revere adopted newer accounting and payment practices and his daybook entries increasingly portrayed monetary exchanges,. The shift from barter to cash transactions spread across America, reflecting the increasing availability of cash, the desire to regularize payment methods, and the increased reliance on wage laborers who also expected payment in cash rather than bartered items.[30]

In spite of Revere's increased production of relatively standard items, he reached the pinnacle of his artistic success under a new style that gained popularity in America in the decades after independence, combining technical mastery and aesthetic beauty. Following the dominance of the ostentatious rococo style prior to the Revolution, fresh excavations of artifacts at Pompeii and Herculaneum inspired English artisans to develop a new "classical" style of silver that emphasized straight structural lines and unbroken curves. Americans equated this new style with Federalist ideals, and the "federalist" (or federal) style of silver featured straight parallel structural lines, beaded or reeded molding, oval and elliptical shapes, and plain, fluted, and urn-shaped finials. The federal style's emphasis upon simplicity and order supposedly represented a return to scientific principles, which might partially explain its popularity in America during the pragmatic self-improvement era that followed the Revolution. The federal style also might have lent itself to standardization more than earlier, more elaborate silverwork. Revere's own skills seemed to fit federal silver better than earlier types: his methodical, measured approach enabled him to integrate and balance the different forms. Unlike in New York and Philadelphia, where there were a large number of competitive and equally skilled silversmiths, in Boston Revere completely dominated the postwar silver scene and mastered the federal style. For example, one prominent craft specialist considers Revere's fluted teapot "brilliant" and "pristine," adding, "In its sophistication and purity of form, it is one of the masterpieces of design in the classical period." Other experts echo this praise, often describing the 1790s as the pinnacle of Revere's artistic mastery.[31]

The fluted teapot became one of the signature pieces of the Revere shop, an item only produced by Paul Revere and two other Bostonian silversmiths at this time. These objects required less labor than ornate objects such as fluted sugar urns and creampots and made use of the uniform sheet output of Revere's new flatting mill equipment, described below. The fluted teapot consistently sold well and had an enduring appeal. After the war Revere practically owned Boston's teapot market, for example, selling thirty-eight teapots during a period when two other leading silversmiths produced a total of five. The complexity of the tea ritual, number of tea-related items, and demand for these items all increased by the end of his career. Revere produced complete tea sets and additional items to expand an existing collection, recognizing that customers increasingly wished to purchase larger sets of silver objects (*en suite*) after the war.[32]

Revere looked to England for inspiration when designing his silver items. In 1784 he imported an illustrated book of Sheffield pattern designs, and many of his ensuing silver items as well as some of his engraving and scripting (font) styles follow the guidelines of this pattern book or other manuals. This practice not only illustrates Revere's business acumen—adding prestige to his own line of silver goods by emulating a chic foreign firm—but also reveals his lifelong facility with emulation and technology transfer. The Sheffield pattern books were an advertisement of wares and not a manual intended to teach silversmiths to produce their own knockoffs, but for Revere this distinction lacked meaning. A few print images offered all the clues he needed to craft quality items that met the aesthetic guidelines of the new style. He also adapted aspects of silver and ceramic work flowing into Boston from China and France. Even a piece of silver offered sufficient clues to enable Revere to emulate and eventually improve upon the work, a process now referred to as reverse engineering.[33] He freely duplicated, adapted, and combined aspects of different designs in crafting his own forms. Emulation was the essence of his genius.

Technological Advances: The Rolling Mill and Sheet Silver

Although Revere left no written description of most of his techniques or tools, we know of one piece of equipment that revolutionized his shop's production methods, boosted his output, and prepared him for future endeavors beyond silverworking. Starting in or after the mid-1780s, nearly all of his products benefited from the highly uniform output of a silver sheet rolling mill, also known as a flatting mill. Flatting mills existed in the colonies at least as early as 1733, particularly in Philadelphia. Revere purchased one in 1785 and hired a carpenter to install a platform for it. Workers needed a sixteen-foot supporting floor beam to hold the mill, suggesting a fairly large size. Most mills had a wooden frame, two cylindrical iron rollers, wooden handles attached to the rollers, and iron screws to regulate the separation of the rollers.[34] A rolling mill used human power, namely, the arm strength of two apprentices or journeymen, to turn two large rollers. When a silversmith fed a bar of silver between these rollers, the rollers flattened and widened the bar into a sheet. The silversmith then turned the screw to move the rollers slightly closer together, and passed the sheet through again, making it even flatter. Repeating this process eventually produced silver sheets of any desired thickness.

In setting up his flatting mill in 1785, Revere began to exploit the distinction between tools and machines, a division that soon sparked major changes in his life and in America's work culture. A tool such as a hammer or file is an inherently versatile crafting device limited by the skill and knowledge of its user. Tools are usually hand-held extensions of the human body, dependent upon their wielder's physical activity, strength, endurance, dexterity, and coordination. A machine, such as a rolling mill, is a specialized, complex, multicomponent device that ideally produces more standardized output, reduces or eliminates the physical labor involved in a task, or performs operations impossible for humans. While some machines require skilled operators, users do not need detailed knowledge of the product: only the machine's maker requires such knowledge. The tool, in effect, moves from the worker's hand into the inner workings of the machine, and machines encapsulate much of the worker's experience as well. The rolling mill, and all the machinery Revere later employed, allowed him to de-emphasize the role of skilled craftsmen in his operations, giving him greater control over the final product.[35]

In a larger sense, American manufacturers' shift from tools to machines foreshadowed the impending arrival of industrial capitalism, a system that eroded workers' connections to the final product as machinery mediated their work processes. Machinery also heralded the expansion of interchangeability in manufacturing, since machine use (alongside the division of labor) made the production process more uniform and more controllable. Of course, machines could not turn out interchangeable products if the machines themselves contained irregularities, as all early devices did. Revere confronted mechanical shortcomings many times throughout his career as he struggled to repair or adjust machines that failed to operate as expected. Machinery took hold in the public mind and became a far more tangible embodiment of industrialization than the division of labor or other changes to production methods. Following the lead of Enlightenment thinkers, Rationalist writers imposed the metaphor of the machine upon social, religious, political, and other aspects of life, for example, referring to the Constitution as "the machine that would go of itself" or the universe as a well-ordered machine. Increased machine use may have been only one component of proto-industry, but it certainly was a visible one.[36]

Revere's use of the silver-rolling mill illustrates the careful thought and planning surrounding its purchase. Armed with this device, Revere and his workers could easily cut and seam rolled sheet silver to form many shapes, eliminating the time-consuming process of "raising hollowware," which involved painstaking hammering, smoothing, and measurement that rarely produced uniform sheets. One silversmith estimated that an unengraved coffeepot required ten working days to create in 1760, but new methods developed by the late eighteenth century decreased the time spent hammering and working the silver. Silversmiths in England used hammering to flatten ingots into sheets even after manufacturers used rolling mills to roll other metals, in part because low labor costs made hammering a cheap procedure. Some "old-fashioned" silversmiths resisted the use of prefabricated silver or silver-rolling technology due to inertia, preferring to continue relying upon familiar manual processes that required greater effort and skill. Revere did not fall in this category. He initially used his rolling mill's sheet silver for the spouts and handle sockets of his cylindrical teapots, a sign that he still lacked the confidence to form a complete object from sheet silver. He soon expanded his use of rolled silver, and his increasingly complex experiments (often including fluting, a style conducive to the use of silver sheets) revealed the utility of this new technology.[37]

Revere quickly adapted many of his manufacturing methods to better utilize the rolling mill's ability to mass-produce silver sheets. Although he initially used a traditional "butt joint" to produce the seams on his new teapots created from rolled silver sheets, he soon switched to an overlapped soldered joint. This method yields a stronger seal, but more important, soldering is an easier process to master and more appropriate for apprentices or journeyman employees with less training. Revere's combination of rolling technology and construction methods allowed him to increase output, maintain quality, lower production costs, and decrease his own involvement in the shop. This masterstroke immediately benefited his business, and the rolling mill enabled much of his increased output after the war. For example, he produced 410 flatware items before the war and 2,069 after.[38]

The rolling mill also enabled Revere's outpouring of silver-plated items: buckles, harnesses, and assorted harness fittings such as bridles, saddle nails, and stirrups. Plating is the fusing or attachment of one type of metal on top of another. Copper buckles could be plated with a thin coating of silver, for example, to inexpensively add a beautiful silver finish to a stronger copper base. Silversmiths traditionally plated their items by mixing up a quantity of silver paste and applying it to the copper surface with their fingers. But some silversmiths used rolling mills to fuse a copper sheet to a silver sheet by heating the rollers and feeding equal-sized sheets of copper and silver through them at the same time. Revere's rolling mill could certainly serve this purpose, and after 1786 Revere began producing thousands of small plated objects far more efficiently than he could have with the earlier method. Revere's increase in silverplate output and his tendency to refer to his mill as a "plating mill" implies that he did indeed use it in this way.[39]

By 1791, Revere had need of a second rolling mill, either to augment his first mill or for resale to another silversmith (though no record of such a sale exists). In his letter to manufacturers George and William Burchell, he reveals a matter of fact approach toward the illegal export of English technology:

> Gentlemen, Please to make and send me a compleat plating mill, the Rolls to be eight inches long & three & one half inches diameter. I would have the Rolls finished in the best manner, the frame I would have substantial & strong, it is for a Silver Smith & one groce of the best well finished and best tempered cast steel gravers, for copper plate engraving. Pack them securley and send them on board the ship Mary bound to Boston N England Capt Tristram Barnard Master, who will pay your bill on sight. I could wish that that the bill might be made out in such way, as that Capt Barnard might not know what the package contains, only that it contains hard ware goods for Your Humble Servt, P Revere[40]

In this letter Revere ordered a rolling mill of specific dimensions, much smaller than the mills he later used for copperwork. Modern readers might balk at the bold manner in which he asked the British manufacturers to disguise the identity of this shipment from Captain Barnard, but at the time a large number of American tradesmen, merchants, and manufacturers contemptuously disregarded British laws attempting to prevent technology transfer, and English manufacturers such as the Burchells often served as accomplices. Revere initiated this subterfuge either to prevent Captain Barnard from rejecting the contraband or to protect him in the case of discovery. It would not be the last time he engaged in illegal technology transfer activities.

As Revere increased his output by reorienting his shop around the rolling mill he soon confronted the issue of labor management, a sticky subject in industrializing America. In spite of his unbounded enthusiasm for new technology, he had to decide how machinery might change longstanding artisan traditions. The mill opened up exciting and scary new possibilities, forcing him to choose whether to alter employee prerogatives and mold his workers into machine operators, or to continue to educate younger workers while allowing his skilled laborers to set their own pace.

Labor Practices: Combining Old and New

Revere's postwar production increases took place in the midst of nationwide changes in employer-employee relationships. His silver shop's workforce seems to have increased in the years after the Revolution, although a lack of comprehensive records obscures the number, names, and ranks of his assistants. Revere made payments to a number of different individuals for various services, implying that they might have been master silversmiths accepting subcontracted work, journeymen receiving wages, apprentices earning income on the side, or outside laborers performing various tasks. While he continued to train apprentices and interact with other laborers according to tried and true artisan traditions, as the century drew to a close, new market-based labor practices entered his repertoire. Throughout this period he adapted his labor practices and production methods to fit the changing conditions and steadily increased his shop's output as a result.

Revere had supervised his workforce according to traditional artisan guidelines for more than twenty years before the war, and naturally resumed those methods afterward. He continued to train apprentices and assumed patriarchal responsibilities over different aspects of their lives while they stayed with him. He continued taking in and training apprentices, treating them as members of his family, quite literally in two cases. Troublemaking apprentice David Moseley ran away from Revere to begin a life on the sea before Revere brought him back into the fold and continued his training. Moseley eventually married Revere's sister Betsey, but in spite of generous loans of money and equipment from Revere he turned to alcoholism and Revere eventually managed his estate in an attempt to make it more solvent. Another apprentice, Thomas Stevens Eayres, married Revere's daughter Frances in 1788 and attempted to go into business for himself in Worcester and later in Boston, until a debilitating mental illness laid him low. Revere eventually served as his guardian until his death. Marriages of Revere's laborers into their master craftsman's family and substantial amounts of monetary and managerial aid paint a picture of Revere's shop as an extension of his household, governed by a capable and helpful, though also authoritative patriarch. Revere's younger brother Thomas also frequently contracted with Revere to provide engraving and other services, receiving at least thirty cash payments during the 1790s. At one point Thomas ran a shop on Newbury Street in Boston under his own name though it is likely that he actually worked in Revere's shop for much of this period. In other cases Revere offered room and board to his journeymen, a service above the minimum expectations of a common wage laborer. Revere's shop, even as its size increased, continued to resemble a familiar group of trainees and skilled workers.[41] But labor conditions started to evolve, both under Revere's roof as well as in the world beyond.

The Revolutionary War produced permanent changes in the master-apprentice relationship, as societal upheavals overturned centuries of imported European traditions and challenged masters' authority. For example, increased literacy and publicly available sources of scientific and technical knowledge such as "how-to" books enabled ambitious newcomers to learn trades without needing to spend seven years in servitude. In addition, Revolutionary republican and egalitarian rhetoric produced, in the words of contemporary Charles Janson, a "loss of subordination in society" that further encouraged independence among workers. While some apprentices had always terminated their contracts by running away and setting up their own shops, new geographical mobility (particularly to western settlements) and a Revolution-inspired ideology of personal liberty greatly accelerated this flight. In the years from 1783 to 1799, twelve states passed new apprenticeship laws to address these runaways. These laws had little impact on apprenticeship, since no law allowed for interstate enforcement or extradition. As the century progressed, apprenticeship contracts became more specific and money oriented. The workplace also changed: in early times the master labored alongside the apprentices and journeymen, and runaways were fewer.[42]

The interests of masters and journeymen also began to diverge at the end of the eighteenth century. Even though artisans traditionally acted as employers of journeymen, the scarcity of labor and surety of employment in colonial times kept this relationship cordial and mutually beneficial, since journeymen knew they would most likely be masters soon. These two groups typically identified with each other as professionals practicing the same trade. Prior to 1750, many craft shops served as extended households for both journeymen and apprentices, treating them as part of the family and even taking in transients in hard times. The wage labor system gained ground by the turn of the century and put an end to this patriarchal relationship, replacing it with employer-employee conflicts. Masters complained about the fickle migratory nature of journeymen, and journeymen protested their low salaries and inability to amass enough capital to start shops of their own. Workers under the new system endured less paternalistic control from craft masters, but still had to live at the mercy of economic cycles, often losing their jobs in rough times. Population increase and the diminished availability of eastern farmland exacerbated the labor problem by creating even more competition for jobs, as well as a growing number of lifelong journeymen with no chance of promotion. By the end of the eighteenth century, the requirements for independent shop ownership became harder to define because they now included managerial responsibilities, and those raised under the apprentice tradition would most likely remain permanent wage laborers. Worker strikes became more common as journeymen looked to each other when craft masters no longer took their needs into account. Well in advance of the birth of factories, eighteenth-century laborers had already encountered many of the elements of proto-industry, such as wage labor, detailed contracts, high worker mobility and turnover, job insecurity, and distance between workers and managers.[43]

Although Revere supervised the training and activities of apprentices and journeymen throughout his career, only after the war did he truly begin to think of himself primarily as a manager. He shared his managerial duties with his oldest son, Paul Jr., thereby freeing more of his time for other projects. After guarding the family property during the British occupation of Boston at the start of the war, Paul Jr. served in the army until 1782 and then briefly went into business for himself. Unfortunately he never equaled his father's skill, judging from his own surviving silver pieces. By 1783, he returned to his father's shop and quickly assumed a managerial position, trusted by his father more than any other colleague. By November 1793, the Reveres referred to their business as Paul Revere and Son to reflect Paul Jr.'s growing role.[44]

As Revere devoted more effort to mercantile activities and other endeavors he decreased his personal involvement in the production of each silver piece. By the late 1780s, his products almost exclusively relied upon standardized patterns and procedures, allowing him to delegate more work to apprentices, journeymen, and even master craftsmen paid with cash, goods, or services. Some of these apprentices and journeymen used the rolling mill to produce large quantities of silver sheets and therefore served a less skilled role to augment the labor of others. Other employees worked sheet silver into finished forms, and based on the explosion of spoons and harness items we can imagine that much of this work took on a standardized feel. Variations in the quality of Revere's shop output throughout the 1780s and 1790s indicate that many workers of different skill levels had a hand in the production of final products. In some cases silver items have inconsistent thicknesses, engraving styles change from piece to piece, or seaming techniques show signs of irregularities.[45] Revere's stamp appears on all products leaving his shop, obscuring the extent of his personal effort in each case. While Revere engaged in some subcontracting before the war, by the end of the eighteenth century subcontracting had become common in his shop.

Revere's employee interactions reveal a mixture of old and new labor practices. It is impossible in many cases to determine whether someone served as an apprentice, a journeyman receiving a wage, or a worker in a different shop performing subcontracting work—in fact, it seems likely that these rigid titles started to lose their meaning by this point. By the end of the eighteenth century, Revere's silver shop still featured apprentices and journeymen trained in the artisan fashion, working at their own pace on a variety of tasks. However, the expansion of his labor force and addition of a rolling mill helped him increase his productivity and division of labor, moving him ever closer to standardization and mass production. This hybrid mixture of craft and what would later be called industrial goals and methods places Revere's shop, even at this early point, into the transitional category of proto-industry. He freely combined the most promising of the new methods with the most reliable of the old ones, and thereby improved his operations without creating a large upheaval for himself or his workers.[46]

Revere's willingness to add new techniques to his artisanal operations may have placed him near the leading edge of late eighteenth-century business practices, but the rest of his countrymen followed closely behind. Many factors collectively undermined the artisan's traditional socioeconomic position, and the proximity of these changes to the Revolution is not coincidental. The Revolution created new practical and ideological opportunities for nearly all members of society that helped unveil the age of capitalism. Old-fashioned concepts such as "sacrifice for the greater good" and "remain within your preordained societal role" changed under the influence of liberal republican ideology, which posited merit and not birthright as the determinant of rank in the new nation. Individual citizens now believed they had the right to act in their own commercial interests without social restrictions, and those who succeeded could then use their wealth and corresponding prestige to acquire a public office and help their country as well. The new wartime and post-wartime market opportunities furthered these beliefs by rewarding some of the most prominent speculators and risk takers, such as Paul Revere. Or so he hoped.[47]

Following the crippling recession that doomed Revere's attempts to thrive as a merchant, major economic improvements took place in the late 1780s thanks to the successful ratification of the Constitution and subsequent establishment of a stronger national government. The improving economy depended in no small part upon Secretary of the Treasury Alexander Hamilton's brilliant fiscal policies. The federal government's power to tax, exclusive control over the right to produce coinage or currency, and firm stance in favor of the full repayment of all debts helped to restore domestic and international confidence, stabilize inflation, increase the money supply, and produce a firm economic basis for development and growth.[48]

In spite of the solidification of America's new economy, many aspects of the new society impaired the craft tradition. For example, apprenticeships no longer represented the only way to learn production methods and technologies because innovation also resided in machine shops, technical manuals, and eventually in factories. The prominence of cash wages made apprenticeships appear less rewarding to young men who wondered why they had to spend up to seven years working for free when one could earn wages while learning on the job. As a result, the apprentice tradition changed, coming into closer alignment with the more market-based society that had already taken root. Ideals of individual enterprise and the universal drive to increase one's capital gradually edged out the former paternalistic family-oriented mutuality of craft shops. Masters no longer considered apprentices and journeymen extended members of the family or household, treating them instead as business associates who received cash wages for their services.

America's changing economic, social, and political institutions fostered the beginning of labor movements. Apprentices and journeymen who previously categorized themselves "vertically" according to their crafts now began to organize "horizontally" alongside other apprentices or journeymen in different trades. Early journeyman organizations usually formed to protest short-term problems such as declining wages, and disbanded when the aggravation faded. Disputes between masters and journeymen were also disputes between current and future masters, and both sides shared common ground in the awareness that they would one day be colleagues. As capitalist practices directed attention from traditions onto profit maximization, this solidarity diminished and labor unrest grew. Master craftsmen also struggled under the new social and economic conditions. Large numbers of craftsmen became wage laborers, others continued operating their small shops despite the erosion of craft traditions, and an elite few made the transition to the world of manufacturing and business ownership. The fate of master craftsmen in this new regime depended on many factors, including access to capital, managerial skill, and the ability to master new machines and technical processes. The decrease of paternal, family-oriented businesses and the widespread availability of currency also affected relationships between shop owners and their customers. Old-fashioned networks of trade exchange, barter, and personal credit gave way to modern commerce when shop owners eventually asked customers to settle transactions with cash at the time of purchase.[49]

In response to this growing fragmentation of master artisans, many cities experienced a proliferation of craft-oriented libraries and mechanics' associations attempting to advocate for common interests. These groups soon reflected the changes to craft institutions. The Boston Mechanic Association, later renamed the Massachusetts Charitable Mechanic Association, immediately became one of the most prominent artisan groups. In 1794 the mechanics of Boston petitioned the Massachusetts legislature to charter a corporation to license apprentices and prevent masters from hiring runaway apprentices. The merchant interests feared that this measure could increase labor costs and lobbied to defeat it in the legislature. The petitioners organized themselves into the Boston Mechanic Association, a voluntary organization with coercive powers. The eighty-two founding members, representing a wide cross section of master artisans and shop proprietors, unanimously elected Paul Revere their first president in 1795, illustrating his reputation among this segment of Boston's population.[50] The association's official annals (published in 1892) recount an engaging anecdotal version of the story of the association's founding, emphasizing Revere's reputation and temperament:

> Trouble growing out of the apprenticeship system was the incipient cause of the movement. The boys would not abide by their indentures, and unprincipled com-

petitors in business seduced them from their allegiance by promises of larger pay and better service. So great was the evil for the then small town, that Col. Henry Purkitt (the maternal grandfather of our much respected fellow-citizen, Henry P Kidder, the generous banker) inserted in the "Columbian Centinel" a notice for the first meeting to consult on the subject. It is a part of the traditions of the times that when Paul Revere saw this notice, much to his surprise, he hurried up town from his North- End residence to inquire of the printer who had dared to take upon himself the responsibility of calling a meeting of the mechanics of the town without first conferring with him! In truth, the gallant colonel had, from the days of the tea episode and the ride to Lexington, been recognized as the foremost man, *par excellence,* of the mechanics of Boston. Colonel Purkitt did not hesitate to acknowledge his action, and to suggest that it was arranged that he (Colonel Revere) should preside at the meeting which was called. This information had a soothing influence upon Colonel Revere, though it was really intended Edward Tuckerman, a South-End baker, a man of great intelligence and public spirit, should be the first president of the association; indeed, he was nominated for the office, but he respected the generous feeling in the community towards Revere, and said he did not care to take the office of president. So Revere was chosen the first officer without opposition, and Mr. Tuckerman became vice-president,—both serving four years, and both retiring from their positions at the same time.[51]

Although the association initially intended to serve as a quasi-guild to regulate apprenticeships, it had trouble collecting its relatively high dues (a one-time registration charge of $2.50 and quarterly fees of 25 cents, both of which soon doubled) and enforcing its requirements. Massachusetts eventually granted the association an act of incorporation and passed a law to regulate apprenticeships in 1805, but by this time the age of artisan traditions in most trades had drawn to a close. Apprentices had largely become paid employees and master craftsmen had bigger problems to worry about than their continued loyalty. The Charitable Mechanics Association expanded its mission statement to include goals such as "the diffusion of benevolence" (realized via a $40 one-time payment to the families of deceased members, and other payments to members in dire financial need) and "the encouragement of improvements in the mechanic arts and manufactures." Its roster eventually filled with well-educated pro-manufacturing businessmen.[52]

The evolution of the Charitable Mechanics Society in many ways mirrors the development of its first president, because Revere aspired to more than craftsmanship. During his silver years his abilities and goals never seemed to converge. Ironically, the very practices that enabled him to succeed at innumerable technical and entrepreneurial challenges had barred him from the colonial gentry by tarring him with the stigma of manual labor. Even after societal expectations shifted toward the appreciation of diligent employment, Revere's standards and goals remained unchanged. He wanted to become a gentleman, to have influence, and to serve society in a meaningful manner. While he initially used silverworking as a way of temporarily financing his mercantile ambitions, by the late 1780s silverworking became his principal route to financial success and he worked hard to improve his shop's income and prestige through practices such as mechanized silver rolling, dividing labor, standardizing his product line, and changing his own role from artisan to manager and owner. Although he failed to redefine his societal position, the recipe for his eventual triumph originated in this early period. The answer lay in changing the rules: instead of using manufacturing as the means for generating income that might lead to societal service and prestige, it had to become an end in itself.

CHAPTER FOUR

To Run a "Furnass"

The Iron Years (1788–1792)

November 3, 1788, was a good day for Paul Revere. In a letter to "Messrs. Brown and Benson," the proprietors of the Rhode Island Hope furnace, he announced, "We have got our furnass a going & find that it answers our expectation & have no doubt the business will do exceedingly well in the Town of Boston."[1] This furnace represented the culmination of years of research and experimentation, but also marked a bold departure from the preexisting patterns of his life. The daily routines of his profitable artisan trade, expensive and artistic silver items, merchants who sold him small packages of silver bars, engraving tools and curved hammers and leather aprons that collected every stray speck of silver dust . . . all of this now passed to the capable hands of Paul Jr. Revere's world would soon revolve around a huge, blazing oven, massive bars of cast iron, sand molds, and a wider clientele. Perhaps, if everything worked according to plan, he could broaden his operations and achieve the wealth and societal status that he still craved. A new beginning.

New beginnings in technical fields were neither cheap nor simple during the late eighteenth century due to crippling shortages in the two most essential commodities for establishing a manufactory: capital and technological expertise. By November 3, Revere had succeeded at the difficult first step of building the furnace. His available financial resources from silver shop profits and family connections proved sufficient for this project, and he solved the intellectual challenges thanks to a combination of his silverworking experience, technical aptitude, and a series of successful fact-finding missions. He still had to make his investment pay off by recruiting knowledgeable laborers, mastering casting processes, arranging dependable supplies of raw materials, and ensuring that his goods found buyers. He had a long road before him, but even he could not imagine how far it would lead him and his family.

Between 1788 and 1795, Revere learned not one but three applications for his furnace, starting with the casting of utilitarian iron objects for sale to consumers, and eventually leading to bell and cannon casting. Each field exposed him to a new network of practitioners, suppliers, and customers, and by repeatedly interacting with these communities and learning their practices, Revere participated in America's ongoing technology transfer. Between independence and Jay's Treaty of 1794, the industrial and economic relationship between Britain and America continued to follow colonial patterns because of America's postwar economic weakness and dependence. Scarce labor supplies, older technologies, and the falling value of export goods produced a trade imbalance and American vulnerability to British mercantile tactics such as credit retraction and product dumping. Many entrepreneurs like Revere identified the potential local market for manufactured goods and searched for help with the technical processes so highly guarded in France and England. They found this help in various ways, often without crossing the Atlantic.

Revere did not record his reasons for investing a large portion of his available assets in an iron furnace and striking out in this new direction. His rudimentary understanding of iron-casting processes and experience with local markets probably gave him confidence that he could master the manufacturing process and turn a profit. His unsuccessful career as a merchant failed to provide him with wealth or prestige, but did allow him to assess the buying habits of Bostonians and identify potentially lucrative product lines. In particular, his hardware store operations taught him much about the local iron market and local iron producers. Something must have impressed him, since his weighing of pros and cons led to his furnace.

But profits and costs represent only part of the picture. Revere's records strongly suggest another reason why he took a bold step into a new field: he tackled the challenges of iron casting because he loved learning new things.

Revere's career consists of repeated shifts from an established trade into a new one, and in each case his correspondence and records speak volumes about his startup procedures while de-emphasizing the ongoing older practices still being used in his shop. At some point after mastering each manufacturing application he turned it over to one of his sons so he could focus on the next one, at which point his documentation of the older enterprise grows sparse. This pattern reiterates several lessons from his silverworking career: Revere loved technical procedures, thrived on challenges, grew even more adept at networking as he grew older, and greatly enjoyed corresponding about metallurgy and metalworking in his free time, absorbing technical and market information from those in the know. However, his restless pursuit of technical knowledge reveals a lack of satisfaction with any one metallurgical operation. He hoped that casting might provide income, prestige, and a doorway to the upper echelons of public service, and not just add a new craft practice to his repertoire. His perspective began to shift as his operations and payroll grew, particularly after the government became an important client in his later years.

The man who decided in 1787, at the age of 52, to invest precious time and money in this new field was ready for a change. Apart from a few intriguing dabblings into gunpowder and cannon casting during the war, Revere's only other opportunity to learn a completely new technical field happened decades before when he learned the "art and mystery" of silverworking at his father's side. The apprentice system offered the opportunity to absorb the knowledge, skills, and culture of a trade from an established practitioner, only "graduating" upon the completion of years of work and demonstration of proven abilities. The artisan educational system had the backing of the entire society, and although this tradition began its decline at the end of the eighteenth century, it still represented the gold standard. Revere therefore faced intimidating challenges when he struck off in a new direction, as he had to learn the new trade, establish a reputation, and integrate his old and new identities.

Paul Revere's experiences as he learned the iron-casting business between 1787 and 1792 offer a wonderful example of the crucial phenomenon of *technology transfer,* a term typically used to explain how nations or societies learn from the example of more advanced ones. Revere attacked this challenge on a smaller scale throughout his self-directed apprenticeship in the iron trade, taking the initiative to experiment and forge valuable connections with "mentors" among the ironmasters at the Hope furnace in Providence, Rhode Island. As he intentionally set aside the rich history of silverworking he unwittingly adopted a new package of equally rich traditions befitting the age of iron, and he came to understand the importance of capital, technology, labor, and environmental inputs in successful manufacturing operations.

Iron from Antiquity to America

Revere's shift from silver to ironworking mirrors the technological progression followed by the metalworkers of antiquity. Many early societies first worked with precious metals such as silver and gold because these metals were attractive to the eye and could occasionally be found in a pure state in large surface veins or riverbeds. Gold and silver have a high degree of malleability and ductility, and workers could easily shape them with simple physical processes such as hammering, requiring only a straightforward set of tools. Precious metals usually served ornamental purposes and had a high degree of symbolic value due to their rarity and beauty. Iron, one of the most abundant metals on Earth, contrasted with these precious metals in nearly all aspects, lending its name to the vibrant "iron age" period of prehistory.

Many metals such as iron occur in either a "native" state, where the metal is relatively uncontaminated by other elements, or as an ore combining the metal and other elements within a quantity of rock. Metalworkers in the earliest human societies of the Near East such as Sumeria lacked the ability to convert iron ore into usable metal, because iron melts at the extremely high temperature of approximately 1535 degrees C, well beyond the reach of early furnaces. Early craftsmen did create a small number of iron implements from native iron sources, which usually came from iron-rich meteors. In fact, the Sumerian word for iron meant "heaven-metal," while Egyptians used the term *black copper from heaven*. Archaeologists and scientists can identify meteoric iron with some degree of certainty because it is the only form of iron to contain trace quantities of nickel. This nickel also helped it to resist rust, explaining why some of these ancient iron objects still exist today.[2]

Widespread human use of iron had to wait until technology provided the means of separating iron from rock pieces and unusable elements chemically bonded to the iron, a process called smelting. Current consensus indicates that the first iron smelting took place in the Near East, most likely in northern Anatolia or near the shores of the Black Sea, around the end of the Bronze Age between the fifteenth and twelfth centuries BC. As with many major technological advances, iron smelting probably began as an accident. Copper melts at a lower temperature than iron, and coppersmiths unintentionally smelted small quantities of iron in their ovens around the early second millennium BC.[3] Enterprising metalworkers fashioned these byproducts into valuable iron jewelry, and in the nineteenth century BC one ounce of iron was priced the same as 40 ounces of silver. The high melting temperature of iron prohibited the larger-scale routine manufacture of iron, but resourceful ironworkers eventually overcame this limitation and even found ways to harden at least the outer skin of worked iron to make it stronger and harder than bronze. The Hittites are often credited with these ironworking breakthroughs, and the dissemination of iron technology accelerated between 1200 and 1000 BC, at which time iron objects became widely used in the Near East and Eastern Mediterranean regions. And thus began the age of iron.[4]

Over the next two thousand years, a series of ironworkers and governments seeking military or economic advancement transferred the technology of ironworking across a continent and an ocean, and finally into the eager hands of Paul Revere.[5] Greece introduced ironmaking to Europe, and iron weapons and implements play a central role throughout Greek history, also appearing throughout Greek literature and mythology. The Greeks passed on ironmaking technology to the Romans, who spread it throughout Europe and even into England. English iron production began a period of rapid expansion in the fifteenth century with the importation of blast furnaces and foreign ironworkers, soon giving England a position of worldwide technological dominance. Deforestation caused the price of charcoal to rise faster than the price of iron in the sixteenth and seventeenth centuries, leading to iron shortages. By the seventeenth century, new ironworks in heavily forested nations such as Scandinavia and Russia, along with larger and more efficient furnaces, stabilized the price of iron. America, the next forested region to receive the transfer of ironmaking technology, nourished Britain's hopes for continued resource sustainability.[6]

The ironworking technology landing on America's shores in the seventeenth century fell into two categories that employed different processes for converting iron ore into usable metal. In the direct, or "blooming," process, workers smelted iron ore on a small scale with simple equipment in facilities called bloomeries. A single skilled worker placed a small quantity of iron ore over a stone hearth and heated it with a charcoal fire, fanned vigorously with bellows to reach higher temperatures. The charcoal fire separated the metallic iron from its waste products in two ways: some impurities chemically combined with carbon monoxide from the fire and escaped as gases, while others physically solidified into slag. Most of the slag melted and drained away from the usable metal, and the remaining slag became hard, brittle lumps throughout the "bloom," a spongy mass of iron. The ironworker hammered the bloom to shatter and separate the brittle slag, leaving only the iron. This bloomery process directly produced bar iron (also known as wrought iron), a fairly pure substance consisting primarily of metallic iron with a small percentage of slag particles. Bar iron was tough and malleable, easily worked into different shapes and by far the most versatile type of iron in use during the colonial period. Iron workers could fuse two pieces of bar iron by heating them to a white-hot temperature and hammering them together, and craftsmen further refined bar iron goods by chiseling and filing them after they cooled.[7] Early ironworkers understood the advantages of bar iron over most other metals, but attempted to discover an alternate production method that would allow them to produce it faster, cheaper, and more efficiently. The solution to this dilemma involved the use of a technological and managerial innovation known as the blast furnace.

The blast furnace method of iron production was also known as the "indirect" process because it required two steps to produce bar iron. In the first step the blast furnace converted iron ore into an intermediate substance called pig or cast iron. Workers filled a gigantic stone chimney twenty or more feet high with alternating layers of iron ore, charcoal fuel, and a calcium-rich "flux" material such as limestone that chemically combined with the slag to help separate it from the iron. Once ignited, this mixture burned for days or weeks at a time, fueled by new inputs of ore, charcoal, flux, and a stream of air pumped via water-driven bellows. Dense molten iron pooled in a crucible at the bottom of the furnace and workers periodically poured it into sand molds, where it hardened into pig iron. Pig iron contained small amounts of carbon and silicon impurities that entered the iron from the charcoal and other substances in the stack. The presence of these and even less-desirable impurities determined the quality and final properties of the end product. Pig iron's heat resistance and hardness made it useful for large or heavy objects such as cauldrons or fireplace backs, but its inevitable brittleness prevented its use in tools or items subject to impacts. Pig iron lacked malleability even when hot, and once it left its molten state it could only be reshaped by being re-melted, which took place in facilities called foundries under the supervision of specialized "founders"

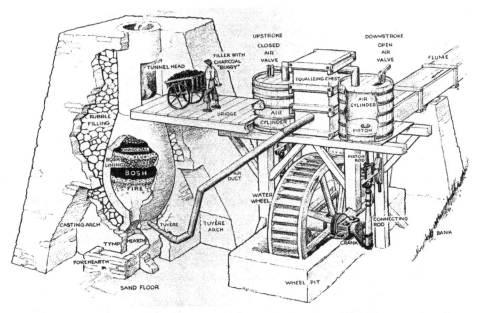

Figure 4.1. Sketch of a blast furnace. From Edwin Tunis, *Colonial Craftsmen and the Beginnings of American Industry* (New York: Thomas Y. Crowell Company, 1965), p. 151. This cutaway sketch shows the components of a typical eighteenth-century American blast furnace. The waterwheel on the right drives air-pumping equipment (often including a bellows not shown here) that fans the fire inside the "bosh." Workers load large quantities of charcoal, iron ore, and flux materials in alternating layers from the top of the furnace, keeping the furnace going for days or weeks at a time. Liquid iron sinks to the bottom and workers periodically release it through the forehearth into a sand pit, where it cools to form iron slabs or "pigs." Paul Revere depended on the output of New England blast furnaces, remelting and casting the iron slabs into utilitarian forms such as fireplace backs or window weights.

such as Paul Revere. Skilled founders could design intricate molds and produce detailed objects that most smiths had trouble matching.[8]

Blast furnaces represented only the first half of the indirect ironmaking process, because their pig iron was less versatile and functional than the bar iron made in bloomeries. In order to produce bar iron, blast furnaces often formed symbiotic links with other ironworking institutions called fineries. Skilled finery ironworkers (or "finers") converted pig iron into bar iron by heating the cast iron in a fanned charcoal fire. The heat and draft melted the pig iron and oxidized the carbon and silicon impurities, which escaped as gases. The finer then hammered the cooling lump of metal to separate the purer bar iron from the impurities in the liquid slag.[9]

Blast furnace operation depended upon managerial as well as technical skills. By contemporary standards all ironworks required large quantities of coordinated labor, but blast furnaces, the largest type of ironworks, required twelve or more workers to feed and tap the furnace all day and all night, every day, for most of the April through November working season. Larger furnaces produced sixteen to twenty tons of iron a week, or around eight hundred tons per year. Blast furnaces required intensive organization and discipline, since one missed step in the process could cause costly setbacks and injuries. As one of the earliest forms of large-scale business in America, blast furnaces fostered the separation of a managerial role from the work of the skilled labor pool. Skilled ironworkers needed intelligence, perceptiveness, and stamina in order to interpret subtle clues and gauge invisible chemical transformations for long shifts under trying conditions. Competent managers ideally possessed some of the same technical skills as well as the organizational and leadership abilities needed to coordinate the workers and deal with problems as they arose.[10] The blast furnace's amalgamation of technological and managerial challenges helped inaugurate a new world of manufacturing endeavors: the large technological system.

In addition to physically constructing and running the blast furnace, an ironworks operator also had to establish mechanisms for continual imports of raw materials, a financial infrastructure ensuring regular payments to creditors and employees, legal or political support, and many other societal factors. Blast furnaces and other complex industrial operations, known as "large technological systems," do not exist in a vacuum, and their technological components depend upon numerous connections with the larger social, political, and economic environment. Large technological systems collect physical artifacts, organizations, intellectual property, legislative support, and natural resources in order to solve problems with increased efficiency. Entrepreneurs and inventors take a leadership role in the early years of a system when it needs to grow rapidly, and managers take over when systems reach a mature state. These managers often attempt to reduce the potential for worker errors through deskilling, routinization, and bureaucracy, and growing systems incorporate external resources such as raw material sources in order to minimize problems such as price fluctuations or shortages.[11]

A blast furnace owner could not carry on operations without considering local customs, market conditions, laws, tariffs, price of competing products, skilled laborer expectations, and many other factors. Similarly, blast furnaces impacted their social and cultural settings by affecting the local labor market, altering the price of goods, forming relationships with military purchasers, sponsoring different political figures, and so on. As blast furnaces grew more prominent many manufacturers received an early taste of the importance of integrating technological challenges with social, financial, and other concerns. In this manner the age of industry gained strength.

Revere chose to enter the ironworking field as a founder, a sensible decision given his circumstances. Bloomeries only made sense in frontier areas close to iron ore sources and far from competing sources of iron products, certainly not the case in Boston. Blast furnaces operated on a massive scale well beyond Revere's means and required considerable capital investments; a large, trained labor force; reliable inputs of iron ore, charcoal, and waterpower; and knowledge of complex chemical processes. A foundry depended upon a process Revere had already learned, the art of heating and casting metal into molds, and as a smaller endeavor it lent itself to artisan ownership and management. Revere might not be in a position to start his own technological system in the late 1780s, but by joining the already established ironworking network he increased his income while teaching himself how to think beyond the scope of an artisan's workshop.

By the time Revere entered the foundry business, America had close to two centuries of ironworking experience under its belt. The story of large-scale American iron production began in Virginia in 1619, when the Southampton Adventurers assembled 4,000 pounds sterling and 80 workmen in Falling Creek, Virginia. This attempt to establish the first colonial ironworks came to a sudden, less than triumphant ending when Indians killed the workmen and destroyed the facilities prior to the start of production. John Winthrop Jr. led the colonists' second ironworking attempt in 1646, when he constructed and oversaw the Saugus ironworks (also called Hammersmith) in Lynn, Massachusetts. Although it closed in 1668 due to disappointing profits and legal difficulties, Hammersmith collected enough expertise and labor over the twenty-two years of its operation to lay the groundwork for spin-off ironworks in other colonies. By the 1680s, this technology had spread to the mid-Atlantic colonies, and the first forges in Maryland and Pennsylvania appeared in the early eighteenth century. By the Revolution, ironworks could be found in every colony except Georgia. While all American ironworks benefited from abundant supplies of wood and waterpower, iron ore was far more abundant in the mid-Atlantic region, with Pennsylvania rapidly asserting its dominance over the field. New iron manufactories often started with a bloomery because of its low startup and operating costs, and later added a blast furnace to produce cast hollowware and pigs. After the blast furnace attained steady production, owners could easily convert the bloomery to a finery and possibly even add a slitting mill, a water-powered cutting device that sliced iron into long, thin strips that could be made into nails.[12]

British investors initially funded and operated most colonial ironworks, following the mercantilist mindset. In other words, the mother country intended to process America's abundant natural resources in English workshops in order to solve its employment and manufacturing shortfalls, and then ship finished products back to the colonies for profitable sales. The earliest colonial ironworks, sawmills, and other technical endeavors attempted to fulfill this plan by focusing upon the processing of raw materials. The earliest ironworks also produced negative social impacts in their local communities: one might imagine the consequences of large groups of male laborers congregating in environments characterized by hard work, liquor, and profanity. The resulting social disorder and societal diversity stood in contrast to early New England's cohesive religious atmosphere.[13]

Americans expanded their ironworking in the 1720s after Britain's disintegrating relationship with Sweden led to restricted Swedish iron imports. Colonial iron helped make up this deficit, and skilled laborers from earlier ironworks provided vital expertise. Although many of these new ironworks sold their output to British merchants, others intended to produce for local markets whose rapidly increasing populations and desire for a higher standard of living fueled a voracious hunger for iron products. At this point the mercantile theory of colonial manufactures had worn thin, since colonial ironworks started producing larger quantities of finished products that competed with goods from the mother country. When British representatives eventually caught on to America's growing iron sector, Parliament passed the Iron Act of 1750 in an attempt to force a return to mercantilist ideals. This legislation removed all import duties from American pig and bar iron in order to encourage the export of raw materials to English manufacturers, but forbade the construction of American slitting and rolling mills, plating and triphammer forges, and other advanced processing works. This law did not have much effect because colonists offered iron regulations the same response they gave to other unpopular British policies: contempt and skillful evasion. The blatant construction of new finishing facilities continued and even increased until the Revolutionary War, fed by an ever-growing population demanding iron goods. American ironworks in the 1770s employed approximately 8,000 men and produced 30,000 tons of pig and bar iron a year, approximately 15 percent of the world's output. The American colonies ranked as the third largest iron-producing nation.[14]

The Iron Act did have one impact: it succeeded in infuriating American ironworkers, many of whom became extremely active in the Revolutionary War. Prescient almanac publisher Nathaniel Ames affirmed America's attachment to iron in 1758, writing, "This Metal, more useful than Gold or Silver, will employ millions of hands, not only to form the martial Sword and peaceful Share alternately; but an Infinity of Utensils improved in the exercise of Art, and Handicraft amongst Men."[15] The British government drastically underestimated the size of the colonial ironmaking industry and its potential value to the colonists in the Revolution. War between the colonies and Britain both helped and hindered the iron industry: wartime demands engendered large ordnance contracts from federal and state governments, but invading armies destroyed furnace equipment and created crippling manpower shortages. In 1785 visiting Swedish engineer Samuel Gustav Hermelin observed that the war destroyed some furnaces and left others idle, while also dispersing or killing a number of the nation's already scarce supply of skilled laborers. Lord Sheffield, a spokesman for British commercial interests, concurred with Hermelin, reporting that Britain would face no American competition whatsoever on a list of articles including copper sheets and utensils, bar steel, and "iron and steel manufacture of every kind."[16]

American ironworkers developed a new business model called the "iron plantation" in the late eighteenth century, most common in the mid-Atlantic states. An even larger variant of the blast furnace, iron plantations occupied thousands of acres of land and integrated operations such as agricultural food production for workers, forest management to ensure sustainable fuel supplies, and diversified ironworks. Plantations heralded modern industrial organizations in many ways, such as the large-scale use of wage labor, collection of all aspects of production (e.g., blast furnace, forge, finery, slitting mill) in one facility, advanced transportation networks, and complicated market analysis and accounting systems. However, the plantations also retained old-fashioned elements such as communal on-site housing, a mix of agricultural and industrial work for the laborers, and paternal family-based ownership.[17] In the same way that Revere and other artisans blended traditional practices with some of the hallmarks of the upcoming age of industrial capitalism, iron plantations also played a transitional proto-industrial role that placed them in, or between, two worlds.

The history of America's iron industry serves as a microcosm of the nation's larger struggle to close its overarching technological gap with the Old World, a particularly frustrating task because of Britain's stern prohibition of any form of knowledge transfer. Americans initially implemented the exact ironworking processes practiced in Europe and Britain. After some early failures, American ironworkers started adapting these processes to react to the unique resources and economic conditions in America. For example, America's abundance of wood encouraged many American ironworkers to continue using charcoal fuel even after Britain upgraded its operations to use alternative fuel sources.[18] In spite of ready access to natural resources, American ironworkers lagged behind their British counterparts throughout the colonial and Revolutionary eras and faced severe technological, managerial, and financial impediments that made iron quality and ironworking profits inconsistent at best.

Ironworkers fared poorly in the earliest postwar years, suffering the effects of harsh foreign competition and unfavorable economic conditions. In contrast with American iron's decline during the war, British ironworks thrived, increasing their efficiency and overall output by converting to coke and coal fuels as well as steam power. The postwar influx of cheap, high-quality British iron forced Americans to cut their prices and profits accordingly. Also, the postwar recession compounded by inflation and the lack of currency affected all business ventures in America. These conditions improved in the late 1780s, when new protective tariffs and a booming economy enabled ironworks to prosper.[19] And whether by coincidence or design, Revere chose to enter the field at this exact point.

Revere the Founder: Climbing the Iron Learning Curve

Revere had no way of knowing that ironworking was only the first of his new manufacturing adventures, involving different metals, new equipment, and unknown fabrication processes. He did know that these first steps, building what he referred to as a large "air furnace" and learning to use it to cast iron objects, involved an intimidating learning curve that would surely tax his financial and intellectual resources. Revere's successful entry into this first new field highlights his many assets and provides a valuable case study of America's turn-of-the-century technological transfer mechanisms.

Technology transfer arises from an imbalance of knowledge between nations, exacerbated by the less advanced nation's immediate needs. The relative scarcity of cutting-edge machinery, machine operators, and relevant technical know-how offers an almost insurmountable barrier to new manufacturers attempting to close this gap. At the end of the eighteenth century Britain led the world in technological refinement, manufacturing output, and economic strength. Colonial America's technology might meet the modest needs of fringe provinces devoted to agriculture and natural resource harvesting, but America needed more advanced manufacturing to avoid a return to economic colonialism after the war. Most historical studies emphasize the central importance of emigrating skilled laborers, particularly from Britain, in early American technology transfer. This generalization contains some truth: America lay in Britain's long technological shadow and early Americans held British (and particularly English) technical skill in the highest regard. Immigrants often publicized their English training and native essayists wrote at length about the need to encourage the emigration of skilled workers in order to correct America's technological deficit. America's technology transfer benefited from a constant stream of English and European artisans emigrating into Philadelphia, New York, and Boston, as well as the importation of numerous high-tech tools or machines.[20]

The actual technology transfer process adopted by many American firms transcended the hiring of British immigrants. American firms could not always enlist relevant foreign experts and had to find other ways to solve their technological problems, such as developing in-house innovations, trading information between shops, and applying the lessons of one field to another. George Washington showed an understanding of the multifaceted nature of technology transfer when he asked Congress to help America ensure its lasting independence by "giving effectual encouragement as well to the introduction of new and useful inventions from abroad as to the exertions of skill and genius in producing them at home." This call to action may have been a step in the right direction, but government and business leaders still had to find a pragmatic way to make technology transfer happen. The diffusion of technology generally follows a four-step approach: demonstration of the potential of a new technology, establishment of a pilot manufactory, diffusion of the new

technology to similar facilities, and modification of the technology by different operators to suit local conditions.[21] Revere's experiences, beginning with
his ironworking, illustrate the mechanisms of technological diffusion to new
facilities and, to a lesser degree, the modification of technology to suit local
conditions. These latter steps of the overall technology transfer process, while
not as glamorous as the role played by the first pioneer inventor to develop a
new technology, are equally important in the establishment of large technological systems and networks of practitioners. Revere's interactions with other
metalworkers portray a fairly open technological community in turn-of-the-
century America, with free exchange of advice even among those who, under
other circumstances, might become competitors. His example also demon-
strates the importance of technological transfer from one field to another,
through channels other than hiring skilled laborers, and in pursuit of lessons
beyond the merely technical.

Revere began preparing for his new foundry in 1787, although he may have
started the ball rolling at an earlier point without recording it. Even though
America lacked technical expertise, managerial skill, and investment capital,
Revere could access enough of each commodity to proceed with good speed.
Despite the extremely different material properties of silver and iron, Revere's
silverworking knowledge helped him master the iron-casting process within
a few years. Revere's craft upbringing bequeathed a general understanding of
metallurgy as well as specific skills such as the ability to melt silver uniformly
in a furnace, prepare small molds, cast molten silver into molds, verify the
consistency and homogeneity of metal products, prepare alloys combining
different metals, and use heat and hammering to reshape metal objects. Now
he needed to replace his understanding of silver's properties—in particular, its
melting point, malleability, and cooling rate—with an equivalent understanding of iron. Revere could also draw upon the casting work he performed for
the Massachusetts government during the Revolutionary War when he helped
Louis de Maresquelle cast cannon at the Titicut forge in 1777. Although the
records do not indicate whether Revere ever worked with iron at Titicut (most
cannon at the time consisted of brass or bronze), he certainly received extensive practice preparing large molds and operating a furnace. Silverworking
and brief cannon-casting experience gave him a head start, but he surely had
much more to learn.

Technical and scientific texts offered a promise of assistance that usually
amounted to nothing. Until 1840, American artisans without reading knowl-

edge of French and German could find little, if any, technical documentation geared at basic ironworking principles. Scientific research also failed to enlighten the would-be ironworker: pragmatic craft knowledge and expertise could not be readily gleaned from the prevalent scientific theories or principles of the day, and artisans' practical craft experience ran far in advance of scientific metallurgical knowledge until the late nineteenth century. Chemists began investigating metallurgical principles in the 1700s and started becoming valuable to ironworks by the mid-1800s, although most artisans resented and rejected the intrusion of theoretical scientists into privileged craft practices. In spite of science's limited role in craft activities, Revere never turned down any possible pathway to useful information and made several early attempts to master the scientific basis of this new field. In November 1788 he asked a colleague to send a copy of Richard Watson's *Chemical Essays*, a set of writings published between 1781 and 1787 for people with minimal chemistry experience.[22] Revere also spoke with Dr. Benjamin Waterhouse, Harvard University's first professor of mineralogy, who remained well connected in British intellectual circles after his education at Oxford. Although we have no record of their conversations, Dr. Waterhouse later described Revere as "the only man (in America) in 1794 who appeared to know anything of the discrimination between ores and the seven metals."[23] While these readings and discussions about chemistry probably provided some useful information, at no point do they ever appear as a decisive factor helping him to overcome problems. In the late eighteenth century a vast gulf separated science and practice, and Revere lived squarely on the practice side of the divide.

Well-established ironworkers clearly offered the most relevant information for anyone trying to enter the field. Every detail—from the moisture content of sand molds to the best clay for lining the furnace—had some impact upon the final product, and experienced professionals could quickly help a newcomer focus on the important issues, avoid stumbling blocks, and learn the tricks of the trade. Fortunately for Revere, most ironworkers in America freely shared knowledge with each other. This openness may have surprised someone whose artisanal education and membership in the Freemasons included the injunction to hide valuable trade mysteries and secrets. Artisanal secrecy descended from British traditions that treated human skill and knowledge like tools and blueprints, valuable commodities to be hoarded for the future well-being of the empire or as a way to preserve a guild's monopoly. The British guarded their ironmaking industry with extra vigilance, as it was a key military asset and one of the great sources of English wealth. Parliament passed a law in 1785 prohibiting the export of ironmaking tools as well as the "seducing" of workers in these industries. A detailed list of non-exportable items included "all forms of rolls, anvils, hammers, molds, presses, or models or plans of such equipment." These new restrictions continued earlier policies, such as a 1718 law prohibiting the emigration of skilled artisans from Britain.[24]

American ironworks for the most part presented a different picture, one in which owners "let neighbors and children stop by their smithies, forges, and furnaces to watch and learn." The open, community-based ironworking field encouraged information exchange and cross-fertilization from other trades.[25] Such fluidity had far-reaching implications for Revere, who would not have succeeded as an ironworker without advice from current practitioners. Revere adopted an equally generous attitude toward knowledge sharing in the years to come, although his letters also reveal an underlying patriotic sentiment. He viewed the discovery and dissemination of new procedures as a benefit to his country, as illustrated in a letter he received from Stephen Rochefontaine in 1795.

> Mr. [name missing] a French Gentleman who will deliver you this, is charged by government to make use of a new method for casting iron guns . . . will you be so kind to help him in doing it yourself or by providing him all the workmen and assistance in your power? I am so well acquainted with your fondness for scientific discoveries that I am persuaded that it is enough to point out the Gentleman who may be useful in enlarging the knowledge of this country to be certain that you will afford him all sort of assistance and meet him with a hearty welcome.[26]

Rochefontaine, a former French military engineer who emigrated to the United States during the French Revolution, served as an inspector of ordnance for the U.S. Army. He represented an invaluable stockpile of technical and military experience in the new republic as well as a potential ally. Although he does not appear frequently in Revere's surviving records, he showed some familiarity with Revere and felt comfortable imposing upon him for this favor. In the years to come a number of other technicians also requested advice from Revere, and as far as the records show, Revere always obliged.

We must not exaggerate the quantity of altruistic information exchange in the young republic. Different firms competed with each other and guarded their most important secrets, and Revere would not let patriotism jeopardize the survival of his own business. Political rhetoric and the theoretical background of patent law offer the prevailing belief that technical and scientific advancement should simultaneously aid the individual as well as the nation: patents enabled inventors to profit from their ideas, but after a time the new invention entered the public domain for widespread use. More pragmatically, metalworkers in general and Revere in particular seldom suffered from excessive competition given the small number of practitioners who operated in such a large market. On the other hand, all metalworkers often had to endure inexperienced laborers, equipment breakdowns, technical hurdles, and external threats such as cheap British imports. In America's industrial infancy, practitioners gained far more by sharing, cooperating, and educating each other than they lost.

Nicholas Brown became a surrogate ironworking mentor to Revere, and happily shared the fruits of his decades of experience. The prosperous Brown family frequently used portions of their mercantile profits to found and operate manufactories in Rhode Island. For example, Nicholas's younger brother Moses gained everlasting fame by investing in Samuel Slater's pioneering textile mills, and the Browns also ran a spermaceti factory that made them the leading colonial-era candle makers. They built a blast furnace in 1765 in a town on the Pawtuxet River called Hope, and christened their endeavor the Hope furnace. In a direct parallel to Revere's own startup procedure more than twenty years later, the Brown brothers sent agents to the prosperous Pennsylvania ironworks; hired workers from some of the furnaces in Connecticut; and purchased pipes, molds, and other equipment from Massachusetts when they set up their facilities. Their completed ironworks employed up to seventy-five men at a time, although at least half of them directed their efforts toward wood gathering, often on the woodlots of cooperative neighboring farms. After a rocky two-year learning process marked by great difficulty in locating capable founders, the operators learned to produce steady quantities of reasonably high-quality iron. They considered building a forge to produce bar iron but had to give up the idea when they could not find any skilled laborers to supervise the process. Similarly, after their attempts to manufacture hollow goods produced inconsistent and low-quality wares, the Browns restricted themselves to pig iron production and sold their output to a wide range of New England forges and also to London. During the Revolution the Browns added cannon casting to their repertoire, providing hundreds of cannon for the government and private interests, and Nicholas Brown joined the younger George Benson in 1783 to form Brown and Benson, the firm that later worked with Revere. While setting up his iron foundry Revere frequently contacted Brown for advice, visited him in person at least once, and included questions in correspondence when arranging iron shipments from the Hope furnace. Years later, Brown helped him learn the bell-making process as well.[27]

Nicholas Brown's learning process serves as an important counterpoint to Revere's experiences, indicative of the owner-operator (or businessman-artisan) divide. While Brown entered a new field by allocating funds, establishing business relationships, and, most important, hiring experienced operators and managers, Revere started his new endeavor by teaching himself the art of ironworking much as he had learned the silverworking trade in his teens. But Revere's reach had grown in the years since he polished silver at his father's side, and his entrance into the iron-casting trade included the same hiring, purchasing, and organizational challenges faced by Brown and other industrial capitalists operating on a larger scale.

After deciding to have a go at the foundry business, Revere first had to set up a line of funding for research and construction costs. His lucrative silver shop financed most of these expenses but because he had not perfected his recordkeeping practices by the 1780s, the finances pertaining to the foundry's inception present a confusing and incomplete picture. In particular, numerous withdrawals labeled merely "To Cash" certainly account for many of the furnace's funding costs, although Revere also used this heading for personal cash withdrawals. This method of mixing shop expenses with personal or other uses of cash was widespread at the time, reflecting the simple truth about the typical intermingling of home and professional life. Without question, the silver shop's profits served as his primary funding source.

Revere's well-to-do Hitchborn cousins also provided essential funding, again obscured by vague notations that hide the specifics of their working relationship. Samuel Hitchborn, the Harvard-trained lawyer and gentleman, allowed Revere to use some of his Lynn Street property for the foundry, located in the north end of Boston just a block from several small wharves. Hitchborn might have charged some rent on this property, but we know for certain that he sold it to Revere on June 28, 1792, commenting, "I have this day executed a Deed of a certain piece of Land situate in Boston on which there is an air furnace to Paul Revere of said Boston Esq."[28] In addition to the use of this land, Samuel Hitchborn also paid more than 9 pounds for coal and carting on two occasions, and more than 37 pounds for iron on three occasions. Revere's other cousin, his former apprentice Benjamin Hitchborn who now worked as a silversmith, also made 9 cash payments to Revere throughout this startup period for a total of more than 66 pounds. Samuel and Benjamin might have been repaying old debts to Revere, in light of the numerous large silver purchases they both made from him throughout the late 1780s. Or they might have loaned some extra cash to their cousin when his own resources were insufficient. A third possibility is that Samuel and Benjamin could have served as silent partners in the foundry, treating these payments as investments that would eventually pay dividends. We cannot know the answer, but we do know that in a time of need Revere's family offered frequent financial aid.[29]

On the flip side of his balance sheet, Revere started listing furnace expenses and startup costs in a revealing ledger that began in March 1787 and covered approximately twenty-one months of activity.[30] Revere's expenses express his priorities in a way that correspondence could not. Unsurprisingly, this ledger tells the story of the four principal commodities that made eighteenth- and nineteenth-century manufacturing firms run: capital, environmental resources, labor, and technology. The entire ledger is a testament to the central importance of capital and technology: Revere's business depended upon his ability to outlay all of this money prior to receiving any income, and every expense served the technological goal of foundry construction. Environmental resources appear in these early expenditures through numerous purchases of raw materials: he bought sand, wood, nails, clay, and up to twenty tons of stones at a time, and paid transportation charges for many of the larger shipments. He purchased most of his construction supplies from local merchants, shop owners, and landowners whom he often lists by name, in contrast to the wider array of distant merchants who provided raw materials once the furnace started operating.

Revere's primary startup expenses also include more than 29 pounds of wages for 10 part-time employees. The records do not always specify the tasks performed by each individual, and it is often unclear whether Revere's workers spent their time constructing, repairing, modifying, or operating the furnace. A spurt of activity took place in April 1787 in which three of the men whom Revere classified as "labourers" performed all their work, and another in July when the majority of the brickwork took place under the supervision of Mr. Richmond, the highest-paid employee, listed as a "brick maker." Revere paid for brick molds, labor, as well as four and a half weeks of Mr. Richmond's board (the only instance in which he paid board) in July 1787. To aid the July construction spurt, Revere rented a windlass, a crank-operated machine used to lift heavy weights, as well as five different purchases of "Drink" costing almost 3 pounds—an artisan-like perk to offer one's contractors. Revere's class views affect the way he referred to his workers: some appear as "Labourer Simmons" or "Joseph Brown, Labourer," while others receive a first and last name, such as Paul Reed. He uses "Mr." to address most merchants and property owners, while he offers the "Esquire" title to a select few, such as his Hitchborn cousins.

Revere billed a total of more than 148 pounds to his furnace account during this startup process, not including cash received from the Hitchborns, which might add more than 59 pounds. This significant investment offers the most tangible proof of Revere's commitment to his new endeavor, such a striking departure from either the silverwork that grounded his career or the mercantile activities that fueled his dreams. Time and success soon proved the foresight of his investment. Research visits to forges and ironworks, readings and correspondence, Revolutionary War and silverworking expertise, and innate mechanical aptitude compensated for Revere's lack of a formal metallurgical education, and he had his furnace running in a short time. These research efforts paid off in more ways than one: not only did ironworking yield great profits in the short term, but casting became a permanent component of Revere's industrial processes. He used this furnace for many applications in the years to come, often in support of his boldest endeavors.

When Revere began his silverworking career he had completed many years of apprentice training, inherited a fully stocked shop, benefited from his father's reputation and professional contacts, and received assistance from family members serving as his apprentices. Entrance into the iron-casting trade featured none of these shortcuts, which makes his rapid journey from research to daily furnace operation all the more impressive. On a typical day Revere might purchase supplies of pig iron from merchants or blast furnaces; help workers load his forge with charcoal and iron; look over his orders and decide which goods to produce; supervise the creation of sand molds and the casting of molten iron into those molds; interact with customers; balance his account books; deal with correspondence; and think about new ways to expand his business and improve his methods. Revere could never have learned the new trade as quickly as he did, and might not have succeeded at all, if he did not have such a strong basis of technical and managerial skills upon which to draw.

Iron casting led Revere a large step closer to the ideals of industrial capitalism. He, as well as many other former artisans or diversifying merchants, sought their own ways to adapt to America's growing population and changing markets, often involving the expansion of their businesses and the adoption of new production practices. Although Revere dealt with intertwined financial and technical challenges each day, we can make more sense of his new operations by dividing them into the four categories that defined America's new industrial paradigm: technology, labor, environmental impacts, and capital.[31]

Technology: Equipment, Production Methods, and Products

Revere never explicitly described his actual ironworking procedures, but we can infer his techniques from common contemporary practices alluded to in his correspondence. Revere's foundry depended most upon the large oven, which he often called a furnace, that he used to melt the pig iron.[32] This oven included a large, brick-lined hearth that contained a charcoal fire fanned with a hand- or foot-powered bellows if workers needed to increase the temperature. Most hearths included a chimney, a fuel and ash pit, and a movable crucible that could be used to hold smaller quantities of metal in or out of the hottest part of the fire.[33] Revere eventually used his oven to melt thousands of pounds of metal at a time—large cannon or bells exceeding a ton—but his first oven could have been a bit smaller before undergoing modifications.

The interior of the furnace had to reach the melting point of pig iron, around 1150 degrees C. After inserting large slabs of pig iron into the hearth, the founder stoked the fire until the metal liquefied. A small tap at the bottom of the furnace allowed him to safely draw out small amounts of molten pig iron, and a spigot or canal enabled him to direct the iron flow into different molds or receptacles prepared in advance. For meticulous work he used a ladle to pour small amounts of iron into the mold as gently as possible. After the item cooled the founder broke off the sand fused to the metal and touched it up with a file. Skilled founders could prepare incredibly detailed and elegant items directly from the casting.

One of the trickiest aspects of a founder's job was the preparation of patterns and molds. Patterns are wooden or metal replicas of the item to be produced. Founders kept and reused patterns for long periods of time, and the pattern's details and proportions had to exactly match the final object: any error on the pattern would mar countless cast items. The mold began as a mixture of sand or clay, carefully mixed from specific ingredients to achieve the optimal moisture and density. Founders packed the sand and clay composition into a flask, a hinged wooden frame that produced an indentation in the shape of the desired object. The founder carefully opened the flask without dislodging any sand, removed the pattern, resealed the mold, and poured iron into the hollow indentation.[34]

When casting larger objects that did not require as much detail, the founder prepared a pit of sand on the ground and pressed a pattern into the sand, creating an indentation to hold the molten iron. The finished item had a level top but the face in contact with the molded sand might contain intricate details from the pattern. Revere used this process, for example, for the casting of stove backs, iron plates placed in the back of fireplaces or hearths to reflect heat into the room and protect the brickwork of the chimney. The stove back usually had a design on the surface facing the room and a rough and flat surface facing the bricks.[35] Years later, when Revere cast even larger objects in this furnace, he probably used an output channel to direct the molten metal into a huge outdoor sand pit.

Even a skilled founder could botch the process at many critical points. The sand's consistency and density had to allow the sand to hold its shape when molten metal flowed into it while remaining loose enough to allow dissolved gases to escape. If dissolved gases could not escape as the iron cooled, the bubbles remained in the iron and made the metal porous. The founder also had to pour iron in a steady manner to minimize dislocation of the sand and ensure that the indentation filled completely. Iron shrank as it cooled, so the mold had to be slightly larger than the final object, but a cavity could form in the center of the object if the metal cooled unevenly. Slag, graphite flakes, or dirt contained in the iron or gathering in cracks in the mold might also enter the iron and weaken it. These errors had a large impact upon ironworkers' reputations, and the highly practical nature of American society guaranteed that all Americans could critically assess the value of all items they purchased, including the quality of the metal composing them.[36] Revere's research and conversations could only go so far in the mastery of such a complex and easily spoiled process, and the only guarantee of proficiency came from on-the-job experience. In this way the foundry learning process paralleled an artisan's apprenticeship, designed to give a practitioner years of experience under an expert's watchful eye before allowing him to work on his own. As a 53-year-old man attempting to juggle this new foundry alongside his silver shop and occasional merchant endeavors, Revere did not have seven years to spend practicing the trade. As a result, we might understand why he bought so much iron and charged it as a startup expense: errors would be frequent, at least for a while.

Many of Revere's expenses related to his ongoing efforts to improve his furnace's functionality, an understandable activity considering his ever-expanding production and product line. For example, significant purchases of lumber, nails, and carpentry work in September 1792 probably involved structural changes to his building. He also bought a new bellows in 1793 and invested in new patterns that allowed him to expand his range of products. Revere recorded one expense for "Altering patterns," several fees for the transport of patterns, one for "pewter for patterns," one mention of "iron plate & hooks for mould," and five purchases of "patterns for stoves and bells" between 1792 and 1794. Revere's eleven purchases of sand helped him make more molds, but might have met other needs as well. Similarly, two purchases of clay could have been used for construction or repairs to the furnace, molds, or for making clay models to be used in casting. He continued making research trips, now to the towns of Hanover, Norton, Providence, and Watertown. The purpose of these trips is not listed, but the Providence trip certainly involved a visit to his friend and advisor at the Hope furnace. As with many new manufacturing endeavors or technological systems, Revere's operations had the most flexibility in their earliest stages, and his detailed accounting of purchases and activities speaks to the number of important decisions he had to make, any of which might move his young establishment down a new path.[37]

Even in these early years when he had so much to learn and so many technological hurdles to overcome, Revere confronted the exciting ideal of standardization. An experienced founder could produce virtually identical sand molds from a high-quality model, and meticulous attention to the heating and pouring of the metal minimized or eliminated most casting errors. In theory, all products resulting from the same pattern should be close to interchangeable. In practice, however, furnace operators found it nearly impossible to control one key ingredient of standardized output—consistent raw material inputs—in light of the state of blast furnace technology. Ironworks produced different types of pig iron depending on the alloy of iron ore added to the furnace and the cooling rate. The major strains included "white" pig iron that was too hard to cut with chisels, softer "gray" iron, and intermediate "mottled" iron. Many ironworkers who lacked any scientific understanding of how carbon and other elements changed the properties of iron blamed all their problems on the quality of their iron ore, hoping to get rich quickly by exploiting pure ores requiring little or no work. While the quality of the ore certainly played a role, the stability of the manufacturing process also impacted the results, and failures to produce consistent high-quality iron frequently resulted from organizational and technical errors.[38]

Revere's records up to 1788 reveal that he sold cast iron window weights, grates, fire backs, and stove components during the early period of his furnace operation, all large and uncomplicated items befitting a novice practitioner. By 1792, he produced a wider range of products that now included smaller and more complex products. He recorded the highest sales of iron boxes, chimney backs, stoves, window and sash weights, and "coggs," a catch-all category that might describe iron connecting pieces or grooved fasteners. The records also contain various descriptions of stoves and related stove products, including Franklin stoves, large and small stoves, ovens, frames for ovens, stove backs, and "dogs," which referred to andirons used to hold up firewood logs in a hearth.[39] Not surprisingly, these functional items were highly standardized, created with simple molds that any worker could quickly press into casting sand.

Revere's other furnace sales represented less popular items often intended for highly specific uses. Some pieces, such as pig iron bars, blocks, and plates, appear to be resold raw materials. Revere may have melted pigs into smaller and more usable shapes for resale, or may have once again acted as a merchant, buying from the blast furnaces and reselling for a profit. Other sales, such as forge hammers, anvils, iron or friction wheels, swages, press plates, iron molds, gudgeons, and furnace covers, illustrate that Revere still supplied the rest of the Boston metalworking community. In preparing these items, Revere continued learning about the tools of the trade and kept his ear to the ground in the growing network of manufacturing professionals. It comes as no surprise that he continued branching into new fields when one considers his love of technical learning and his connection with so many skilled practitioners.

Labor in the Post-Artisan Mode

Revere increased his departure from the artisanal apprentice tradition by employing between six and ten full-time wage laborers in his foundry. He paid a wage according to the number of days they worked, although he occasionally continued his earlier practice of paying board in addition to wages. In addition to furnace operators, he hired people to build and repair equipment in the facility, carry materials, or perform related activities such as blacksmith work. Everyone served different roles as needed in the still-informal communal shop.

As part of the transition from artisanal to industrial managerial practices, Revere developed three different methods of tracking his employees' wages and hours. The expense ledger includes numerous cash payments to employees along with occasional memoranda about their rates of pay or days worked. He supplemented the ledger with assorted receipts that he now organized into a receipt book instead of throwing them into a box with other loose papers as he had done in his early silver shop. These receipts contained more detail than the ledger, including pay rates, number of days worked, occasional descriptions about the type of work, and tallies of the total pay owed to the employee before and after each salary payment. The third employee record is a one-page narrative describing each laborer's comings and goings.

The ledger is the most complete record containing entries for eight employees: Solomon Oliver, John Freelove, "Henry a black man," Stanley Carter, Nelson Miller, Whitney, William Story, and Capt White. The records illuminate the flexible nature of work in general and manufacturing in particular. Solomon Oliver was by far Revere's most frequently paid employee, with twenty-nine entries in Revere's ledger. Revere paid him nearly every week between May and October 1793, implying that he might have worked as a caster or in some other fundamental capacity throughout the entire working season. Oliver's salary started at 3 shillings a day, and increased to 4 shillings a day on September 1, 1793, not as high as the 5 or 6 shillings a day earned by several others. Zebulon White's receipts show that Revere also made fairly complicated arrangements with some of his workers to suit their skill sets and work patterns. On November 14, 1793, Revere paid him for 2 types of work: "77 days work at the furnace for my self" at a rate of 6 shillings and 8 pence per day; and "95 days work at [furnace] for John Freelove" at 5 shillings per day.[40] The former charge is the highest rate paid to any of Revere's employees, probably indicating that White was the most skilled laborer. The second memorandum implies a relationship between White and Freelove, in which White received or managed Freelove's wages. Revere's customized treatment of each of his employees makes perfect sense at this time: as a small manufacturer dealing with a specialized workforce he had no need for or conception of a standard contract, preferring to negotiate terms that pleased each individual while keeping the iron flowing.

The mysterious listing of "Henry a black man" represents one of the few mentions of race throughout the entire collection of Revere's papers. Henry appears twice in the ledger: once listing a fee of 1 pound, 11 shillings, 6 pence for 3 weeks of work, and also an expense of 2 pounds, 5 shillings, 1 pence for 3 weeks of his board. Henry received a smaller weekly wage than other employees, but this discrepancy might reflect his race, a different rate for less skilled tasks, or a compensation for the payment of board: the wage disparity vanishes if we combine Henry's wage and board into one charge. Other than one other listing for a payment of a worker's board (to a man named Elizer Homer, about whom nothing else is known), Henry is the only recipient of board in these records, marking a nearly complete break with the apprentice tradition in a short time. Paul Revere's old friend Thomas Wadsworth also alluded to Revere's racial views in a 1785 letter. Wadsworth moved from Boston to South Carolina and was happy in his new setting, though a bit guilty when he imagined Revere's response: "I have land enough for a very fine Plantation and can Slave it so as to put me in Independence. Suppose you will ask where it will put my Slaves I answer I have nothing to say on that subject now only they are black and therefore ought of right to be Slaves but don't ask me any more of those questions."[41] Wadsworth's tone seems to imply a familiarity with Revere's antislavery views, but in the absence of Revere's response we can only speculate. A vocal fan of merit-based social status such as Revere might very well oppose slavery, the ultimate form of hierarchical subjugation.

Revere paid most of his employees at the end of each working season, indicating that, like most early manufacturers, he served as a form of bank during the interim. In the early nineteenth century most New Englanders bartered a variety of goods and services all year long and settled accounts with coin about once a year, and employers such as Revere played a key role in these credit networks.[42] As an example, on September 13, 1792, he paid Miller, Story, and Carter for eight to sixteen days of "air furnace work." Story and Carter earned 5 shillings a day each. All three of these men chose to receive 1 pound and let Revere hold the remainder of their earnings. Miller eventually received ninety-five days' pay for "work at times since May" on December 2, 1793.[43] If Revere had a larger operation with a more diverse product line, as in the iron plantations to the south, he might have allowed employees to set up credit accounts and buy some of his products for their own use without ever needing to convert wages into cash. Revere's workers obviously trusted him with their wages, and his thorough records tell us that he took this responsibility seriously.

Although Revere's employment patterns approximated a regular work schedule, the expectations of late eighteenth-century skilled laborers required many concessions. His records include one extremely dense page of attendance and absentee information in paragraph form:

> Mr Miller came Thursday the 23 at night. Mr White & Mr Story came Saturday the 25 at night. Freelove came Monday night 27.
>
> Capt White Mr Miller Mr Story went home Saturday mor'g June 15. Mr Miller came back Tuesday 18 at night Capt White & Mr. Story came back Saturday 22d @ night. Mr Carter came Friday morning 28, went home 29 @ night—Capt White & Mr Story went home Saturd 29 at night.[44]

Unlike the careful spreadsheet-like lists of expenses or cash receipts, Revere used a free narrative in this attempt to track employee attendance. The attendance page describes the work habits of Captain White, Mr. Story, Mr. Miller, Mr. Carter, Stephen Metcalf, Mathew Metcalf, and Freelove, who is listed without any title.[45] Revere employed between one and five of these workers at a time, usually a minimum of three. The workers started and stopped working at seemingly random dates and times, occasionally beginning at night or on a Saturday, and ending in the morning or on a weekday. The duration of each employee's labor varied considerably throughout the 146-day period between May 23 and October 15: Miller, White, and Freelove worked approximately 70 to 75 days, Story worked approximately 48 days, and Carter and the Metcalfs worked between 20 and 25 days. The furnace operated in 3- to 5-week shifts, always followed by at least a week of rest. These worker routines mirrored the work patterns of a blast furnace despite the fact that Revere's operation did not depend on the seasons like blast furnaces did, for example, when they shut down in winter due to waterpower shortages. Perhaps Revere adopted a blast furnace schedule because he learned the trade from Brown and Benson and he decided to emulate their schedule. Alternately he might have worked whenever he received regular shipments of pig iron, which naturally followed the rhythm of blast furnace life.

Workers received benefits other than some control over the dates and times of their work. Revere charged seven rum payments (two of them for "6 days"), two beer payments, and one payment for "Liquor—7 days" to his furnace expenses. This notation attests to a steady liquor allowance for his workers, since rum and liquor could now be measured in units of time. The steady flow of liquor appears in most, if not all, early craft and manufacturing shops whose skilled workers expected certain privileges in exchange for their services, and many owners or workers grew accustomed to such treatment from their own apprentice and journeyman terms. Heavy liquor use on and off the job united early working men across trades, nations, and skill levels. One medical analysis published in 1804 calculated that the average workingman drank eight gills of hard liquor a day: two gills (the equivalent of four small drinks) before breakfast; three gills between breakfast and the completion of the noontime meal; and three large drinks typically taken at 2:00 p.m., in the middle of the afternoon, and upon leaving work at 6:00 p.m. This daily liquor, totaling a quart of hard alcohol, cost approximately 4 shillings, or half a dollar, a fairly substantial sum. The article concluded that although this level of drinking had financial and long-term health implications, in the short term most workers were able to work at their jobs while drinking at this level without showing signs of drunkenness.[46]

Revere's overall employment picture combines elements of discipline and freedom: for the most part his employees all arrived at the beginning of one major work period and all left at the end, but they definitely had the freedom to alter their schedule. At least one worker took occasional days off to go fishing, and all the workers adjusted their arrival and departure times by a few days whenever the furnace started or stopped.[47] Revere's laborers received a reasonable set of perks in exchange for their efforts: wages in line with their skill levels, steady liquor on the job, and control over their workload and work dates. As much as he might have wanted to dictate his workers' attendance patterns to suit a more predictable schedule, the flexible patterns portrayed in the records tell us that labor negotiations were a two-way affair. Revere and his workers both had a say, and growing streams of output imply that this arrangement suited all parties.

Raw Material Availability and Environmental Impacts

The switch from silver to iron brought Revere in contact with a new assortment of customers, suppliers, and raw materials. A silversmith dealt with merchants, other silversmiths, and upper-class clients when buying silver and selling pricey silverware, but a foundryman interacted with blast furnace operators, colliers, other metalworkers, and anyone who needed iron goods, watching tons of iron and charcoal enter and leave his shop. The price of silver depended upon tariff, mercantile, and political factors, while the prices of iron and charcoal primarily depended upon their local abundance. And while a silversmith's activities had an impact upon his clients, suppliers, and fellow practitioners, an iron founder's labor left its mark upon the face of the land itself.

Neither Paul Revere nor anyone in early America thought about the environment as an entity, let alone one facing endangerment from human activities. Beginning with the first New World colonists, Americans focused upon the specific local conditions, such as forests or rivers, that might aid or hinder their struggle to survive and prosper. One of the main compensations for America's isolation and lack of European civilization was the abundance of natural resources, such as land and timber, valuable commodities in light of England's dense population and almost total deforestation. Colonists saw forests as infinite and inexhaustible and viewed other natural resources in almost the same light. Postwar Americans gradually realized the finite extent of these resources, leading to variations in prices, occasional shortages, competition, and eventual conservation. Manufacturing firms of the late eighteenth and early nineteenth centuries had a growing impact upon their natural surroundings, leaving imprints such as dammed or polluted waterways, slag heaps, and air and noise pollution. Paul Revere's evolving environmental practices and attitudes mirror that of the nation as a whole, although he always stayed one step ahead of the multitude. While he had minimal resource needs during his early career as a silversmith, he became increasingly dependent upon raw materials as his endeavors grew in number and size, starting with his ironworking practices.

Although he purchased large quantities of coal, limestone, and clay for his furnace, the lion's share of Revere's raw material efforts and expenses centered upon iron. In a pattern that foreshadowed his upcoming copperworking experiences, Revere and his purchasing agents cast a wide net in search of cast, bar, and "old" recycled iron, often making purchases in Rhode Island and New Jersey as well as Massachusetts to meet the various needs of his manufactory. Pig iron directly fed the furnace, and served as his primary and most fundamental purchase. Bar iron cost more than pig iron, reflecting the extra labor needed to refine the pig iron into this stronger and tougher metal. The fact that Revere purchased bar iron implies that he may have performed or subcontracted blacksmith operations in addition to his casting, reshaping the bar iron into tools and common household products. "Old" iron commonly referred to a variety of recycled iron goods Revere purchased from merchants or accepted from customers as a credit, though he disliked its dubious and inconsistent quality. Revere purchased a large quantity and variety of raw materials between 1792 and 1794, including at least 32 pounds spent on coal, 11 shillings for arsenic, 36 pounds for copper, 29 pounds on tin, and an unknown amount (probably more than 50 pounds) on iron.[48] Raw materials for iron casting and all his other endeavors cost a lot of money, but natural resources had many other impacts upon his operations, and Revere influenced his environmental surroundings in many ways as well.

With the exception of silverworking, all of Revere's endeavors repeatedly slowed or stopped their operations because of raw material shortages. A lengthy correspondence record illustrates his endless quest for iron. He started searching for iron sources more than a year before he announced his foundry's operation. J. Blagge, his purchasing agent, quoted pig iron prices at the Alsion and Batson furnaces (10 pounds and 9 pounds, 10 shillings per ton, respectively) in November 1787, but reported that both furnaces had ended their yearly production and could not sell any goods until they restarted in the spring. Brown and Benson wrote Revere in April 1788, offering six tons of pig iron at $28 per ton, although they had trouble shipping it to Boston and asked if he could pick it up in Marblehead.[49] Revere tried his best to procure a regular supply in the November 3, 1788 letter to Brown and Benson: "Mr. John Brown when in Boston informed me your Furnace was to go soon. I should be glad that you would ship as soon as possible ten tons of Pigs . . . We are desirous to have a constant & regular supply of Piggs from your furnace, & in order to do it we think there cannot be a more effectual way than by interesting you, or some of your gentlemen owners of the Hope furnace, in our furnace, for that end we are willing to sell either one quarter or one third of it."[50] Revere would not offer to relinquish one-third of his newly finished furnace for anything less than a critical component of his operations. The Hope furnace produced fairly consistent output and had a solid reputation by this point. Revere could have secured a permanent iron supply if Brown and Benson accepted this offer, but they turned him down. Revere wrote another letter to Brown and Benson in September 1789, repeating many of the details and more formally requesting up to one hundred tons of pig iron a year, a staggering quantity of metal in a time when most blast furnaces produced less than one thousand

tons annually.[51] In spite of his foundry's small size, Revere adopted the perspective of a large technological system because he hoped to minimize uncertainty by expanding his operations to incorporate and control environmental
factors such as raw material sources. Revere disliked uncertainty and correctly
identified iron shortages as a continuing thorn in his side. As bad as they
were, supply uncertainties grew far worse when he later worked with copper,
a scarcer material than iron.

Like all manufacturing operations, Revere's foundry impacted its natural
environment through the raw materials and pollutants that entered and exited his premises. The pig iron from Brown and Benson and other sources, as
well as charcoal, stone, sand, and wood purchases, all had an impact on the
local ecosystems and landscapes from which they were taken. And Revere
affected his immediate environment in at least one documented way, by producing a large plume of smoke from his burning charcoal. J. Callendar's 1788
copper plate engraving of Boston includes and almost highlights a plume of
smoke emanating from Revere's foundry.[52] In a very real sense he changed the
Boston landscape.

Even though Revere's records delineate the foundry's raw material usage, we
cannot connect his many activities to specific environmental consequences.
Revere's foundry operated alongside thousands of other small craft establishments in the Northeast. All of these manufacturing impacts competed with
other uses of New England's then-abundant natural resources, particularly
fuel use and agricultural land clearing, which eliminated more forestland each
year. Revere contributed to deforestation, air pollution, and the depletion of
local and distant ore supplies, but his direct effect upon the environment cannot be quantified: each of his raw material suppliers also made sales to many
other customers and Revere's environmental impacts had plenty of company.
Direct impacts, however, are not the whole story. Revere's continual orders of
pig iron, when added to the purchases of other founders like himself, provided
a reliable local market for blast furnace owners who wanted to avoid the unreliability of overseas shipping and competition from British furnaces in distant
markets. Revere's presence in the network of blast furnaces, foundries, fineries,
blacksmiths, and other local metalworkers enabled the entire community to
supply the needs of the growing American market more efficiently, and therefore, he indirectly contributed to increased iron production and all of its side
effects.

By supporting blast furnaces Revere did contribute to deforestation, because blast furnaces did more damage to the wooded environment than any other existing operation. Blast furnaces used about an acre of old growth wood or a larger amount of secondary growth wood each day to make charcoal, a fuel that burned hotter and faster than wood. Total wood usage doubled if the ironworks included a finery. The high sulfur content in coal made it difficult to use, and plentiful forests enabled American ironmasters to continue working with charcoal until the rising price of wood encouraged conversion to coal-based ironworking practices, such as puddling in the 1830s. Even so, charcoal fuel still accounted for nearly 100 percent of all American pig iron production at that time, and charcoal still accounted for 87 percent of iron production by 1860. In contrast, British ironmasters had to switch to coal as soon as the 1770s to remain in business, enabling a renewal of England's beleaguered forests.[53]

Despite blast furnace operators' seemingly wasteful wood practices, they recognized some of the perils of deforestation to the extent that they affected operations. American federal and state authorities in the eighteenth and early nineteenth centuries never attempted to regulate natural resource use and gave ironmakers free rein over their environment. Even though business owners usually focused their efforts upon maximizing labor efficiency because they identified labor as the scarcest resource, blast furnace operators often pioneered woodland management techniques. Most operators attempted to procure twenty thousand acres of timber per furnace, assuming that they consumed one thousand acres each year. This strategy allowed every harvested portion of the forest to re-grow for twenty years before using it again. Unfortunately, erosion and other problems occasionally inhibited the second growth, and when it did appear it often contained smaller amounts of poorer quality wood. To mitigate wood shortfalls, some ironworks initiated rudimentary forest management policies. Notations in a small number of ironwork ledgers indicate that some owners paid to fence, protect, and transplant "sprouts," or new saplings, that started to re-grow from previously harvested areas. While forward-looking, blast furnace forest management was not surprising: if a blast furnace ran out of wood, it closed its doors.[54] Ironmasters adopted these resource protection measures out of the most pragmatic reasons, because they wished to avoid causing irreparable harm to vital ingredients of their business. In this sense they heralded the conservation mindset that gained prominent advocates at the start of the twentieth century when America's growing population and industrial development visibly endangered many natural resources.

Although he did not operate a blast furnace, Revere impacted the environment by helping to expand the growth of the metalworking community and increase its use of natural resources. Environmental limitations impacted him in return. Raw material scarcity played a much greater role in his foundry activities than in his silverworking trade. As a silversmith, he purchased silver from overseas merchants or local customers, and used small amounts of wood, charcoal, and other materials. His production depended more upon the affluence of the local population than upon the availability of materials. The opposite circumstances applied to ironworking. Revere never alluded to any problems obtaining sufficient fuel for his operations (although fuel prices rose steadily), but iron shortages became a continual impediment. Seasonal patterns of iron production forced Revere to stop his production during the winter months and he constantly asked merchants and purchasing agents to locate new iron sources. He often resorted to purchases of "old" iron that he melted down and reused in spite of its questionable quality. Customers and large clients paid some bills in old metal, a practice that expanded when Revere shifted to copperworking. From this point forward, metal availability and quality played a primary role in all of his managerial and technical decisions, and he increasingly acted in a capitalist-industrialist manner, treating the environment as a commodity and limitation. This trend reached its apex years later, when he operated a rolling mill complex in Canton.

Capital Concerns: Sales, Profits, and Management

In 1788, for the first time in his career, Revere found himself in the position of starting a new technical business from scratch. Thanks to a profitable silver shop and an unspecified arrangement with his well-off cousins he avoided the pitfall of limited investment capital that plagued the vast majority of new endeavors in postwar America. But financing his foundry's construction hardly ended his pecuniary challenges. He still had to manage the operations, balance day-to-day problem solving against longer-term strategic choices, and simultaneously wear technical and managerial hats. Many of the prominent startup manufacturing operations of this era succeeded due to fortunate alliances between technical practitioners and moneyed entrepreneurs such as

Samuel Slater and Moses Brown in the Rhode Island spinning mills, Robert Fulton's and Robert Livingston's work on early steamboats, and Paul Moody and Francis Lowell in the Lowell textile mills. In keeping with the theory of large technological systems, Revere played the combined role of an inventor-entrepreneur by overseeing all aspects of the new business, technical and otherwise. Even if Hitchborns fronted some of the money and skilled employees performed most of the work, success or failure, with all of the consequences, rested almost solely on his shoulders.

Revere's sales information, while sketchy, paints a picture of early financial success. One of his ledgers records payments (income) and expenses (costs) for most of the year 1793. The lower estimate of income from his sales during his 1793 foundry operations is slightly more than 393 pounds. His 1793 expenses total slightly more than 311 pounds, yielding Revere at least an 82 pounds profit for the year, and possibly more.[55] Revere's strategic efforts as manager of his shop centered on this fundamental equation, and for the remainder of his career he attempted to raise his income as much as possible while lowering his expenses.

The very existence of separate credit and debit tallies reveals Revere's growing interest in accurately measuring his financial performance. The foundry represented another giant step in his proto-industrial progression from informal barter and credit exchanges to monetary accounting. His silver shop began as an eighteenth-century craft operation, but like most American businessmen in this period he started adopting capitalist techniques in response to the expanding market economy. Revere's workshop and early foundry operations involved an even greater reliance upon cash, wage labor, and written contractual agreements. He started writing receipts for many services and these receipts invariably included the monetary value of all services rendered. He also changed his accounting and recordkeeping technique upon opening his furnace. His early silver shop records are far more confusing than the surviving furnace records: for example, he freely commingled credit and debit accounts, listing expenses and sales together.[56] With the beginning of the furnace, Revere clearly separated his expenses from his sales and income via a double entry bookkeeping system.

Revere gained familiarity with the sales and pricing patterns of metal goods from his mercantile experience. Starting in 1783, his all-purpose hardware and general goods store sold metal items such as candlesticks, kettles, saws, files, window weights, hinges, iron nails, locks, and buckles, as well as items used in a metalworking shop such as bellows, melting pots, and "moulding sand." Although he probably did not make many of these items at first, the retail business placed him in contact with merchants, local blacksmiths and coppersmiths knowledgeable in their arts, as well as the local clientele, a much larger group than the generally well-off patrons of his silver shop. Revere had lengthy and complex transactions with local producers, involving a reciprocal exchange of goods, services, cash, and debt. As a result, he gained an intimate awareness of the local demand for different items, the range of available products, typical wholesale and retail prices, and the practices of the metalworking community such as the use of the most popular metalworking tools. This experience also allowed him to inspect the construction details of different items and chat with their makers, which must have helped him learn the technical processes as well.

Furnace records illustrate the seasonal nature of iron production and sales. The working year began with slow sales in March, which increased by the end of April. The high production season extended from May through September. Business dropped off in October and November, and operations then ceased until the following spring. Incomplete customer cash payment records reveal a lag of several months (averaging around sixty days) between product sale and payment. Customers usually started paying their bills in June and July, the cash inflow picked up in August, and the largest number of payments arrived in September and October.[57] This cycle reveals the close ties between Revere and the networks that supported him. His production cycle relates to that of a blast furnace, which ceased operations in winter when rivers froze and in summer when rivers ran low. The payment cycle also coincides with a typical harvest season, since most farmers sold their crops at some point in the fall and used the proceeds to pay their debts and make purchases for the next season. As Revere shifted his operations from silver into iron casting, he started falling out of touch with merchants' patterns and grew more connected to natural rhythms of water flow and agricultural production.

Although Revere served in a managerial capacity when he directed his silver shop workers, the iron trade required significantly more complex purchasing strategies, market research, and labor supervision. Revere observed the practices and cultures of different ironmaking establishments, the largest of the technical communities and some of the largest managerial concerns in America, throughout his ironworking career. Ironworks introduced many managerial innovations as well as some of the first rigid labor expectations, and the industrial discipline that spread throughout textile and other factories in the mid-nineteenth century had roots among colonial furnaces and foundries.[58] The overlay of individual skilled labor and rigorous supervised discipline grew increasingly prevalent in Revere's shop, although as we have seen, he still allowed his men to enjoy many of the privileges and freedoms of skilled laborers. His managerial practices reveal, as do so many of his other operations, how Revere kept a foot in both worlds, bringing in or devising his own new policies to deal with the growing scale of his operations while often falling back on the tried and true artisan traditions that guided his own training.

Ironworks, from the smallest foundry to the largest iron plantation, represented a vital step in America's proto-industrial transition, as they retained ties to rural and pre-modern traditions and practices while also necessitating the adoption of continuous flow processes, the designation of a distinct managerial class, labor discipline and regimentation, and the concentration of large quantities of investment capital.[59] Paul Revere joined this movement for practical reasons, as he expected to quickly learn the business and profitably meet a market demand. But he soon thought about manufacturing and management from a perspective even farther removed from that of the skilled practitioner. Revere's iron career brought him into a new world, across a point of no return. The days of artisanal skilled labor in his silver shop lay forever in his past, to be replaced by an ever-increasing scale of production. In the years from 1787 to 1792, when he focused on setting up his foundry and worked primarily on iron goods, he engaged aspects of large technological systems, technology transfer, and environmental impacts that later characterized the age of industrial capitalism.

Revere's rapid foundry success resulted from fortuitous timing, innate technical aptitude, thorough research, and the casting experience he gained from silverworking. He came to realize that the foundry oven melded the characteristics of tools and machines: it required skilled labor and could be used in a flexible manner to produce different products, but an expert could produce consistent output by following a standard set of production practices. During his iron-casting career Revere taught himself and his employees to use the foundry oven in an even more machinelike way, optimizing its use until they could produce highly standardized output. He gained great experience in the technical aspects of this challenge and his confidence served him well in the years to come.

Paul Revere had already begun moving toward standardized output while working in his postwar silver shop but he truly embraced this ideal as a founder. As a silversmith, Revere produced works largely custom-made for each individual customer. Although silverwork styles imposed certain common forms and conventions on all pieces, customers prized many items for their uniqueness. Revere and other silversmiths started rethinking their trade after the proliferation of fairly standardized utensils and common items, aided by the output of silver-flatting mills, but the highest calling of all silversmiths remained the prestigious production of unique silver plate. Ironworking imposed entirely new values and goals upon the producer: from blast furnace operators to founders and finers, the measure of a skilled ironworker was his ability to produce uniform metal and standardized objects. The use of molds and patterns imposed an ideal upon the founder in particular: the best objects duplicated the model used in casting, and all deviations from it decreased the item's value and perfection. Iron founding changed Revere's standards as well as his methods.

Revere could not control the managerial aspects of standardized output at this time, nor could anyone. The iron network, while strong, had a long way to go before iron production took place with any regularity. American pig iron had a fairly consistent quality by this point but its scarcity forced him to turn to undependable "old" iron. And even if he had arranged a reliable iron supply, his laborers still expected to be treated as craftsmen, with control over their work schedules manifesting in ways that disrupted regular foundry operations. These complications did not particularly concern Revere because standardization did not yet exist in the technical vocabulary. His version of standardization essentially enabled his shop to produce more objects cheaper and faster without requiring his own labor. In this sense he succeeded. For his next endeavor, standardization was not even a possibility.

CHAPTER FIVE

Bells, Cannon, and Malleable Copper (1792–1801)

In 1792 the leaders of the New Brick Church convened a meeting to discuss a matter of some urgency. Their church bell, transferred from its previous home in the Old North Church in 1780 after somehow surviving the wartime dangers of the British occupation, had developed a substantial crack. The damaged bell could no longer be used for services, and, in fact, could only be rung in the case of fire. The membership of this meeting included thirty-five church officials and major donors, including Thomas Hitchborn, Samuel Hitchborn, and Paul Revere.

Paul Revere's presence at such a meeting reflects a lifetime of diligent church membership and service. Even as a 13-year-old boy he took his religion seriously and equated it with public service, signing a formal compact with six friends who joined him as bell ringers. Two years later he risked his father's wrath to attend a controversial sermon at a different church. In spite of this youthful rebellion, Revere followed his father's lead by remaining loyal to the New Brick Church as long as he lived in Boston and he served on numerous church committees between 1788 and 1800.[1] As with his other activities, Revere's church membership afforded him the opportunity to network with colleagues, drum up potential customers, serve in prestigious positions of authority, and play a visible role in his community. But even if the church offered occasional benefits he never exploited it, and the majority of his charitable church work or religious comments in his letters and speeches demonstrated his piety without yielding any tangible reward. His youthful respect for religion had only increased by this later stage in his life, and he took matters such as the 1792 church bell committee to heart.

The primary purpose of this meeting involved a subscription drive to generate the funds needed to pay for a new bell. In a moving show of solidarity, the group voted that any member then present who failed to pay the amount of his promised donation would not be allowed to ever hear the bell. In a more serious vein, these men had to determine the proper course of action in acquiring a new bell. We can imagine 57-year-old Paul Revere half-listening to proposals to ship the old bell to England for recasting while quietly calculating the price and availability of bell metal, the size of his own furnace, and the process for making what, to him, would be a gigantic mold. He knew a thing or two about bell making from earlier readings and discussions, but the field remained abstract to him, both intimidating and intriguing. Technical challenges abounded, and he had to wonder about the optimal thickness; the ratio of height to width; the proper proportion of copper, tin, and other metals; and so many other issues great and small. Bell casting would be difficult, but not impossible for a man who had already mastered the art of casting small silver items and large iron ones. Could he afford to take on this responsibility, or more accurately, could he possibly refuse such an exciting challenge?

Revere accepted the job at that very meeting and immediately set off to work. Despite the complexities of bell making and the lack of a supporting network along the lines of America's close-knit ironworking community, Revere's broad range of skills prepared him quite well for this new task. Most important of all, he did not need to modify his furnace for this new trade. He added 412 pounds of new copper and tin to the approximately 500 pounds of material salvaged from the cracked bell, and to help with the casting process he visited other foundries. By the end of 1792 he had succeeded, in a manner of speaking. The bell's tone received poor reviews, such as Reverend William Bentley's private observation that "The sound is not clear and prolonged, from the lips to the crown shrill." Furthermore, alone among the more than one hundred bells his shop produced during the period of his active involvement, it contains visible creases and imperfections and soon showed many signs of wear, a sign of his inexperience.[2] And thus began Revere's latest enterprise, not with a bang, but with a shriek.

Even this inauspicious beginning made Revere proud, with good reason. His was the first bell cast in Boston, as attested by its inscription "THE FIRST CHURCH BELL CAST IN BOSTON 1792 BY P. REVERE." This same bell received an extended tribute from the Reverend Edward G. Porter almost a century later:

> Few bells have such a record as this. It has hung on three conspicuous churches, either in its original or enlarged form. It has summoned six generations of worshippers to the sanctuary. It has tolled for the dead, and awakened the living from their morning slumbers. It has opened the daily market, announced the hour for lunch, called the hungry to their dinner, and the weary to their beds. It has broken the stillness of the night by its dread alarm of fire. On momentous occasions it has rallied the citizens to meet in defence of liberty. It has sounded the tocsin of war, and rung merrily on the return of peace. It has assisted in the patriotic celebrations of the Fifth of March, the Seventeenth of June, and the Fourth of July. Truly such an active and faithful participant in the affairs of Boston during so long a period of our history deserves a place among the famous bells of the world.[3]

Reverend Porter's reverential listing of the tasks performed by Revere's bell and his placement of the bell into the center of local history personified it with a level of respect accorded to few other technologies. This encomium could just as easily serve as a memorial to a human being.

Bell casting opened doors that soon led to new endeavors. Sitting in that meeting room, volunteering to learn the bell-casting business, Revere completed a paradigm shift that began when he first established his foundry. He now saw his "air furnace" as a multifaceted machine that could cast items from different metals or even alloys of metals, in sizes far larger than anything he previously attempted. Following this realization and armed with growing confidence, Revere embarked on a host of new metallurgical adventures in the years following 1792. If he could make the leap from iron implements to bronze bells, surely he could transfer his bell experience to cannon casting. And after working with large objects consisting primarily of copper and tin alloys, he had no reason to fear a shift to smaller items made solely of copper. In the wrong hands confidence becomes a great liability, but in this case it was well placed. By 1795, Revere's product line had grown so large that some experienced observers found it hard to believe any American could really do all of that.

Revere's rapidly increasing experience coincided with local and nationwide economic and political developments. His timely shift from iron to copper alloy products possibly resulted from an assessment of his location and an understanding of his raw material needs and limitations. New England's iron sources paled in comparison to sources in the rest of the country, particularly in the mid-Atlantic states. When small New England furnaces could not supply his needs he had to pay expensive transportation fees. While he could (and did) continue to produce cast iron implements on a moderate scale, he had more trouble expanding his output and making his mark on the field when saddled with exorbitant operating expenses. In contrast, America possessed only scarce supplies of copper. The ideal site for a copper manufactory was a location near a major port city that could provide supply and demand, surplus labor, and mechanical expertise. Boston was one of the three best locations in the nation.

The economic climate also favored his new endeavor, as the credit contraction and depressed markets of the mid-1780s had faded to an unpleasant memory. The establishment of a stronger national government headed by George Washington in 1789, as well as Alexander Hamilton's proactive array of aggressive economic policies, provided the nation with much needed stability. This stability offered a firm platform for steady economic improvement in the late 1780s and 1790s, fueled by internal factors such as population growth and the growth of urban centers that expanded America's market economy, as well as external factors such as rising foreign demand for American goods and shipping. The new nation now benefited from stronger political leadership, solid national credit, and increased circulating capital.[4] It needed these assets in the coming years, as a series of international crises required the creation of a military establishment practically from scratch. Many Americans, including Revere, took advantage of the strong economy by enlarging their operations to satisfy the growing national or international demand for manufactured goods. His wonderful ability to learn new trades was vital once again, but as his operations matured the technological challenges were rivaled by the need to secure sufficient amounts of investment capital, manage and coordinate a growing body of skilled workers, and react to rapidly changing market conditions. The successful eighteenth-century manufacturer continually integrated technical, entrepreneurial, and managerial aptitudes as Revere had done for some time.

Revere's decision to enter the bell-casting field also marks a turning point in his perception of his career. While he enjoyed the status of a skilled craftsman before the war, in the years that followed he hoped to use technical work as a springboard to a position of societal service, at which point his sons could take over the workshop. By 1792, his merchant career had clearly failed to take off, and while there is no reason to believe that Revere lived an unhappy life or disliked his silver and iron trades, he also must have understood that he had not moved any closer to the upper classes. Bell making started out as a chance to help his church out of a bind while adding a lucrative new product to his repertoire, but this new activity soon reinforced an important lesson: manufacturing had its own rewards. Revere now had the opportunity to craft beautiful items that proclaimed the glory of God while also serving practical societal needs. Bells also made a statement about America's growing manufacturing competence, all the more powerful due to the absence of local bell casters. Revere could serve his religion, his society, and his bank account at the same time. This advanced high-profile technical work unknowingly started a pattern that defined the remainder of his career. Over the next few years he produced cannon, bolts, and spikes for his government, and in so doing helped it to overcome the great technological gap separating it from potentially hostile foreign powers. Throughout the 1790s the practical and symbolic importance of technology became increasingly clear to Revere, his community, and the American government. Beginning with that first bell, he realized that the makers of quality items played their own vital role in the history of the new nation.

Becoming a Bell Maker: An Art and a Science

Early metalworkers undoubtedly noticed the sound-making potential of different alloys whenever they banged a newly made bar or sheet and observed its vibrating tone. The earliest bells, intentionally produced for that purpose, probably appeared in China, as they are mentioned in various myths and legends there dating to the twenty-ninth century BC. In the Western world, Greeks on the island of Crete produced bells at points between 2000 and 1800 BC, with bronze bells in Athens dating to the sixth century BC. These early bells often carried mythic overtones, as various legends associated their peal with the ability to separate truth from falsehood or punish wrongdoing. As different societies gained metalworking experience, particularly in the field of casting, they learned to produce larger bells capable of being heard at farther distances. Christian churches started using large bells in Western Europe and England around the eighth to tenth centuries AD, becoming both prominent and prevalent by the eleventh century. Large bells encouraged architectural changes in churches, such as the development of high front gables, soon known as bell towers, which extended the range of their music even farther and summoned the most distant townspeople to services.[5]

Although bells served many different purposes, the church bell played the largest role in early American society. Looking down upon the world from majestic towers at the tops of churches, often the largest structures in places like religious New England, church bells maintained the highest possible visual and auditory profiles and illustrated the symbolic power of technology. Church bells served a variety of named functions each week: for example, the "Gabriel" bell woke the community; the "Sermon" bell summoned parishioners to church services; the "Pardon" bell symbolized the pardoning of sins in the middle of some services; and the "Passing" bell announced the death of a community member with three sets of three rings for an adult male, followed by one ring for each year of his age. Bells also served as important instruments of general communication, used for fire alarms or to proclaim momentous news. Some deacons left bell ropes hanging outside their churches to allow community members to ring them in emergencies. Church bells held such an esteemed societal role that they often inspired complimentary verses, either published or inscribed on the bell itself. Many bells cast by Revere and others carried the inscription "The Living to the Church I Call, and to the Grave I Summon All." An anonymous writer humorously outlined the connection between church bell and parish in 1816 when Revere's largest bell made its debut in the stone tower of King's Chapel:

The Chapel Church
Left in the lurch
Must surely fall;
For Church and people
And bell and steeple
Are crazy all.
The Church still lives,
The Priest survives,
With mind the same.

Revere refounds

A bell resounds

And all is well again.[6]

The complexity of the manufacturing process grants each bell a unique personality and, one hopes, its own special allure. Metallurgists often describe bell making as both an art and a science, partly because minor variations in every aspect of the procedure enable workers to produce a nearly infinite range of final products. The quality of a bell's sound primarily depends on factors such as the type, quality, and proportion of metals and the shape and size of the mold. Increasing the amount of copper, decreasing the percentage of tin, or adding small amounts of other metals affects each bell's final attributes. Bell making also requires an understanding of geometry and general mathematics, to enable the bell maker to scale a general pattern to different sizes without altering the bell's complex acoustical properties. When struck properly, bells emit a musical chord consisting of main note called the "fundamental," a "nominal" note an octave higher, a "hum" note an octave lower, and several "partial" notes in between. The topmost parts of the bell emit the higher notes and the low hum note emerges from the lowest part, or lip, of the bell. Therefore, each alteration of the ratio of height to width, the thickness at any point, or the overall size of the bell completely changes its volume and tone. A skilled bell founder can pitch a bell in different musical keys based upon its size and dimensions, and can alter the bell's sound during the tuning process by removing metal to diminish some of the partial notes while maintaining the proper intervals between the entire chord. The rare bell able to produce a desired sound without tuning is known as a maiden bell, the holy grail of bell founders.[7]

Revere's records illustrate the difficulty in scaling a bell mold to a different size. One 700-pound Revere bell measured 32 inches diameter across the bottom, 27 inches high, and 17 inches across the top. In comparison, the 2,437-pound bell at King's Chapel in Boston, the largest ever cast by Revere and Son, measured 49 inches across the bottom, 36 inches high, and 27 inches across the top.[8] Revere struggled to optimize his bells' shapes and weights throughout his career, as illustrated by occasional customer complaints or severe miscalculations of some bells' final weights.

Bells seemed to invite paradoxes, as their sounds had to be simultaneously loud, clear, and beautiful. No two church bells sounded the same, and especially in a large town such as Boston hosting many churches, the sound of one's bell was well known by members and nonmembers alike, representing that parish in the eyes and ears of society. Revere the artisan, maker of beautiful objects, worker of metals, and lover of all things technical, immediately and irrevocably succumbed to the enchantment of bells. He embraced the many intricacies of bell casting from 1792 through the end of his life, producing drawings, recipes, and sample bells in his drive to unravel the perfect combination of metals and the optimal bell production process, as did countless bell makers before him.

Bell casting first introduced Paul Revere to the challenges of working with copper, by far the largest component of bell metal. Copper shares some of silver's characteristics such as its high degree of malleability, but also offers some of the functionality of iron. Unlike Revere, early human societies worked with copper well before iron because they could more easily smelt it into a usable form. Most copper occurs in copper ores that combine metal, oxygen, and other elements in a rocky mineral. The visual distinctiveness of these ores and the existence of surface deposits in many parts of the world undoubtedly aided the early adoption of copper. Malachite, for example, is a fairly common copper ore known for its beautiful green color. The first copper smelting took place in the Sinai desert in southern Israel, approximately around 3500 BC, perhaps as an unintentional byproduct of pottery firing. Copper smelting is a physical and chemical process requiring heat above 1084 degrees C as well as a chemically "reducing" atmosphere rich in carbon but lacking in oxygen. These conditions are hard to create accidentally, but definitely exist within a pottery kiln.[9] Copper mines and artifacts have been found at various European sites dated between 4000 and 3000 BC, with a general diffusion of the technology from southern to northern Europe and west to East Asia. Common copper items of the ancient world included agricultural tools such as sickles and plows, carpentry tools, and ornaments.[10]

Copper, while harder than metals such as silver, still proved fairly soft and of limited utility in items such as blades, weapons, and armor. Early smiths could harden copper through hammering and heating, but this ran the risk of making the metal brittle and unusable. Copperworking technology underwent a dramatic improvement when ancient metalworkers began experimenting with alloys, or different mixtures of metals, in search of improved physical properties. After unsuccessfully experimenting with copper-arsenic alloys, metalworkers settled on the combination of copper and tin, known as bronze.[11]

Bronze offered many advantages over copper, such as increased hardness and strength, fewer bubbles produced during cooling, and a lower melting point (around 950 degrees C) that made it easier to work. Bronze first appeared in the Near East and Mesopotamia around 3000 BC, and caused an explosive growth in overall metal production and mining, with bronze weapons and armor quickly gaining ascendancy over their competitors. Bronze remained the preferred metal alloy for more than one thousand years and gave its name to a dynamic age of human history.[12] The rise of iron, discussed in the previous chapter, eventually reduced the general demand for bronze goods because of iron's greater strength and hardness. However, bronze continued to serve wherever one needed a lighter or more ductile substitute for iron. Because of the acoustical properties of bronze, bells became one such application.

Bronze alloys have many intrinsic advantages for cast items such as large bells. A tough metal, bronze can withstand the shocks and impacts that bells receive throughout their lives. It also melts at a relatively low temperature, which allows the caster to ensure that it melts in a uniform manner, not hardening until all the metal has time to fill the mold. Bronze is fairly soft, especially in comparison to iron, which makes it much easier to manipulate during the tuning process. Bronze's softness also relates to its elasticity, which allows the bell's vibrations to last a long time and propagate over great distances. Finally, bronze resists corrosion and is lighter than iron, essential qualities for items hoisted into high towers and exposed to the elements.[13]

The composition of bell metal remained a matter of great inconsistency at the time that Revere contemplated his entry into the bell-making trade. Most bells consisted of copper alloys, but the ratio and range of metals in these alloys depended on tradition e ach bell maker's preference, and the local abundance or scarcity of each constituent. Although modern experts define bell metal as a variant of bronze consisting of 75 percent copper and 25 percent tin, many bell makers experimented with small additions of other metals to improve the bell's sound. Some bells include quantities of zinc, which combines with copper to form brass, an alloy with similar properties to bronze. Most bell makers also added minute quantities of silver or even gold to their bells largely out of the superstition that this would produce a sweeter sound. Contemporary bell makers kept descriptions of the proportions of different metals, heating times, and temperatures purposely vague, either to guard trade secrets or to acknowledge the fact that they failed to follow one consistent recipe. One definition stated that "Bell-metal is a composition of tin and copper in due proportion; which has the property, that it is more sonorous than any of its ingredients taken apart." A study of fragments of a Revere bell shattered by lightning illustrates that he followed the prevailing wisdom concerning bell composition: his bell consisted of approximately 77 percent copper and 21 percent tin, with small amounts of lead, arsenic, zinc, nickel, and silicon, and a trace of silver. The lesser ingredients probably resulted from impurities in the copper, although Revere probably added the silver intentionally, a hopeful nod to the prevailing tradition.[14]

Bell making remained a high-tech trade in Revere's time, and all large bells in America had to be ordered and imported from England. The first and most famous bell produced in America, the Liberty Bell, originated in England but cracked upon its first striking. Philadelphia's Pass & Stowe foundry recast the Liberty Bell in the 1750s and needed two tries to produce one with an acceptable sound. The number of early bell makers, while undoubtedly small, is obscured by the fact that several foundries, such as Revere and Son, probably cast a small number of bells alongside their normal activities such as blast furnace operations, iron casting, and other metallurgical work.[15] In this way the early American metalworkers formed a loose professional network that transcended any single material, product, or process: most skilled workers who attained mastery in some aspect of metalworking offered advice or aid to the others. Upon receiving the contract to cast a new bell for his church, Revere would have attempted to contact anyone who could offer pertinent advice, especially local experts. He almost certainly learned some of the details of bell casting from Aaron Hobart of Arlington, Massachusetts, one of the few Americans who understood the principles of bell founding in the early 1790s. But Revere's bell-making preparations predate his 1792 contract, as revealed in earlier correspondence directed at metallurgical research.

Revere first investigated the possibility of casting bells, at least briefly, several years earlier. Nicholas Brown wrote to him in October 1789, responding to an earlier letter in which Revere asked many questions about bell making. Brown had witnessed the bell-casting process at his furnace, although he did not qualify as an expert. Brown informed Revere that a recently cast bell cost more than 60 pounds sterling and consisted of 60 pounds of copper and 35 pounds of block tin, a deviation from current and eighteenth-century metal proportions. This letter makes repeated mention of Brown's papers on the subject of bell making, such as "I set to overhauling the file of papers about recasting our meeting bell," and "I found authors differed about the loss & proportion of metal," providing another illustration of the research that all metalworkers commonly performed and the importance of any available expertise.[16] Even without Revere's original letter to Brown, it seems clear that he had toyed with the idea of bell making since 1789.

In late 1791 Revere's interest in metallurgy took on a heightened and more scientific intensity. He began a correspondence with Doctor Lettsom (which he misspelled "Lestrom"), a London scholar. Revere's questions to Dr. Lettsom illustrate his advanced knowledge on many aspects of metallurgy as well as his practical curiosity concerning related topics. Before writing Dr. Lettsom, Revere tested a sample of tin from a recently discovered Massachusetts source. After describing the sample as "1/32" heavier than his sample of Cornwall block tin, Revere theorized that his ore sample was not "divested of the crude minerals which it is commonly mixed with," possibly implying that it was what the miners call stream tin. He requested Dr. Lettsom's opinion, as well as samples of "shade, stream, and mine tin" from Cornwall or Devon "in their crude state." Revere sent Dr. Lettsom some samples of minerals found in the area and promised to continue doing this every spring and fall, "for I am realy selfish in the cause, for I doubt my abilities in chemistree, and am sensible that I shall git a true estimate of all that I shall send you." Dr. Lettsom responded in August 1792, and sent Revere the tin ores he requested. He also identified Revere's samples and mused about the establishment of a mineralogy school at Harvard. In March 1793 Dr. Lettsom responded to another letter and identified a new batch of minerals sent to him.[17] This abstract research did not directly relate to Revere's work; he needed to cast the copper and tin, not learn to identify different ores. This correspondence does illustrate his interest in the field, and increased his overall understanding of metallurgical processes. In addition, the ease with which Revere obtained valuable advice, feedback, and physical samples from a well-known scholar in a different nation clarifies the early scientific and technical networks of the English-speaking world. Unlike the American politicians who refused Revere's requests for appointed positions, Dr. Lettsom saw merit in this American stranger's scientific questions and took the time to respond. The lack of professionalism allowed the small number of interested amateur scientific practitioners, typically self-funded gentleman-scholars, to make useful contributions such as Revere's samples and observations. Revere must have received a bit of a thrill when a noted scientific expert and writer took the time to converse with him as an equal.

Revere's metalworking experience, in this case his iron foundry work and his earlier silverworking, again proved directly relevant to the new field he wished to enter. Armed with his iron-casting knowledge, he now faced several new learning objectives: learn how bronze differed from iron throughout the heating and cooling process; practice the casting of larger objects; acquire some bell molds; and experiment with acoustics. Because copper and silver have some common properties, his silverworking career offered guidance about how to cast copper alloys and how they differed from iron. For example, iron hardens when it cools quickly, while copper and silver soften.[18] Revere's general understanding of metal casting allowed him to focus on updating and honing his knowledge rather than spending time learning the basics. He addressed his highest priority, the acquisition of bell-related molds and patterns, with several purchases recorded in his ledger as "core moulds," "patterns of crown of 3 small bells," and "patterns of crown of large bell." He also purchased sand, clay, and other patterns that might have applied to either iron casting or bell making.[19] All of Revere's correspondence, discussions, and equipment prepared him to enter this trade, but nothing could replace the value of experimentation and personal experience. Complaints about Revere's first bell remind us that practice makes perfect, and the lack of complaints regarding his second and later bells attests to his continual improvement.

Revere's first bell contract with his own parish gave him the opportunity to learn the basics of the bronze-casting process in a supportive environment. Although the bell's shrill tone and surface imperfections did not prevent him from getting paid, he clearly had ample room for improvement. Revere and his workers stepped up to the challenge and eventually mastered the details of one of the most complex and unforgiving metalworking processes practiced in eighteenth-century America.

Bell makers first created an inner core that modeled the inside diameter of the bell. They accomplished this by digging a hole in the ground, building a hollow pile of bricks in the center, covering them in a special mud, and using a pattern to pack the mud into the precise shape of the interior of the bell. Second, they applied a mixture of tallow and wax to the outside of the core, creating a perfect wax model of what the finished bell would look like, including lettering or designs. Third, workers applied numerous coatings of "bell mud" on top of the wax, creating the shell, a model of the outer surface of the bell. The shell pressed right against the wax model and hardened against it, conforming to its exact shape. At this point the workers had created a mud core resembling the bell's interior, a wax middle layer that modeled the final bell, and an outer coating of mud on top of the wax. A workman then lit a fire inside the hollow brick structure at the center of the core to melt the wax and harden the mud: the core and the shell now resembled the inner and outer surfaces of the bell, separated only by air. The shell was then hardened with additional fire, and covered in sand or loose soil to prevent it from bursting. Casting experts then poured molten metal into the space between the core and the shell. After it cooled, which would require a day for bells weighing up to a ton and up to a week for larger ones, workmen removed the bell and trimmed any casting irregularities. Holes, pockets of air, or cracks in the core or shell could ruin the bell at any stage of this process. Finally, craftsmen began the laborious process of polishing and tuning the bell. They painstakingly smoothed the inner and outer surfaces, and removed small quantities of metal at different points to prevent partial notes from interfering with the tone of the fundamental note.[20]

Although Revere and his workers grew more proficient and confident with practice, the bell-making procedure did not lend itself to standardization. Revere's correspondence and records show that he probably used existing bells as models for new ones: he refers at points to the weights and dimensions of some of Boston's church bells when discussing a new contract, and his bell sales include clusters of bells with similar weights that probably followed the same general pattern.[21] Each bell required its own mold that did not survive the casting process, and every step of the mold-forming and casting process needed careful attention and adjustment. For example, the mud used to pack the bell mold had a special formula that optimized its moisture content, density, and other properties, recorded as a recipe in his 1793 memoranda book: "The mud for thickness of Bell one part horse dung one Sand & one part Clay For Navel & Cope 6 parts horse dung 4 Sand & 4 Clay & some Cow Horn." Revere's crew had extensive experience with the preparation of smaller molds for iron casting, but bell molds occupied many cubic feet and had to receive hundreds or thousands of pounds of molten metal.[22] Each mold required days to prepare, and therefore an error during the casting process became costly.

Revere practiced this new trade alongside his iron-casting work, getting double use out of the foundry oven he had already constructed while taking advantage of his workers' familiarity with metal casting. According to Reverend Edward Griffin Porter, who captured various recollections about Boston history in his 1887 book *Rambles in Old Boston*, Revere used the backyard of his

Figure 5.1. Lowering the shell and forming the bell core. From Denis Diderot and Jean le Rond d'Alembert's *Encyclopedie: Recueil de Planches, sur les sciences, les arts libéraux et les arts méchaniques aves leur explication* (Quatrieme Livraison; Paris; Briasson, David, LeBreton; 1767), vol. V, "Fonte des Cloches," plate II; as cited in Charles Gillispie, ed., *A Diderot Pictoral Encyclopedia of Trades and Industry*, vol. 1 (New York: Dover Publications, 1987), plate 115. Beginning in the 1750s, Denis Diderot's multivolume *Encyclopedie* illustrated the equipment and processes used in many French trades. This illustration depicts several stages of the bell-making process that some founders still followed in Paul Revere's time. On the left, workers (not pictured) lower the outer shell onto a finished inner core. On the right, two workers use a mold to produce another core. Once the shell and core are properly aligned, bell founders pour molten metal into the space between them to form the bell. Other founders used wax to initially fill the space between the shell and core, and melted the wax with a fire before pouring the bell metal.

nearby Charter Street house to display and sound some of his finished bells. This text states that Revere's property

> was about sixty-four feet wide and one hundred and forty feet deep, containing a
> large yard in the rear, where bells were often placed for inspection after being cast
> in the foundry. Purchasers would come to hear them sounded; and boys would

often gather round out of curiosity. One of their number remembers being present with others on such an occasion, when they were probably in the way; for Mr. Revere pushed them aside with his cane, saying: "Take care, boys! If that hammer should hit your head, you'd ring louder than those bells do."[23]

One of these boys was probably his son Joseph Warren. Born in 1777, Joseph Warren had some silversmith training but also spent time in the foundry, which he eventually supervised. Older son Paul Jr., by this point an adult in charge of most aspects of the family silver shop, also helped out in the bell foundry. He apparently preferred bell casting over silverworking since he left the family business to run his own bell foundry in 1801.[24] Revere ran each of his businesses as a family concern, illustrated by the use of his own house as a bell showplace, and in the case of something as audible and fascinating as bell casting, the business became something of a neighborhood draw as well. Revere's open shop policy remains consistent with the practices at many other establishments, some of which offered invaluable aid to him with each new endeavor.

Alongside his growing technical expertise, Revere's marketing savvy increased. He devised a one-year warranty policy to aid this new trade, initially offered on all church bells:

> This bell is warranted for twelve months accidents & improper usage excepted; and unless it shall be rung or struck before it is placed in the belfry, or tolled by pulling or forcing the tongue against the bell, by a string or otherwise.[25]

This simple statement illustrates a major shift in early retail policy, in which Revere backs each item with a guarantee of quality. His shrewdness comes across in the "accidents & improper usage" exception to this warranty, because he did not want to spend his shop's time and money fixing breaks caused by common user errors. The most frequent mistake receives explicit mention in his policy: many bell ringers rang their bell in the easiest manner possible, by grabbing a rope attached to the bell's tongue and smashing it into the side of the bell. This improper ringing technique prompted Joseph Warren Revere to write, "with a yard of twine I would undertake to break every Church bell in Boston."[26] English bells, including ones made by Revere, should have been mounted on a large wheel that, when turned, swung the side of the bell into the dangling clapper or tongue. The tremendous weight of these bells made the installation process somewhat challenging, especially for churches that had not previously owned one, and Revere's warranty also takes care to exclude accidental damage because these bells would not survive a long drop or similar mishap. Upon selling each bell Revere probably discussed the details of their use and upkeep with the buyer, and this warranty helped him underscore the most important points.

A second note elaborates several of his pricing policies:

Price of Bells

All bells of 300 lbs and over are to be warranted as Church Bells, and the same discount made for cash. All under 300 lbs, or all sums less than one hundred dollars, 60 days credit or one pr cent to be discounted for cash.[27]

Revere extended a warranty only to church bells and the largest school bells and also offered a cash discount of 1 percent. The cash discount may appear trivial by modern standards, but represented another new step for Revere, as he realized that the continued extension of credit hurt his business. Departing from complex silver shop dealings in credit, barter, and cash, he now attempted to move to a fully cash system.

Revere finished his first bell in 1792, and sold five the following year. He made three types of bells, with different sizes befitting their different uses. Church bells, by far the largest and most complicated, weighed more than 500 pounds and occasionally exceeded a ton. Schoolhouse bells usually weighed between 100 and 500 pounds, though rarely more than 300 pounds, and ship bells weighed 100 pounds or less.[28] Table 5.1 presents the bells made by Revere's foundry until 1810, the last full year of his employment, as recorded in his own painstaking tally in two different ledgers in his family papers.[29] He probably ended his personal involvement in the bell-making process earlier than 1810, but this table records the entire production line completed during his active career, including church, school, and ship bells.

Table 5.1 reveals repeated fluctuations from year to year, with the fewest bell sales in the first three years and the highest number of sales between 1798 and 1808. This irregularity resulted from Revere's other responsibilities as well as the changing economic climate. Following the startup period of the first three years, the low output of two bells in 1799 probably reflects Revere's preoccupation with federal contracts for cannon, bolts, and sheeting, while the economic recession most likely caused the 1804–1805 lull.

Bell making certainly had the potential to bring in good money, but Revere

Table 5.1. Revere's Bell Production and Weights

Year	Number of Bells	Total Weight of Bells (pounds)	Average Weight of Bells (pounds)
1792	1	912	912
1793	5	1,643	329
1794	1	673	673
1795	5	3,096	619
1796	4	3,481	870
1797	5	2,799	560
1798	10	5,302	530
1799	2	1,199	600
1800	8	6,691	836
1801	7	6,011	859
1802	9	9,153	1,017
1803	6	5,276	879
1804	4	3,517	879
1805	3	2,640	880
1806	12	12,553	1,046
1807	5	5,379	1,076
1808	9	6,736	748
1809	5	5,975	1,195
1810	2	2,086	1,043
Total	103	85,122	826

could not count on it for a steady income: he never sold a large number of bells in a given year and the demand fluctuated too much. Each bell yielded a considerable degree of profit, in keeping with its status as a finely crafted item. Revere's bell price changed over the years but usually remained between 42 and 50 cents per pound of bell weight: typically 45 cents (or 2 shillings, 8 pence) per pound for church bells, and 50 cents per pound for smaller bells. Revere typically charged around 50 cents a pound for pure copper items such as bolts, spikes, and sheets, which made his church bell price something of a bargain, possibly reflecting an economy of scale when casting larger items.[30] Without a record of Revere's labor costs we cannot determine the profit that he realized on each bell, but as with all of his high-weight products, bells generated large income surges with each sale. He surely appreciated this addition to his product line, and for reasons beyond the monetary.

Unlike his increasingly standardized silver-rolling and iron-casting activities, bell making was a more artistic activity requiring immense amounts of skilled labor and customized attention for each piece. In this case a step backward in terms of mass production still equated with progress in other areas, such as high-profile visibility and distinction for the Revere shop. Even Revere's earliest bell, with a tone that made some listeners squirm, represented a manufacturing achievement for Boston and a triumphant accomplishment for its founder. This first bell still rings twice a year, on Good Friday and Christmas Eve, in the St. James Episcopal Church in Cambridge. More than one hundred other bells made in the Revere shop survive to this day, praising the skill of their maker with each chime.

Cannon Founding and Government Contracting

In 1794 Revere made the next evolutionary step in his career by entering the cutting-edge field of cannon casting. One might be hard pressed to imagine a larger ideological shift: shortly after learning to cast bells that created beautiful music, transmitted vital information, or summoned townspeople to church, Paul Revere began receiving commissions from state and federal war departments seeking the most destructive weapon of the eighteenth century. Without hesitation the almost 60-year-old entrepreneur plunged into this new field, perhaps because his military background helped him understand the strategic importance of cannon, or possibly because he realized that both church bells and cannon served societal needs in a prominent manner. This new trade dominated his shop within a few months.

Cannon casting was the quickest and easiest technological leap of Revere's lifetime, as attested by his speedy mastery and the dearth of written materials in the Revere Family Papers pertaining to his learning process. Cannon casting used almost identical raw materials and equipment as bell making: throughout the history of European warfare innumerable invading armies tore down the church towers of their enemies and recast bells into usable artillery. In addition to Revere's experience with the casting of large bronze and iron objects, he also had some exposure to cannon-specific techniques and tools, such as cannon molds from his earlier work with Louis de Maresquelle at the Titicut furnace in 1777. In addition, Revere had a working familiarity with the use and limitations of different types of cannon from his service as an artillery officer in both the French and Indian and Revolutionary wars. We can safely assume that he carried out his typical research process by consulting several of his knowledgeable colleagues, obtaining some molds, and experimenting until he mastered the new technology. As with the bell-making trade, cannon casting had not reached the state of maturity in America to enable a large or connected community to form, forcing Revere to learn complex processes largely through trial and error. In Europe, however, cannon casting held an exalted status as one of the fields that ushered in the era of centralized nation-states.

The first cannon originated in ancient China, as did so many other technologies. Chinese alchemists developed the first usable gunpowder around the ninth century AD, and developed metal-barreled cannon to use this gunpowder at some time in the twelfth or thirteenth century. Early cannon resembled metal vases that fired arrow-shaped projectiles, and this technology soon spread through the Middle East and into Europe. While cannon are depicted in some European texts and played a small role in several early engagements, they first became vital strategic weapons in the Hundred Years' War in the fourteenth and fifteenth centuries when France used large numbers of cannon to defeat and evict invading English armies, often by shattering formidable static defenses such as castle walls. Modern observers would recognize many aspects of these early cannons: French armories cast a tubular cannon body in a single piece instead of attempting to weld together a series of bars; bronze and brass became the choice construction materials due to their favorable material properties; and spherical cannonballs often made of stone became the dominant form of ammunition. The rapid spread of cannon led to a European arms race and a "second bronze age" in which the nations able to lay hold of reliable copper, tin, and zinc resources gained a strategic advantage, even after England improved its ability to cast iron cannon in the sixteenth century.[31]

Cannon technology took a great leap forward in the 1460s and 1470s thanks to pioneering metalworkers in France and Burgundy. Size became the critical design constraint: cannons to date had been immense and often immovable devices, useful when they could be cast near the site of use but not suitable for field or mobile needs. A series of complementary developments enabled cannon casters to greatly reduce the amount of metal they had to use while also increasing their weapons' overall destructive power. Advances in gunpowder manufacture led to the use of small grains or corns of powder that produced bigger explosions, and iron cannonballs replaced stone ones. Along with the use of mobile gun carriages to facilitate the transport of heavy weapons, the newly mobile array of siege weapons reconfigured the power structure of Europe, giving attacking forces a new advantage over defensive fortifications. In addition, large and centrally organized countries reaped the biggest rewards because only they could raise enough funds to support constant purchases of new weapons. The Senate of Venice gave words to this military paradigm shift in 1498, declaring, "the wars of the present time are influenced more by the force of bombards and artillery than by men at arms." Lighter-weight cannons also functioned well on ships, which now became potent mobile weapon platforms. The considerable weight of large numbers of cannon ran the risk of making ships too top-heavy, but this could be mitigated by cutting gun ports in hulls and placing heavy weapons closer to the waterline. King Henry VIII of England financed the first of this new class of warships, which inaugurated a naval revolution that eventually led to European, and specifically English, dominance of the high seas. England's and Europe's advanced metallurgical technology became a crucial advantage in Renaissance-era warfare, with numerous contemporary reports (supported by modern chemical analyses) praising the strength and durability of Western bronze and iron in comparison to more brittle and overly bulky ordnance captured from the Middle East and India.[32] Britain guarded its cannon-casting technology even more strictly than its other technological secrets due to its strategic importance to the empire's worldwide interests. While the American colonies benefited from access to British arms throughout the colonial years, the ability to create these weapons remained entrenched in the mother country, which had major impacts upon the Patriots' ability to carry on their Revolutionary struggle.

In June 1776, then-Colonel Henry Knox proclaimed, "The business of casting Cannon and making fire arms is of infinite importance to this Continent and cannot be too much encourag'd." Although colonial American gunsmiths developed impressive skills in the art of making and repairing muskets and rifles, cannon casting remained a virtually unknown technology until the Revolutionary War. By the start of the war, Americans lacked significant numbers of cannon as well as the means or knowledge to make them. British troops carefully guarded cannon stockpiles, and merchants owned the only other cannon for use on their ships. Different colonies, beginning with Massachusetts in 1774, sought to purchase cannon for defense against British attacks. Early purchases and captured stockpiles succeeded in arming some colonial forts and militias with a confusing array of different-sized cannon, but the Patriots soon realized they had to learn to make cannon and ammunition if they wanted any hope of mounting an effective defense. Numerous ironworks had the theoretical capacity to produce cannon, provided they could learn how.[33]

During the Revolution, various founders took on this challenge. Beginning with the Hope furnace in Rhode Island, the Salisbury furnace in Connecticut, and the Reading and Warwick furnaces in Pennsylvania, American foundrymen turned out increasing numbers of cannon. They had to produce iron cannon at first because of copper shortages and their unfamiliarity with copper casting. They soon learned the many disadvantages of iron: because iron is less elastic than bronze or brass, metalworkers had to make iron cannon larger to prevent them from bursting but this made them too heavy to move, more suitable for static defenses than battlefield action. Early American cannon had an unfortunate tendency to burst during testing or use, often with fatal results. In the sober words of George Washington, "It is a melancholy Consideration that in these cases we suffer more from our own Artillery than the Enemy." Still, by the end of the war many hundreds of American-made cannon had been used or placed in fortifications.[34] American Revolution–era ironworkers borrowed from existing casting technology, publications such as European "shot tables," European foundrymen such as Louis de Maresquelle, and trial-and-error experimentation. Revere adopted a similar educational strategy in 1794, though by this time researching new metallurgical fields had become quite familiar to him.

The cannon-casting process resembled bell or iron casting in many ways, but a cannon undergoes significantly different stresses over its operating life than just about any other object. Cannon casters over the centuries optimized their products to maximize strength and durability and minimize the overall size and weight of the cannon. In addition, casters standardized the production process as much as possible to reduce casting errors and guarantee that ammunition, itself a highly standardized product, securely fit the barrel without jamming. Critical steps in the production process involved selecting the proper alloy, understanding the casting process, and learning to properly finish and test the final product.

Most early cannon casters preferred to work with "gunmetal," an alloy of bronze consisting of approximately 90 percent copper and 10 percent tin. American founders often substituted brass for bronze until a special form of gunmetal consisting of 85 percent copper, 5 percent tin, 5 percent zinc, and 5 percent lead eventually supplanted other alloys. Bronze remained the preferred material for cannon casting because its low melting point enabled castings to be more reliably uniform, it resisted rust, and it could be worked relatively easily after being cast into final form. The ease of casting and greater elasticity of bronze also made bronze cannon less likely to burst than iron cannon, and when they did burst the gun tubes usually split without hurling deadly metal shards into the gun crew.[35]

The cannon-casting process changed little between the Middle Ages and post-Revolutionary America, and even though Revere never described his own procedures in writing we can confidently assume that he followed the prevailing practices. First, the founder created a mold from baked clay. This mold contained three parts: the cannon barrel's exterior, which usually included decorations such as coats of arms; the breech, or closed back portion of the barrel; and the core, a plug in the shape of the interior of the barrel that allowed the caster to produce a cannon with a hollow middle. Some founders reinforced their molds with iron bars to better withstand the stresses of casting. Second, the workmen secured the mold in a pit and filled it with molten metal. After the metal cooled, the workmen removed it by breaking the mold. Finally, the caster finished the cannon by carefully drilling or boring out additional metal to enlarge the hollow center, ensuring its proper size and uniformity.

The traditional cannon-casting and finishing process left room for improvement. The core portion of the mold rarely remained in the appropriate location in the exact center of the barrel while molten metal poured in all around it. Casting difficulties led to many non-uniform cannon with irregular interior dimensions. The French first developed the "solid bore" casting method to correct these problems in the 1740s and 1750s. Cannon casters could cast the barrel of their cannon in a single heavy piece without a hollow core, and then drill out the bore with a powerful boring machine that usually made use of water or steam power and hardened iron drill bits. As a result of the accurate placement and uniformity of the center shaft, barrels more reliably fit the diameter of the cannonballs, reducing the risk of jamming and the loss of explosive force due to gases escaping around the edges of the ball. This new degree of safety allowed casters to reduce the barrel thickness and overall weight of the cannon. Improved cannon-casting technology spread to other countries, reaching England in the 1770s. Revere adopted this process, and subcontracted the drilling portion of his work at least in 1795 and 1796.[36]

The high failure rate of colonial and Revolutionary cannon, both during the proving process and under battle conditions, illustrates the incomplete American mastery of casting technology. Many colonial founders initially had a lot of difficulty making sound castings that could withstand the stress of use, or the even worse stress of "proving," the process of testing cannon for flaws. Revere described this process to a friend in 1795: "In proving my ten I lost one. Lt. Major Lillie was imployed by the Secretary to prove them he filled the Chamber full of powder, the shell full of sand and then Clayed the space between the shaft & Cillander, one of them recoiled 39 feet & all the rest above 30."[37] Proving typically involved discharging a cannon under unusually rigorous conditions, for example, by using triple the normal quantity of gunpowder, discharging the cannonball into the ground, or somehow plugging the barrel. Presumably, if the cannon survived this test it would endure ordinary operating conditions. Many cannon did not pass this test, leading Revere and other founders to complain (to no avail) that it represented unrealistic and pointless expectations.

As late as December 1795, Secretary of War Timothy Pickering complained, "The casting of cannon has not been attended hitherto with the expected success." He also stated that the government recently hired a "French gentleman" to overhaul the casting and boring process, and mentioned ongoing efforts to hire a "complete cannon founder" from Europe.[38] This letter may refer to the same French gentleman whom Stephen Rochefontaine directed to Revere in 1795, as quoted in the previous chapter. Revere's startup challenges echo the novelty and risk surrounding American cannon casting in general: even the so-called experts had a hard time turning out consistently excellent weapons.

Perhaps the major difference between bell and cannon casting involved operational management. Bell casting took place on a much smaller scale, with a maximum of twelve sales in a given year, and therefore Revere could afford to wait for a contract before searching for metal. Cannon casting had more in common with Revere's ironwork because it took place on a large scale and required more raw materials and a higher emphasis upon steady throughput. Fortunately, his skilled laborers stepped up to this new challenge thanks to the training they had received when working on iron casting. Likewise, Revere had personally acquired helpful managerial experience developing reliable supply sources and tracking his income and expenditures. All systems were go.

Revere initiated his cannon-casting career by contacting his old friend Henry Knox, now conveniently positioned as America's first secretary of war, and offering to produce artillery for the government. He must have seen cannon casting as an opportunity to combine some of the best aspects of his earlier products: cannon equaled or exceeded the size of bells and therefore commanded high prices, and founders usually contracted for large sales involving many cannon at the same time. They embodied great craftsmanship as did his silverwork and bell casting, but they also served a utilitarian goal, all while helping his country cement its hard-won independence. Unfortunately, the full picture of cannon casting included negative aspects as well. Cannon casting involved frustration and setbacks due to the complex process and unreliability of most of Revere's clients, particularly the federal government. But in spite of these impediments, Revere's cannon-founding career certainly offered more pros than cons as it provided him with high-profile work that greatly expanded his workshop's reputation, income, and metallurgical competence.[39]

As with the bell-making process, cannon casting relied heavily upon skilled labor and personal judgment. Even experienced cannon founders had trouble getting the metal to solidify in a uniform manner, and a small misalignment during the boring process produced an off-center shaft more likely to burst or misfire. Revere and his workers had to climb a steep learning curve, and his first contract proved the most troublesome of his four client relationships between 1794 and 1800. The federal government hired him in July 1794 to cast ten howitzers, large army field pieces that fired six-pound cannonballs with an eight-inch caliber. Both the War and Treasury departments corresponded with Revere, occasionally in a repetitive or confusing manner, reflecting the unclear division of authority and responsibility in the early executive branch of the government. Over the course of this contract Revere and the government representatives all made mistakes that delayed completion of the cannon and the ensuing payment.[40]

Government officials took an interesting initial step when they divided the howitzer contract between Revere and James Byers, a supervisor at the federal armory in Springfield, Massachusetts. The government wanted twenty howitzers, and asked each man to produce half of them for identical contract terms. In addition to spreading the risk between two facilities and producing the cannon twice as fast, this enabled a comparison between the two foundries to help determine the competence and efficiency of each. Indeed, this caused Revere some problems later on when his wasted metal stood in contrast to Byers's better results. Although under other circumstances Revere and Byers might have seen each other as rivals, they quickly learned that they had similar personalities and technical interests and faced a common set of manufacturing challenges and frustrating bureaucratic hurdles. At this early point in the cannon-founding field, having a local colleague familiar with your trade outweighed the potential loss of business from competition. Revere and Byers chose to cooperate from the outset by exchanging molds, patterns, and advice, and they eventually became close friends. Letters between Revere and Byers display an increasingly friendly tone and the inclusion of personal inquiries in addition to business matters. For example, Revere begins his letters with "Friend Byers" and makes comments such as "My best regards to your good lady."[41]

The vast majority of problems with Revere's first federal cannon contract centered on raw material procurement, again, an unappreciated harbinger of what Revere and his son faced repeatedly in the years to come. Copper and tin were scarce in early America and the quality of metal was often suspect. Revere initially bought raw materials with a specific product in mind, as he had during his early bell castings. In other words, if he knew he needed ten tons of copper for a set of cannon he would buy precisely ten tons. As his shop grew he gained confidence in his ability to procure contracts that would use up any metal he could lay his hands on, and consequently his raw material purchases became more regular, subject to merchant prices and availability. His shop maintained a stock of metal that he could apply to any project, and he confidently used his own funds to expand this stock whenever he could—a striking step toward large-scale industrial practice. His ledgers contain one furnace stock inventory performed by two of his sons in 1799, presented in Table 5.2.[42]

By 1799, Revere had more or less eliminated his iron casting, as shown by the absence of iron in Table 5.3, to emphasize bronze and copper products. Gunmetal became his highest priority at this time, and the separate listing for gunmetal either represents an alloy that he had already created, or a stock of "old" gunmetal from a cannon or other source ready for recycling. Revere's 410 pounds of bell metal likely refer to a broken bell awaiting recasting. The large weights of copper and tin that he kept on hand enabled him to create more bronze or produce pure copper goods, and the brass may have referred to either brass or bronze since practitioners often used these terms, as well as the alloys they described, interchangeably. The fact that Revere dealt with thousands of pounds of tin and did not have any zinc on hand implies that what he called brass was really bronze.

Raw material stockpiling represented a logical response to the larger quantities of metal he used in his ever-increasing contracts. In 1795 Revere tallied all his metal expenditures for the federal howitzer contract at the government's request. He bought 38,032 pounds of copper and 4,712 pounds of tin for 10

Table 5.2.   May 1799 Furnace Inventory

Item	Quantity (pounds)
Gun & Brace Mettal	4209
Copper	2114
New Spikes & Nails	180
Bell Mettal	410
Coarse Brass	1284
Fine Brass	556
India Tin	1912
Worn Tin Melted	3115
Brass Wire	574

howitzers, of which he sent 15,473 pounds of copper and approximately 1,800 pounds of tin to Byers in Springfield at the government's request. Revere's final howitzers weighed 16,917 pounds, and although he had some leftover copper and tin at the completion of the contract, he still wasted as much as 8,200 pounds of metal, an enormous amount.[43] Metal shortages had a dire impact on Revere's operations. Both Revere and the government initially estimated the weight of the finished guns to be 1,500 pounds each, when in reality they weighed between 1,633 and 1,750 pounds each. Even though Revere's cost estimate factored in 16 to 20 percent extra metal, "which is very great," to account for errors and waste, he still fell short of the mark, and as a result he procured less metal than he needed to carry out his work.[44] This miscalculation, among other issues, led to copper and tin shortages that slowed the rate of production while he scrambled to purchase additional materials. Revere also lost one howitzer during the proving process, forcing him to recast it and use even more metal. Since the government agreed to reimburse him for all metal purchases, this wasted metal did not come out of his own funds. But it did not benefit him either: Revere's salary depended on the weight of the final product, a fee of 17 cents per pound of finished cannon. This fee had to cover his labor, fuel, and other incidental charges, which increased each time he began a new casting or labored to correct an error.

In addition to slowing his operations, Revere's wasted metal greatly concerned the government. Although the War Department's first contract with Revere agreed to provide him with all needed metal, government officials quickly asked Revere to take on this task. In this case Revere's purchasing network proved itself more effective than the government's. He ended up writing many letters to various merchants, cobbling together enough small purchases of copper, tin, and "old" metals to supply both himself and Byers in Springfield. The large percentage of waste reflected the quality of the metal as well as his inexperience. The government limited him to a price of 1 shilling, 2 pence per pound of copper and 1 shilling, 8 pence per pound of tin, which forced him to buy inferior material.[45] Added restrictions on his allowable shipping costs forced him to send the highest-quality copper to Springfield because its density made it the cheapest to transport. In contrast, Revere had to work with the bulkier recycled copper and brass implements that had to be melted and recast. In addition, reused copper contained small pieces of iron that either accumulated in the furnace or entered the final product as imperfections. In justifying his larger waste percentages in comparison with Springfield, he felt the need to defend his honor and judgment: "I was as carefull and as prudent as if the metal belonged to my self & what is deficient was wasted." He finished his howitzers by January 1795, but the government questioned his excess metal usage and delayed paying his contract until May 1796, more than a year later, when the different departments convinced themselves of his honesty. He grew more efficient as he mastered the casting process and wasted far less in future contracts.[46]

Revere's first federal contract reveals the inexperience present on both sides. Revere took on too many responsibilities and accepted constraints that excessively limited his operations. He could not procure high-quality metal at the price the government set, and wasted time and money working with inferior "old" copper. He also, understandably, had not mastered the casting process as quickly as he expected. His burst cannon and high amount of wasted metal triggered a government review process that delayed his reimbursement more than a year. The government's division of authority between the War and Treasury departments also complicated all transactions, as Revere often received two sets of identical questions from Secretary of the Treasury Knox and Commissioner of Revenue Tench Coxe, while his own concerns often went unanswered. By the time he finally received his payment, the extra labor for the burst cannon and the lost use of his money during the long delay reduced his profit.

Federal contracting had its upside as well. Revere's second federal contract involved the first American attempt to cast carronades, light, short-range naval cannons that splintered ship hulls in an attempt to kill enemy sailors. Henry Knox asked Revere to board the French frigate *Concorde* while it was docked in the Boston harbor, and had him sketch these weapons. Revere took his research seriously and wrote Knox: "By post I send you the draft of the Charonade. I have endeavored to make it as plain as possible, as in several things it differs from other guns. I enclose directions translated from a French printed paper. For the use of the seaman, the officers speak very much in favor of the Charonade."[47] Revere had to lobby the government for years until it agreed to fund his project in 1798, requesting ten carronades for the *USS Constitution*. This contract proceeded more smoothly than the earlier one and in addition to generating profits it helped keep Revere in touch with influential government officials. Revere was far from the only person to find federal work a mixed blessing: many reputable ironmasters who sold their goods to federal government armories thought the government's "sometimes arbitrary military specifications" counterbalanced the high prices and large sales volumes it offered.[48] Revere endured these difficulties with great persistence and energy, and thereby increased his experience while situating himself for future government work.

Once he started working for the federal government Revere found it easy to attract many other clients. Proven cannon casters were rare in the young republic and word spread quickly. The Massachusetts state government immediately became his most frequent client, beginning in April 1794 when he won his first contract to cast twelve brass "three-pounder" cannon. Three-pounders were small cannons frequently used by infantry regiments, so named because they fired three-pound cannonballs. Revere completed this first job in October 1794 and by June 1795 he claimed to have made a total of more than thirty three-pounders for Massachusetts, with pending orders for ten more. He cast at least sixty-eight cannon for Massachusetts between 1794 and 1800, and continued well into the 1800s.[49]

Revere's reputation rapidly extended to different states through recommendations from military and government officials familiar with his work. James Lawrason of the Alexandria Artillery Company in Virginia ordered one six-pounder, two three-pounders, and three carriages from Revere in July 1794 upon the recommendation of the Philadelphia superintendent of stores.[50] And on May 10, 1798, William Rhodes and Nathaniel Fischer of Rhode Island contacted Revere about his brass artillery after hearing him highly recommended: "We the undersigned being appointed by the Honorable House of Representatives of the State of Rhode Island for the purpose of seeing whether Brass Field Artillery can be procured in this country and to report the expense, we understand by Col. Corliss that you have cast some at your foundry which is equal to any imported."[51] A favorable comparison to imported cannon was lofty praise. The letter then requested Revere's prices and terms of payment for twelve four-pounders. Revere responded on May 16, advising Rhode Island to buy three-pounders instead of four-pounders: most customers eschewed four-pounders because their ammunition looked similar to three-pound balls, thereby causing "very great dificultys in time of action" when artillery suppliers confused the two.[52] The Rhode Island delegation wisely took his advice. By this time he had more experience in artillery matters than most non-military clients, as illustrated by Table 5.3, which summarizes his early production of different artillery pieces.[53]

With at least eight cannon sales each year, Revere had found a steady and lucrative line of work. But other than his seamless ongoing business relationship with Massachusetts, he encountered difficulty with every contract he accepted. After experiencing the frustration and financial hardship caused by payment delays, Revere implemented a generous credit policy and explicitly stated his terms at the start of each new contract. But no solution seemed equal to this challenge, and each non-Massachusetts contract ran afoul of major misunderstandings about each party's obligations.

Revere's interactions with the Alexandria Artillery Company soured almost immediately. James Lawrason expressed interest in buying one six-pounder, two three-pounders, and three carriages from Revere in early 1795. He submit-

Table 5.3. Revere's Ordnance Production, 1794–1800

Year	Ordnance Type	Client	Quantity
1794	Three-Pounders	Massachusetts	12
1795	Howitzers	United States	10
	Three-Pounders	Massachusetts	12
	Three-Pounders	Alexandria Artillery Co. (Virginia)	2
	Six-Pounder	Alexandria Artillery Co. (Virginia)	1
1796	Three-Pounders	Massachusetts	8
1797	Three-Pounders	Massachusetts	8
1798	Three-Pounders	Massachusetts	12
	Carronades	United States	10
	Three-Pounders	Rhode Island	6
1799	Three-Pounders	Massachusetts	2
	Twelve-Pounders	Massachusetts	6
1800	Three-Pounders	Massachusetts	8
	Total		97

ted this order in writing before hearing the prices, and Revere started work at once. By July 1795, Lawrason asked to remove the six-pounder from his order because he had trouble collecting his subscription money. Unfortunately, Revere had already finished and engraved all the cannon, making them hard to resell. Furthermore, he complained that some of his subcontractors, "The Carriage maker & Blacksmith," demanded their fees for the work they had performed. Lawrason countered that he never definitively asked for the six-pounder, he assumed Revere would not start without an advance, and he would not have ordered any cannon if he knew he needed to pay so quickly. Lawrason attempted to satisfy Revere with partial payments, and eventually paid in full, after a huge delay.[54]

Now that he recognized this potential ambiguity, Revere clearly summarized his terms at the start of the Rhode Island contract:

> The lowest price I can cast & finish them for is fifty cents @ pound, & if the State Arms are put on them, that will be an extra sixpence. I am willing that they be proved by any person who understands the business; the proving is at the expense of the purchaser. Those three pounders I have cast for this State weigh from 412 to 420 lb. each . . . My terms of payment will be one half down when delivered & the other in ninety days. I believe the carriages +c may be procured upon the same terms. I am not certain that I have copper sufficient for twelve pieces should therefore be glad to have some copper procured for which I will allow twenty cents pr pound.[55]

These policies attempted to pass some unpleasant responsibilities to the client. For example, Revere gladly credited the account for the copper cost if this allowed him to avoid the hassle of buying it. Proving also became the buyer's responsibility. And the 50 percent payment upon delivery followed by ninety days' credit seemed a fair compromise, protecting Revere from complete payment delay while allowing the customer some time to raise funds.

Unfortunately, this written clarification did not produce a smooth transaction. William Rhodes and Nathan Fisher, the Rhode Island purchasing agents, asked Revere to cast six howitzers in May 1798, including state arms. By August, Fisher announced he had changed his mind and wanted to buy only four; once again, Revere had already cast all six and placed the Rhode Island arms upon them. Revere won this argument, but the terms of one-half down and the other half within ninety days did not materialize. Of the $1,292 price tag, Revere received only $450 upon delivery at the end of August. The clients should have paid the remainder by November 30, but still owed him $492 as of February 13, at which point the records fall silent. This dispute reiterates the vagueness and lack of commitment in early work agreements, when contractual obligations such as Rhode Island's inability to reduce the size of the order after it was finished remained unclear, and even Revere's ninety-day credit policy seemed too strict for a state government to meet.[56]

Contractual difficulties obviously did not dissuade Revere from continuing his cannon-casting trade. All his expense reports compute a "labor" fee, usually listed as "for casting," based on the weight of the finished product. For his first federal contract Revere computed a labor fee of $2,897.39 that amounted to 17 cents a pound for 16,917 pounds of ordnance plus $22.50 for the engraving of the U.S. coat of arms on the finished pieces. Revere gradually raised his rate over the years, to 20, 22, 23, and finally 25 cents per pound exclusive of the cost of the raw materials. This simple fee, based only on the weight of the final product, hides the many actual operating expenses incurred by Revere, such as the wages he paid his employees, depreciation, shipping insurance, and fuel. Revere's labor fee constituted between 45 and 60 percent of finished contract prices for different small contracts with the Massachusetts government.[57]

Revere's pricing policy represents an interesting simplified method of value determination. Revere estimated all of his costs and aimed for a reasonable amount of profit at the start of the contract, with no attempt to pass on specific costs—other than the cost of metal and customized engraving charges, which always received separate treatment—to the client. For example, recasting a cannon after it burst did not change the weight of the final product, but it did require additional fuel and labor costs that had to come out of Revere's profits at the end of the contract. This fee structure encouraged casters to correct their wasteful practices and ensure quality products to avoid costly setbacks. Revere's frequent increases in his "labor charge" from 17 to 25 cents per pound over a period of 6 years represents his improved understanding of the actual cost of running his business as well as a growing confidence in his product. The pricing policy protected the shop from miscalculations in the weight of the final product because the final weight determined the price. A weight-centered pricing approach reflected the uncertainty accompanying every stage of the process: as mentioned above, even though Revere cast 10 government howitzers from the same pattern, the cannons' final weights ranged from 1,633 to 1,750 pounds. And apart from the federal government, future clients did not even attempt to provide copper or other raw materials, preferring to pass this unpleasant search process on to Revere, which cost him an enormous amount of time and effort as attested by numerous letters to merchants each year.

Two receipts shed light on how Revere interacted with his most highly skilled employees. Nelson Miller acknowledged receiving $24 for 12 days of work for himself and his son ending on June 1, 1795, an extremely high wage reflecting the faith Revere had in him. During his federal howitzer contract Revere wrote a personal letter asking his "Friend Miller" to return to Boston quickly and help him recast the cannon that burst during testing. Later receipts indicate that Miller continued working for Revere for at least two more years and remained one of his highest-paid employees. Revere paid an even larger sum of money to Elib Faxon in August 1795, broken down by task and not by the amount of time he worked. Revere subcontracted some tasks such as artillery boring to Faxon, who owned his own equipment. According to this receipt, Revere paid Faxon for a variety of services between August 1795 and April 2, 1796, that required large equipment and specialized skills.[58] Since Revere had just entered the cannon-casting field, he had not yet purchased bulky and expensive cannon-turning and -boring devices. Two early charges from Faxon correspond to the turning and boring work on Revere's first two contracts (ten howitzers for the federal government and twelve smaller cannon for Massachusetts). Revere counted on Faxon and Miller to perform a range of tasks competently, with minimal supervision, and in Miller's case, on short notice. Revere also had a trusting relationship with both of these men, illustrated by the faith he placed in their abilities as well as their willingness to defer receipt of their wages, by many months in Faxon's case. As Revere grew older and more experienced, he continued changing his role from skilled worker to manager and overseer.

Malleable Copper: Bolts, Spikes, and Technical Experimentation

As his experience and ambitions continued to grow, Revere decided to solicit more work from his largest and most influential client, the federal government. This time he sought a steadier and more lucrative relationship by offering to mass-produce vast quantities of copper bolts, spikes, and other ship fittings. This new line of work departed from anything he had previously learned and forced him to implement new technical processes and managerial techniques, develop new contacts, and greatly expand his scope of operations.

With a low-cost and high-volume product he stepped farther away from the skilled craft paradigm than ever before, but he knew what he was doing just as he understood the financial and other rewards he could attain. And thus he entered the world of malleable copper.

Copper maintained a low profile throughout most of America's history. Early North American settlers, as well as the governments and corporations that sponsored them, hoped to grow rich from gold and silver deposits similar to those located by the Spanish in Central and South America. When these precious metals failed to materialize, prospectors shifted their efforts to more utilitarian resources such as iron and copper. Dutch, French, and British colonists found plentiful quantities of usable iron ore in many colonies but had a harder time locating copper. Unlike iron, copper must be virtually free of contamination before it can be worked, so the rise of copperworking has always been inextricably linked to the ability of copper smelters to remove impurities. Copperworkers often had high expectations each time they started working with ore taken from a new mine: the slow atmospheric oxidation of impurities partially purifies copper ore near the surface. American mine operators learned to perform relatively simple smelting processes to make these high-quality ores usable, and new mines produced the highest profits. After miners exhausted these surface ores, the next extractions of copper contained sulfur and other contaminants that required the far more complex, costly, and time-consuming "pyretic ore" smelting process. This process remained a highly guarded British secret and Cornish miners and Swansea smelters held a virtual monopoly on copper production until the 1830s.[59]

The first sustained copper mining in the American colonies took place in the early eighteenth century using imported labor from Germany and, to a lesser degree, Britain. British law required that all copper ore be smelted in England, but as we have seen, Americans preferred to look the other way when faced with restrictive imperial legislation. Although several primitive colonial refineries briefly satisfied local copper needs, high operating costs, a lack of skilled workers, low copper demand, and frequent imperfections in the final product prevented these ventures from turning a profit. As a result, colonial copper mines shipped nearly all their copper ore to Britain for smelting and fabrication, and Britain shipped finished copper goods back to America.[60]

Many Americans still wished to develop a domestic copper industry to keep some of these profits on their side of the ocean. Mines at Simsbury, Connecticut, produced small and undependable quantities of copper ore from 1720 to

1788, and after 1773 the mine became known as the Newgate Prison when incarcerated convicts served as the primary labor force. The Schuyler mine in New Jersey produced the largest quantity of American copper in the decades after its 1715 opening, but its recorded output fell to between ten and forty tons a year by the 1760s and it closed in 1773. The mine reopened after the Revolution, with minimal output. Investors opened other mines in New Jersey and Maryland but failed to find significant quantities of copper. Almost all of these mines relied upon foreign workers, primarily imported from Germany and Wales, and most colonies discontinued all copper mining by the Revolution. The American colonies produced between one hundred and two hundred tons of copper a year in the 1720s and 1730s, and output fell under one hundred tons a year by the 1750s.[61] Thus the local supply of copper, never high, started its plunge at a time when population increases, not to mention new applications, drove demand ever higher.

American copper demand increased steadily with population growth, and Americans remained dependent on British imports to meet their needs. Years later in 1814, Thomas Cooper, the chair of natural philosophy and chemistry at Dickinson College, lamented in *The Emporium of Arts and Sciences*: "Copper is an article so necessary to us at present for sheathing ships, for making distilling vessels, for vessels used for culinary purposes, for plated ware, for coin, &c., &c., that I hardly know of any manufacture of such importance, after that of iron: and yet we have no smelting work for copper, or any copper mine worked in the United States."[62] Copper imports, usually of pots and cooking utensils, increased from under 20 tons a year in the beginning of the eighteenth century to a peak of 350 tons in 1760. Copper items are superior to iron ones for many cooking tasks because of their heat conductivity and imperviousness to rust, but copper's high cost made iron the metal of choice for most early colonial pots and other utensils. Colonial coppersmiths, typically itinerant laborers who performed a variety of simple patching or tinkering services, were few and far between. These smiths usually required a second trade to meet their meager operating expenses and might double as tinsmiths and braziers who performed other metalwork. Nearly all copper objects at this time were made from sheet copper, which furnaces produced by casting copper into long and thin molds. Coppersmiths spent most of their time shaping cast copper sheets to produce and mend items such as pans, kettles, pots, ladles, skillets, stills, funnels, and utensils.[63]

The disparity between copper's demand and supply became even more exaggerated when America's shipbuilding activities increased, including vigorous military purchasing under the new government. Shipbuilders relied upon a variety of metal fasteners, such as bolts, spikes, and staples, to hold together the timbers of wooden ships. The ocean environment subjected these fasteners to saltwater, necessitating a rustproof metal such as copper.[64] But shipbuilding added new constraints to the copper-making process as well, since metal fasteners also had to withstand the constant strain of a massive ship's weight and motion pulling it in different directions. Itinerant coppersmiths and metalworking artisans across different trades often struggled to understand and adapt to these challenges: the artisan mentality placed a high premium upon tradition and secret knowledge, but the nation's new technological demands required a more experimental approach. Fortunately, this was where Revere excelled.

Technical Experimentation and the "Secret" of Malleable Copper

Revere launched his new career in May 1794 by contacting a government official (whose name is not recorded on the surviving letter) and discussing upcoming shipbuilding activity:

> I understand that there are to be two Ships built in this State, for the General government, and that they are to be Coppered, if so, they will want Composition bolts, Rudder braces, &c. &c.
>
> I can purchase several tons of Copper here, and my works are fitted for such business: Should those things be wanted, and I understand by General Jackson that it is in your department, if you will be kind enough to give me the refusal, you will much oblige me.
>
> I will do them as cheap as any one, and as well.[65]

Revere's networking abilities shine through this letter: he knew about the two ships soon to be under construction before work began, contacted General Henry Jackson and obtained more information plus the use of the general's name as a contact, and fired off a letter to a government official (possibly Henry Knox or a purchasing agent) in order to lock up a contract. What this letter does not say is the fact that Revere had never produced a single bolt or spike or worked with pure copper on any large scale. As usual he had quite a learning curve to climb, but he confidently and correctly assumed he could master this new process.

In future letters written long after finishing this first contract, Revere repeatedly refers to the magnitude of his technological achievement, for example, describing the "considerable labour & expense" he exerted when he learned to produce strong but non-brittle copper bolts and spikes.[66] He offers even more detail in a revealing letter to Massachusetts Congressman Harrison G. Otis in 1800, in which he reflected upon some of his metallurgical successes:

> Before the Frigate Constitution and the other two Ships were built the new merchant ships that were to be coppered were bolted & spiked with cast composition metal (Copper & Tin) which from its being brittle, did not answer the end. When the Copper came from England for the Above Frigate by some Accident a part of the Bolts were too large, I was applyed to by Genral Jackson the Agent, to draw them smaller. I then found that it was necessary that Bolts & Spikes for Ship Building should be made out of Maleable Copper. After discoursing with a Number of Old Copper Smiths, they one & all agreed that they could not melt copper, and make it so malleable as to hammer it Hot. I farther found, that it was a Secret in Europe that lay in but very few Breasts.
>
> I determined if possible to gain the Secret. I have the satisfaction to say, that after a great many tryals and very considerable expense, I have so far attained my wishes, that I did supply the Constitution, for dove tails, staples, nails, &c &c to the amount of 1000 Weight drawn from Copper of my Melting. Since which, I manufactured for the Ships Boston and Essex, upwards of 10,000 weight of Copper into Bolts & Spikes, from Old Copper.[67]

This letter lays out the context for Revere's entry into this new field: his first commission asked him to modify existing bolts and spikes, an easier task than creating new ones from scratch. But when his inquiries among the local metallurgical community revealed the difficulty of this challenge and the complete lack of experience among all technical practitioners, Revere attacked it on his own via "a great many tryals and very considerable expense." He described this process as a "secret," hearkening to his artisan days, when he pledged to learn the "art and mystery" of his trade. Revere also repeatedly compared his product quality to that of British-made bolts and spikes; it went without saying that only an exemplary grade of copper could equal that of the British. Without any clues regarding his learning process, we might imagine that he probably practiced in his foundry until he could produce items similar to British ones. And while learning the secret of malleable copper manufacture, he also gained knowledge of a number of metallurgical principles that underlay these processes.

The real challenge faced by Revere and other American metalworkers concerned the tradeoff between the different, and seemingly contradictory material properties that fasteners needed in order to survive the harsh conditions present in shipbuilding and ocean voyages. Metal fasteners often secured massive wooden beams to each other, and they had to resist being pulled apart or bent by the ship's motion. Resistance to different stresses is a material property known as strength.[68] But strength is most helpful in a material when it is combined with ductility, the ability to deform without breaking. Modern materials scientists describe the combination of strength and ductility as toughness, referring to a material that can absorb a huge amount of energy without rupturing. In other words, a tough material can resist most stresses applied to it (strength), and when extreme stresses finally begin to deform the material it bends without breaking (ductility). Tough metal led to more durable spikes and fasteners that enabled a ship to avoid frequent and costly overhauls and refitting, but contemporary practitioners could not express the product's needs in this manner. Even with respect to iron, a metal used and understood far more than copper in Revere's day, skilled workers did not fully comprehend the ramifications of their procedures or the different characteristics of metals. One industrial archaeologist contends that "Throughout most of the nineteenth century, mechanics thought that the stronger iron was, the better it was. They failed to appreciate the relationship between strength and toughness."[69] To make matters even more complicated, many fasteners received sharp impacts during the shipbuilding process as shipwrights hammered them into cured wood, requiring the additional property known as hardness to resist dents or deformations from blows. Many copperworkers initially attempted to make fasteners out of cast copper, a complete mistake. As with cast iron, cast copper is brittle and breaks or shatters under impact, possessing hardness but not ductility or toughness. The answer lay in an entirely different direction.

By 1783, the British copperworking industry had developed a rolling process to produce "malleable" copper bolts and spikes, far stronger and less brittle than cast copper. The proper combination of cold-working, annealing, and hot-working produces fundamental changes in the microstructure of cast copper, successfully achieving the desired toughness. All metalworkers practiced cold-working when they changed the shape of copper or other metals through hammering or bending. Phrased in a more technical manner, cold-working produces plastic deformation, or a permanent physical alteration in the shape of the substance, such as bending a piece of iron into a horseshoe shape.[70] Cold-working increases the amount of defects and strain within the metal, making it more brittle and less tough: for example, if a blacksmith bends a horseshoe too many times it will snap into two pieces. Britain's real discovery came about when metalworkers combined the standard process of cold-working with annealing and hot-working, in order to mitigate the brittleness problem. Annealing is the process of heating metal without melting it in order to rearrange its atoms into a more perfect structure with fewer defects. This releases some of the metal's strain and brittleness, and thereby restores its ductility and toughness. This process requires careful judgment on the part of the worker: insufficient heating fails to remove the brittleness, while excessive heating, or even prolonged heating at the proper temperature, removes the metal's strength and makes it soft. Hot-working is the procedure of reshaping a metal after first heating it to the temperature range that restores ductility. Skilled coppersmiths fabricated malleable copper by a combination of hot- and cold-working processes, alternately heating, cooling, and hammering the copper until producing the desired blend of qualities. In this manner, copper could become both hard and tough, or in other words, simultaneously resistant to penetration, resistant to breaking, and non-brittle.[71]

Two tasks faced Revere: he had to understand at least the rudimentary theory behind this procedure and he had to learn how to implement it. In correspondence throughout the remainder of his life he correctly uses terms (albeit with creative spellings) such as *annealing, malleability, hard, tough,* and *brittle.* He understood some of these concepts from earlier projects dating back to his silverwork and he probably picked up new terminology during his many visits to different metal shops, but malleable copper forging gave him the chance to synthesize and direct all his knowledge toward a single goal. One clue hinting at the actual technical processes developed by Revere lies in his use of the word *draw* when describing his copper bolt and spike production process. As a part of his earlier silversmith training Revere gained plenty of familiarity with the wire-drawing process. Wire drawing closely approximated several nuances of malleable copper working: too little heating did not restore flexibility to silver wire, and too much heating softened or even melted it. He could not make copper bolts or spikes the way he made silver wire, by pulling them through a series of small apertures, but his silverworking experiences did teach him the concept of working metal to the point of maximum strain before heating it just enough to remove the strain, and then working it some more.

Revere also learned a related technical skill during this period. In a January

1800 letter he mentioned that he used his furnace to refine 1,800 pounds of copper at a time.[72] As noted earlier, copper refining was an extremely complex technology when done properly. Different copper ores could easily become contaminated with other chemicals such as sulfur or oxygen, and British refineries carried out a many-stage process that used different techniques to gradually remove larger and larger quantities of contaminants. Revere almost certainly did not possess the knowledge or equipment to carry out such a difficult operation but he might have imperfectly refined the metal by heating it in a reducing (i.e., oxygen-lacking) environment, which removed oxygen from copper. He might have also used the term *refine* to refer to the separation of copper from other metals in a mixed batch of "old" copper that often included tin, zinc, iron, and other materials. In this case he carefully adjusted the furnace temperature in order to liquefy metals with lower melting points and physically separate them from higher-melting-point substances. Although Revere lacked the ability to properly refine copper, any partial step toward this technology represented a landmark in post-Revolutionary America. Revere's bolt and spike "drawing" process, therefore, probably involved a combination of refining cast or "old" copper into a purer state, forming it into bar shapes, and hammering and annealing it to produce the desired proportions and material properties. As always, he failed to commit any of his metalworking knowledge to paper, instead letting his products do the talking. Sales soon followed.

Revere exhibited his growing metallurgical knowledge in an October 28, 1795 letter to Portsmouth Naval Agent Jacob Sheaf. Revere told Sheaf that he had just delivered fifteen tons of copper bolts to naval agent General Jackson. In his letter to Sheaf, Revere proved that he understood the technical nuances of this work: "should they [copper bolts] be <u>cast</u>, they will not answer for the use they came for, for when that metal is cast in sand, it looses a very great part of its M̶a̶l̶l̶i̶b̶i̶l̶i̶t̶y̶ ̶M̶a̶l̶l̶e̶a̶b̶l̶e̶ Mallebility and is very easily broken. But those which are <u>drawn</u> retain their Mallebility and are as tough as iron."[73] This advice proved timely because Sheaf planned to reduce the size of his own spikes by casting them, a move that certainly produced brittle products. Revere also singled out "casting in sand" as part of the problem. In another letter, written on February 7, 1796, to an unknown recipient, he uses metallurgical terminology with confidence: "Now Sir, any person who is aquainted with the nature of metals must know, that if it will stand the force of a Trip Hammer, after it is drawn to the size & swaged smoth, it must be tougher than at first, because the grain of the metal is finer."[74] Even though Revere may not have used the term *tougher* precisely in its modern form, he still shows a concern and awareness of material properties, made all the more impressive by his discussion of metal grain size, revealing a casual understanding of the correlation between a material's microstructure and its properties. During his malleable copper research period Revere learned to produce quality output, but also strove to understand why metals behave the way they do. This scientific approach set him apart from the majority of practicing artisans and captured the love of learning that guided him through so many metallurgical adventures.

Revere initially had unrealistic expectations regarding his products. In the February 7, 1796 letter cited above he claimed his bolts "acted under the hammer as tough as Iron," adding, "I will risque my reputation, that you shall take one of those bolts and place it across two blocks of Iron and strike with a large Blacksmith sledge, backwards and forwards, three hundred times, before you can break them, and yet they are left Hammer hard."[75] This is a bet Revere could not win. He chose this imagery to demonstrate his bolts' toughness, but even though many contracts attest to the quality of his products, infrequent customer complaints reveal that his bolts were not indestructible.

Revere wrote another letter to Jacob Sheafe on January 7, 1799, in order to defend his spikes and reject a request to take back a shipment. Sheafe reported that he could not use the 616 pounds of composition spikes Revere recently sent him because workers cut the heads too large and left one side uneven. Even though he and his workers had to turn out many tons of bolts and spikes for each contract, Revere had not truly mastered the precision manufacturing and quality control methods needed for true interchangeability. His two-part response reveals much about his view of his products. Concerning the issue of the spike heads and unevenness, Revere noted "that all the English ones have large heads & the French ones have still larger & ar square," the only direct response Revere made to the charge that his spike heads were too large. He did not discuss their unevenness at all. Instead, Revere used this letter to defend the workmanship and quality of his spikes in general, and to discuss the different types of metal used in spikes. He took a didactic historical and scientific approach to the subject:

> I wish to mention, when the Brittish Nation first began to [word missing] other than Iron bolts & spikes into their Ships, they were made of a composition of Copper & Tin, but they soon found that it would not Answer, by reason it was too brittle. They then found they could harden copper; they now make use of

no Composition Bolts or Spikes, but clear copper.—If you wish to distinguish between Copper & Composition bolts or spikes, lay them across two pieces of Iron, bend them backwards & forwards, & you will soon find which breaks first, then look at the grain.[76]

Revere also pointed out that Sheafe erred in describing the spikes as "composition" metal, since he sent only spikes made from "clear copper, drawn from barrs of copper under the hammer in the same manner as iron spikes are made." In Revere's terminology, "composition" copper referred to bronze alloys whereas "clear," "pure," or "malleable" copper referred to homogeneous copper. He added that Sheafe's spikes exactly resembled the 1,500 pounds of spikes he had recently sold to the *USS Constitution,* and "The work men that drove them [the spikes] told me they were equal to the English ones. And what is more no man but my self in the four New England States, can melt the Copper & draw it in to spikes but my self." The bolts were also "such as the Sec'y for the Naval department ordered to be drove into every Vessell building for the United States." Revere frequently resorted to this tactic of defending a specific item or product by citing his body of work and overall experience.[77]

Revere wrote a second letter to Sheafe on February 13 and did a better job of addressing the actual topic. In this letter, Revere admitted that he rushed this shipment of spikes to ensure they all arrived in one package. Therefore, "some of them might not be quite so even as others & their heads a little larger." He devoted the majority of his letter to the charge that his bolts were too soft to penetrate the hull of a ship. This time, he contended that some woods, particularly the prized live oak used on certain naval vessels, were hard enough to break even British spikes. The shipbuilders of the *Constitution* had a similar problem, and overcame it by boring holes before driving each spike. In conclusion, Revere mentioned that he could draw the spikes to a smaller size for $5 per one hundred spikes, and make them strong enough to drive "near as well as iron."[78] Even though Sheafe clearly had a problem with this shipment of spikes, Revere's offer to draw them down, as well as his helpful advice about pre-drilling the hull before using the spikes, presented several helpful solutions.

Although Revere held his technical prowess in high esteem, his self-praise carried a large element of truth. His achievement was, without question, a milestone for America. In his previously mentioned letter to Harrison G. Otis in March 1800, Revere presented a "Short History of that valuable & necessary Metal in this Country," a history of American copperworking that suspiciously resembled a history of his own copper experiences. This letter confirms some of Revere's statements concerning the rarity of copperworking knowledge in America: "[Colonel Joshua Humphreys] asured Mr. Stoddard that there were no person in America that could make Copper maleable so that it could be drawn in to Bolts & Spikes. I have shewn him, and he acknowledges that He was mistaken and gave me leave to make use of His Name."[79] If an expert as qualified as Colonel Joshua Humphreys, the nation's most gifted naval architect, believed that no American possessed the knowledge and experience required to make malleable copper bolts and spikes, Revere's achievement truly separated him from most if not all American metalworkers. Long after dismissing Revere's claims, Humphreys had the opportunity to visit Revere's shop, at which time he observed Revere's advanced copper processes and products and immediately admitted his error. Humphreys' retraction may have boosted Revere's ego in the short term, but this powerful recommendation would have even larger long-term impacts, to be discussed in the following chapter.

Revere's malleable copperworking accomplishments soon elicited public praise. An article titled "The Launch," printed in the May 21, 1799 issue of the *Massachusetts Mercury*, pointed out that the frigate *Boston* was the first ship exclusively constructed from bolts and spikes manufactured in the United States: "We think the publick are under obligation to PAUL REVERE, Esq., for his indefatigable attention to this Branch of Naval Architecture, especially at a time when the British Government has prohibited the exportation of that valuable Article."[80] This public use of "Esquire," a gentleman's title, surely had a powerful effect upon Revere. Many Americans recognized the importance of Revere's accomplishment even though he duplicated, rather than invented, an existing British process. Being first was not as important as staying in the race.

Malleable Copper Contracts and Profits

Revere finished his first assignment for resizing 15 tons of bolts on October 14, 1795, and charged the U.S. government $2,756, which it paid on March 10, 1796.[81] Over the next few years he received more contracts and turned out more output than he had in his earlier endeavors. He was now positioned to provide a relatively high-tech product to a country that needed it desperately.

As an American craftsman and manufacturer, Revere had always competed with British goods, whether silverware, ironware, or bronze bells. He produced fairly limited quantities of most of these earlier products and focused more on meeting the specialized needs of his local market than on combating the pressure exerted by England. Upon entering the malleable copper field, in which he had to produce enormous numbers of bolts and spikes, he fully realized the power of the British industrial juggernaut and confronted America's longstanding love of cheap imported goods. American preferences produced a form of inertia driven by hundreds of years of history: English newspapers and styles drove American tastes; the British pound, shilling, and pence system shaped Americans' conceptions of monetary value; and British textiles, iron, china, glass, and other products had long served as standards of utility and excellence. A new manufacturer such as Revere had his work cut out for him, needing to overcome a general bias in favor of all things British. Worse yet, this bias had a solid grounding in truth, because British goods did offer excellent quality at low prices.[82] Revere employed a methodical strategy in his struggle to overcome British competition. He first negotiated several contracts in 1795 and 1796 with local government officials or merchants in need of bolts and spikes. As his shop mastered these manufacturing processes his reputation grew, and he used his satisfied clients as references and contacts to drum up new business. He eventually compared the quality of his goods to that of British manufactures and benefited from both a preference for local products as well as the patriotic satisfaction of buying American.

As with his cannon contracts Revere charged a price per pound for finished products and incorporated all raw material and production costs into this price, although this rate varied depending on the complexity of the material. For example, in an August 31, 1798 letter to "the Committee building the ship at Mr. Hartt's yard" he offered to prepare spikes, rudder bands, a chain for the rudder, bolts, and cogs for 37 cents per pound for cast copper or 41 cents a pound for malleable copper, a difference of 4 cents a pound reflecting the extra work needed for annealing the malleable copper. With the price of copper hovering around 20 cents per pound, this yielded a "labor fee" of 17 to 21 cents per pound of work, which he used for salaries, fuel costs, other expenses, and his own profit. By the time of this letter, his reputation had preceded him and he frequently responded to letters inquiring about his prices and production speed. By December of the same year, his price for copper bolts and spikes rose to 50 cents per pound and he expressed concern that he could not accept a new contract from Jacob Sheafe because he could not find enough copper to finish his current contract. Raw material shortages once again reared their ugly head, affecting his price as well as his ability to accept new work.[83]

By 1799, Revere had streamlined his pricing policies. One ledger, beginning in September 1798 and ending in April 1799, contained all of Revere's bolt, spike, and staple orders for the *USS Constitution*. At the end of the contract, he computed a total of 8,116 pounds of copper products sold to the *Constitution*, but he did not include an equivalent dollar amount. We can assume that he computed a final charge for the contract based on an overall price per pound, probably 50 cents a pound, which included labor and raw material costs as in his earlier contracts. Taking this price simplification one step further, Revere clearly ignored any differences between bolts, spikes, staples, and other products by this point, and merely combined everything into a single total weight.[84]

One comprehensive ledger describes all of Revere's expenses and income for bell, cannon, and copper work from May 1, 1799, to February 28, 1801. The majority of these activities involved the production of bolts and spikes. The debit column covers a variety of expenses, and the credit column lists payments received from different sources. Table 5.4 outlines Revere's major expense categories.[85]

Although most of Revere's costs are listed as a payment to an individual "for copper" or "for 6 weeks work," some of the expenses listed in the "miscellaneous" category consist of ambiguous payments to individuals that could represent additional raw material purchases or wages. Other miscellaneous expenses include purchases of bricks, wire, and sand, and numerous small withdrawals labeled only as "cash."

Several patterns emerge from Revere's expense list. The constant need for copper—and iron, brass, bell metal, and tin, to lesser extents—dominated his operations and accounted for more than three-fourths of his total operating expenses. He recorded metal purchases ranging from $7 to more than $1,000.

Table 5.4.   Furnace Expenditures, 1799–1801

Expense	Cost (dollars)	Total (percent)
Metal Purchases	16,339.04	76.3
Labor Costs	1,131.15	5.3
Fuel Costs	903.80	4.2
Transportation (Shipping)	83.81	0.4
Miscellaneous	2,949.70	13.8
Total	21,407.50	100

Although the majority of his metal acquisitions are simply labeled "copper," he recorded some purchases of "old" copper, copper stills, cannon, and bells, indicating the need to recycle in order to meet demand. His supply network became more widespread than ever before as he patronized at least forty-six different metal suppliers (and probably more than fifty after the unlabeled entries are considered), as well as numerous suppliers of coal, wood, charcoal, and other goods. Even with this huge number of individuals selling him raw materials he occasionally had to shut down his operations because of copper shortages, so we can imagine the consequences had his legendary networking skills proven less effective.

Revere's income is listed in a credit column that is far less comprehensive than the expenses ledger due to abbreviations and missing pages. Fortunately, he computed subtotals of his income at several points, and these subtotals cover the missing entries. He received $26,450.17 in cash payments throughout this period. Although some of the 61 recorded payments are as small as $2, they averaged more than $430 thanks to 12 large payments of $500 or more, 7 of which exceeded $1,000. Payments from Stephen Higginson and Amasa Davis account for the four largest ones, all for Revere's work on government contracts that each incorporated a range of goods. Most of the payments merely record the name of the payee instead of the item sold, but Revere did make sure to note that five payments corresponded to bell purchases, a product that continued to stand out from all others. His product line continued to increase as he sold his clients copper nails, sheathing nails, flat point nails, braces, pintles, dove tails, composition cleats, braces, eye bolts, chambers for pumps, cast rollers, and chains in addition to bolts and spikes. If his prior experiences serve as a guide, Revere could reproduce nearly any item after the proper study and practice. This period of product expansion and experimentation exposed him to the technology surrounding sheathing copper and probably gave him the idea to attempt to produce that commodity in the years to come.

Revere's gross revenues from May 1, 1799, to February 28, 1801, exceeded his expenses by $5,042.67, yielding an approximation of his profit. Of course, this figure cannot be treated with any confidence or precision, because it depends upon Revere's subtotals and assumes that the beginning and ending of the debit and credit lists correspond to the same transactions. Since many of his clients took years to pay him, and we know his sales volume increased steadily throughout this period, we might expect that his eventual profit for this period could be higher when clients paid some of the larger bills from the end of this production run. This ledger also fails to include hidden expenses, such as Revere's depreciation on tools and machinery or any rent he paid for his property, calculations beyond the consideration of eighteenth-century businessmen. With these disclaimers in mind, if the "profit" is compared to gross revenues, every $1 of sales yielded 19 cents income for Revere. This income, while not hurling him into the uppermost social or economic class, enabled him to continue amassing capital for his final endeavor, which eventually led America into a new copper age.

Adaptation and Networking

In the extremely fluid economic climate of post-Revolutionary America, flexibility, adaptation, and innovation paved the way for great success. The 1790s saw a revival of commerce, and America's growing market economy created large demands for a wide range of products at a time when few domestic producers could compete with British goods. Britain's dominance of most manufacturing fields was an established fact, rarely questioned and in some cases even praised for its impact upon America. But the hazards, costs, and delays of international transport often rendered British goods vulnerable in American markets. An enterprising American manufacturer hoping to compete with these foreign products merely needed the initiative to enter the ring, enough capital to launch a manufactory, a command of the technological and business processes required to establish a production system capable of meeting local needs at accepted prices, and a network able to bring in enough raw materials to keep the fires burning. These requirements set the bar fairly high: most artisans lacked the necessary capital and business acumen, and most merchants lacked technical expertise. Paul Revere could do it all.

By the end of the 1790s, Revere exploited America's favorable economic conditions by helping to fill its seemingly unquenchable demand for malleable copper products. His position as a manufactory owner and supervisor enabled him to use his talents in a way that separated him decisively from the different laborers in his shop. The majority of his responsibilities fell into two categories: technical adaptation and networking.

Revere personally pioneered all technical processes employed in his shop. His adaptability is more evident at this point than at any other time in his life. Between 1792 and 1796, he learned to cast bells and cannon, produce "malleable" copper via annealing and cold working, and fabricate a variety of fasteners such as bolts and spikes, all without any formal instruction or benefit from the emigration of knowledgeable experts—the traditional form of technology transfer. Revere demonstrated his intellect and creativity through his exceptional ability to learn new trades by borrowing from ones he had already mastered. In many ways he embodied the ideal of an adaptive American society struggling to emerge from Britain's imposing shadow.

Revere also derived significant benefits from his formidable networking skills. During his artisan days he used his contacts to drum up silver orders from the members of many organizations, and his networking ability also helped him spread the alarm during his Midnight Ride. In the 1790s networking enabled his workshop to thrive when so many other manufacturing endeavors failed, because he could generate work for his shop and overcome raw material shortages better than most. His location near the docks of a major port city and his numerous colleagues in the public and private sectors helped him identify lucrative projects and offer his services at the right time, often before the work formally materialized. Of course, discovering the work was only half the job. No longer a newcomer to the metallurgical world, Revere had a growing reputation that accurately described his capabilities. He had become a respected manufacturer and he had earned that respect.

Revere's many successes illustrate how a choice subset of post-Revolutionary artisans improved their circumstances, becoming managers and owners of growing manufactories in a time when most other master craftsmen continued working in small shops or as skilled laborers in the employ of others. Profits from a number of endeavors, starting with silverworking and iron founding and soon incorporating bell and cannon casting, provided fixed and liquid capital that empowered Revere to continue trying new applications such as malleable copper work. His furnace, tools, workforce, and property, as well as intangible assets such as experience, supply networks, and reputation, allowed him to experiment in new fields without excessive risk or delay. And steady income from already-mastered occupations gave him the freedom he needed to survive the startup of each new operation. During this early federal period, the high entrance requirements for new technological endeavors discouraged the bulk of the population, but certain members of the middle class still took a chance. However, in his later endeavors Revere needed help to overcome higher startup costs associated with larger and more complex manufacturing processes.

Revere marketed himself to the government during a period of Federalist dominance, marked by international tension and fiscal uncertainty. What the government lacked in competence and experience it made up for with ambition. Indeed, its ambition was probably aided by the inability of most officeholders to recognize the funding limitations of the young bureaucracy. Revere corrected some of the government's deficiencies by confidently proposing new products or methods in letters to military purchasing agents or treasury officials. He did this with the same two goals that drove his entire career from his earliest days of patriotic artisanship. On one hand, he unapologetically served his own interests, zealously learning new trades in hopes of making a profit and augmenting his reputation and social standing. However, he also drove himself so hard because he wanted to help his country increase its manufacturing potential and technical expertise. Patriotism is often hard to prove: ironically, if Revere ever professed his patriotism in too direct a manner one might question his real motives. However, he went out of his way to offer free advice and other assistance in correspondence with key governmental figures such as Byers, Knox, and Coxe, and he offered the government extra services such as the procurement of raw materials for the national armory at Springfield without requesting a finder's fee. His interactions with government officials employed a more respectful tone than his letters to private clients, and in spite of frequent delays and errors he never considered ending the collaboration. Revere's desire to serve his community also reinforced his choice of products during this period: bells, cannon, and fasteners all represented both cutting-edge technology previously limited to Britain and high-profile items that helped local communities or the government function smoothly. Revere's pride in his work mirrors his love of his country, and he finally accepted that his most valuable contributions lay in manufacturing instead of office holding.

National economic growth, turbulent domestic and international politics, public service, and the drive to improve his societal standing propelled Revere into his final endeavor. Changing times produced new demands, which called for new products. An advantageous location, his growing reputation, and a vast stockpile of experience helped open the door to the greatest challenge of his life, rolling copper. And for this he would need government support beyond the founding fathers' wildest dreams.

CHAPTER SIX

Paul Revere's Last Ride

The Road to Rolling Copper (1798–1801)

On New Year's Eve of 1798, Revere wrote an introductory letter to Benjamin Stoddert, America's first secretary of the navy:

> Sir, I understand you have advised the Committee for building the Frigate in Boston, not to send abroad for any thing they can git manufactured in this Country; these sentiments have induced me to trouble you with this letter. I can manufactor old or New Copper into Bolts, Spikes, Staples, Nails &c &c or any thing that is wanted in Ship building; that is cast the Copper into Pigs, draw the pigs into Barrs under the Forge hammer, and then manufactor the Barrs into Bolts, Spikes, Nails, &c &c. I supplied the Constitution with Dovetails, Staples, Nails &c &c. The Frigate Building here has upwards of 5000 lb. of Bolts, & Spikes, allready in her, of my manufactor, & I have supplyed Jacob Sheafe Esq., Naval Agent at Portsmouth with 600 lb. of Spikes for the Frigate building there.—My greatest difficulty is to git old Copper. Could I git a sufficient supply of Copper I would undertake to roll sheet Copper for Sheathing Ships.[1]

Revere's contacts had once again provided a valuable tip: Stoddert wanted to build a strong navy from domestically produced components and actively sought manufacturers who could produce bolts, spikes, and sheeting from malleable copper. Increased military spending fit perfectly into Revere's plans. He had spent the prior three years perfecting his ability to work copper into a malleable state and then draw it into various shapes such as bolts and spikes. The long list of products in his letter to Stoddert reveals his growing technical mastery of large-quantity output for a fairly wide array of products. Revere's letter also displays his confident eagerness to learn to roll copper as long as Stoddert helped him find enough raw materials, the biggest stumbling block at this stage of his business. In spite of the seemingly perfect match between Stoddert's needs and Revere's skills, Stoddert ignored this letter after a consultation with Colonel Joshua Humphreys suggested that Revere's claims were ludicrous. Humphreys was America's foremost turn-of-the-century naval architect and also served as the Navy Department's chief technical advisor, appointed as America's first "naval constructor" in 1794. His knowledge about both the science of ship design as well as the technical details underlying different ship components made him eminently qualified to assess the nation's technological limitations.[2] Humphreys assured Stoddert that no American could produce malleable copper. Humphreys was wrong.

Revere had been producing and experimenting with malleable copper since 1795 and after learning to construct and modify various copper fasteners he investigated the possibility of making sheathing copper for ships' hulls. The road to rolling copper consists of two intertwining narrative pathways: Revere's technological progress and the government's evolving goals and capabilities. These threads converged when he finally made contact with Stoddert, and they remained connected until his retirement. Revere's technical accomplishments continued the self-education that characterized his foundry endeavors, particularly his drive toward standardization. Revere's copper-rolling education differed from all his previous studies, because he now hoped to learn an elusive British process that Americans had yet to master. As the first American practitioner of a new process, Revere moved up the technology transfer ladder described in Chapter Four, climbing from the internal diffusion phase to become the initiator of a pilot plant. Revere did not travel to Britain or record any conversations with British workers, instead relying on other methods closer at hand and more familiar by this point in his career.

Paul Revere's last ride cannot be properly understood without also following the development of the early federal government, still young and inexperienced in 1800 yet embroiled in intense ideological and diplomatic struggles.

In addition to serving as the single largest customer for American businesses, the government cared about the source of its products. A strong market mentality developed in large port towns by the 1790s and private buyers sought the highest quality at the lowest price, which almost always led them to British goods. Certain individuals within the federal government, particularly Benjamin Stoddert, realized they had to think of the future and prepare for a time when volatile international relations might lead to blockades, the cessation of foreign imports, or war. A well-equipped, domestically supplied navy would guard the young nation against these scenarios. Concerns for national security dovetailed with a growing public discourse on the benefits of a domestic manufacturing industry. Influential leaders such as Alexander Hamilton and Tench Coxe crafted careful arguments, government policies, and public shows of support for high-profile manufacturing endeavors in order to help America overcome its significant technological deficiencies. Revere had finally positioned himself in the right place at the right time, and his technological skills could now meet a national need while reaping financial and social rewards. But first he had to attract the federal government's attention, something easier said than done.

The Early Federal Government and Benjamin Stoddert's Navy

The federal government played an increasing role in Revere's business in the 1790s, first as a client for his cannon and later as the purchaser of innumerable bolts and spikes. Government contracting became a major part of his identity as it provided him with income, technical challenges, and the chance to serve his society. Government negotiations soon became even more challenging than the technical aspects of the trade, and Revere's vaunted networking skills offered him little guidance in this new arena. Fortunately for him, the government's reaction to changing international alliances and events provided large long-term contracts that balanced the frustrating delays and errors of the inexperienced bureaucracy. All of these impacts, good and bad, had their roots in the early years of the republic.

The government's deficit of administrative experience owed much to America's colonial status during much of the eighteenth century. Local and colonial governments operated on small, informal scales far from the center of empire, and few colonists had the opportunity to participate in colonial bureaucracies such as the customs service. Some Americans received administrative or managerial education in the army or in larger businesses with many employees, but even this preparation hardly met the needs of postwar America.[3] As much as Paul Revere wished for a more proactive and accomplished government able to commit to large military contracts, he and his contemporaries had ample experience with the negative aspects of central government in the years leading up to the Revolution.

Following the war, Patriot leaders hesitated in establishing a strong centralized source of authority so soon after throwing off the ruling hand of Great Britain. In 1783, following the successful negotiation of a peace treaty with Great Britain, America finally received official recognition of its independence. But independence had few immediate political ramifications because the Articles of Confederation had been ratified in 1781 and the Congress of the Confederation, continuing the membership and mission of the Second Continental Congress, had directed the war effort and served as a federal government since then. The Articles of Confederation created a weak federal bureaucracy to oversee governmental functions, but it was severely understaffed and overworked from the beginning. Without the power to tax or regulate commerce the Congress of the Confederation often took a back seat to the individual states. A group of Federalist statesmen, including George Washington, Alexander Hamilton, and James Madison, championed a new system to collect greater power in the hands of a national government. Passionate political leaders and much of the general population debated the merits and dangers of this new system in state after state during the ratification debates of the 1780s. The proposed Constitution gave the federal government the authority to borrow and coin money, tax, and regulate commerce. To avoid the jurisdictional turf wars and confusing overlaps that plagued the government under the Articles of Confederation, the Constitution prevented states from usurping federal prerogatives such as issuing money or levying tariffs. With the election of George Washington in 1789 the new government finally came to life, and an eager nation awaited signs of the tone it would set, the actions it would take, and the world it would produce.

Three defining characteristics colored all government actions at the turn of the century. First and foremost, the federal government was completely inexperienced: every aspect of it was new and its employees and officeholders had little or no administrative experience.[4] Each decision represented not only the solution to a short-term need, but also a potentially vital precedent that could shape future policies in unforeseen ways. As a result, lawmakers often focused upon the theoretical underpinnings of the Constitution rather than the practical details of daily policy execution, rendering detailed transactions haphazard at best. Revere eventually learned that many government officials knew little about the products they asked him to produce, and even less about the funding mechanisms that ensured he received his fees. Second, the small government had a limited budget and modest goals. The ongoing constitutional debate concerning strict versus loose construction forced all officials to carefully justify and defend their actions, discouraging broad or decisive action. The government's concerns over budgets and strict construction caused all of Revere's contracts to undergo extensive review after he provided his products but before he received his fee, and plunged all operations into chaos every time any officeholder retired or lost an election. Third, the government was intensely partisan, and many newspapers cast the very future of the nation into doubt each time an opposition party threatened to win a majority. Although the partisan nature of government helped Revere win some contracts in the 1790s, it played a larger role after 1800, when the Jeffersonian party of Democratic-Republicans actively opposed his Federalist ideology.

Due to its large size and close ties to the recent Revolution, the army stood in a class by itself, encapsulating all the administrative challenges facing early America. Although Congress demobilized the Continental army after the war, the government reorganized and strengthened the War Department in 1791. The new standing army reached a size of 12,000 men by 1798 in response to fears of a French invasion, a shocking development for a nation that had previously equated large military forces with imperial oppression. Financing and supply problems plagued the War Department from the start, leaving behind a depressing legacy of mismanagement, waste, supply line breakdowns, and undelegated authority. A damning 1791 report singled out the quartermaster's department and private contractors for particular criticism. Alexander Hamilton described the prevailing situation in a colorful 1799 letter to the secretary of war: "the management of your Agents, as to the affair of supplies, is ridiculously bad. Besides the extreme delay, which attends every operation, articles go forward in the most incomplete manner."[5] The government's chronic inefficiency had many sources. The Constitution did not specify whether the War or Treasury Department should purchase and pay for supplies, and the two departments frequently resolved this stalemate by not paying at all. The comptroller of the treasury had to approve all federal funding claims personally, and even small errors in the application process led to rejection and delay. Revere ran afoul of these pitfalls repeatedly in his manufacturing career and his papers contain many letters reminding different branches of the federal government that he had been waiting months or longer for his payment. Revere's experience was far from unique: military contractor James O' Hara, for example, had still not been paid for his 1796 contract in 1801.[6]

Unlike the American army, which dealt primarily with internal issues such as protecting the frontier until the War of 1812, America created its navy in order to address ever-changing international relations with various European powers. Although the Department of the Navy was only two years old when it contacted Revere in 1800, its origins stretch back to the creation of the new federal government and the mercantile and diplomatic policies of the early republic. Naval history presents an unusual paradox. For some European powers, America's navy represented their primary interaction with the government of this new nation, and its successes or failures often embodied America's overseas reputation. The early navy became a vital component of national diplomacy and defense, and its existence and operations engendered continual high-profile attention as well as partisan debate at all levels of government and throughout most American cities and towns. At the same time, a single individual could fully supervise the small navy, making it personal and flexible on good days but inefficient and overworked in hard times. This combination of visibility and compactness explains many of the factors that influenced Revere's copper-rolling contracts and says much about the navy's first chief executive.

The establishment and growth of the Department of the Navy would not have been possible without the financial prosperity and expansion of trade fostered by the Federalists, particularly Alexander Hamilton. Hamilton's policies bore fruit by the mid-1790s, when America enjoyed restored public credit, stable national currency, huge quantities of investment capital, and a thriving merchant marine. America's growing merchant fleet went hand in hand with improved coastal fortifications, the creation of a navy, and an often pro-British foreign policy, culminating in John Jay's controversial treaty with England in 1794. In addition to removing British forts and troops that constrained America's westward expansion, Jay's Treaty (also known as the Treaty of London) allowed American merchants to continue their profitable activities but alienated powerful interests by granting "most favored nation" trade status to Britain and failing to compensate slave owners for captured slaves. This treaty effectively cooled the rising tensions that threatened imminent war between America and Britain at a time when America was militarily and economically unprepared. In the years of peace following this treaty, American exports increased in value from $33 million in 1794 to $94 million in 1801.[7]

Although many Americans, and particularly Democratic-Republicans, protested the unfairness of Jay's generous treaty with Britain, France did more than protest. Viewing it as a tacit alliance with Britain, France declared war on American commerce in 1796 and 1797, and French privateers and warships began capturing merchant ships, particularly in the vicinity of the West Indies. By 1799, America had lost approximately $20 million of mercantile property to French attacks and shipping insurance increased from 6 to 25 percent. When French officials insisted upon a large bribe in order to meet with John Adams's diplomats (later known as the scandalous XYZ Affair), America revoked its earlier treaty with France and retaliated with its own privateers and navy. Thus began an undeclared naval war with France in 1798, often called the Quasi-War. This "war" was unique in American history because of America's lack of clear wartime goals: America did not seek victory or French defeat, the government prohibited privateering against French merchants, and the navy initially avoided French warships except under special circumstances. America only wished to protect its overseas commerce without provoking a larger conflict. The young nation received an early lesson regarding the price of independence; Americans now had to protect their own interests without the backing of Britain's imperial might.[8]

With the increasing complexity of naval affairs, Congress took the controversial step of establishing a separate Department of the Navy. Prior to 1798, the secretary of war supervised all naval operations. James McHenry served as the secretary of war in the late 1790s, and friends and foes alike described him as a bureaucratically inept administrator. His lack of talent aided the birth of a separate Navy Department, since many of his allies hoped he might perform better if his responsibilities diminished. The creation of this new department stirred arguments between Federalists and Democratic-Republicans over the ideal size and strength of a national navy. Federalists supported a strong navy that could defend American shipping interests and generate international respect, while Democratic-Republicans believed the distance between America and Europe rendered a large navy unnecessary and prohibitively expensive. As with most issues during this volatile period, the debate over the navy highlighted diverging ideological views of America's future: a nation with a prominent navy anticipated international interactions and conflicts, while a nation without one expected neutrality and isolation. Amid all the conflict and rhetoric, 1798 also marked the appointment of America's first secretary of the navy, Benjamin Stoddert.[9]

Benjamin Stoddert was the only cabinet member President John Adams did not inherit from Washington's administration, and was by some accounts the president's most trusted advisor. Born in Maryland in 1751, Stoddert belonged to a prominent landholding family and trained as a merchant in his youth, serving as a major in the Pennsylvania militia during the Revolutionary War. He soon received an appointment to the Board of War, which provided him with invaluable administrative experience and political connections. He established and managed a successful shipping firm in Georgetown after the war until he received his appointment as secretary of the navy in June 1798. A Federalist to the core, Stoddert immediately used his cabinet office to promote his vision of America as a powerful, self-sufficient, and commerce-driven nation. Throughout the turbulent years of the early republic, Stoddert's intelligence, leadership, and drive stand out against a horde of less experienced government employees. According to Treasury Secretary Oliver Wolcott, "Mr. Stoddert is a man of great sagacity, and conducts the business of department with success and energy: he means to be popular; he has more of the confidence of the president than any officer of the government." Stoddert's clerk Charles Goldsborough added his own praise at the end of Stoddert's term, saying, "A more fortunate selection could not well have been made. To the most ardent patriotism he united an inflexible integrity, a discriminating mind, great capacity for business, and the most persevering industry."[10]

Upon taking office Stoddert inherited a single four-week-old vessel, several appointed captains whose only experience lay in the merchant marine, two office rooms, and no administrative staff whatsoever. He quickly assembled a staff of six clerks, and better yet, he received the services of three frigates whose construction had begun years earlier. The Naval Act of 1794 provided for the armament of six ships under the Department of War in order to protect American shipping from pirates hailing from Morocco, Algeria, Tunisia, and Tripoli, known collectively as the Barbary states.[11] Republicans opposed to the idea of a standing navy forced the insertion of a clause stating that these ships would be decommissioned if hostilities ceased, and indeed, American ambassadors signed a peace treaty with Algeria in 1795. However, George Washington lobbied in favor of continuing at least a partial armament of naval vessels, and three of the frigates survived. Christened the *USS Constitution, United States,* and *Constellation,* they were completed and seaworthy by July 1798. The term *seaworthy* fails to do justice to these vessels, also termed *super frigates,* which combined the world's most advanced technological design with first-rate materials only available in America. Brilliantly designed by naval architect Joshua Humphreys, these ships were longer, faster, heavier, more durable, and more efficient than any existing ships of their class. They could outfight anything their own size, and outrun anything bigger.[12]

Stoddert faced a shortage of skilled manpower for all aspects of naval operations and personally performed nearly all diplomatic, financial, technical, and logistical administrative tasks, ranging from the development of national policy to the addressing of envelopes. Stoddert's correspondence illustrates his ability to successfully address practical and detailed questions related to timber selection, the number of cannon appropriate for ships of different sizes, the amount of nails needed during construction, and many other issues.[13] Although administrative tasks occupied an enormous amount of his time, the manpower deficit gave him free rein to shape the navy as he saw fit and allowed him to handle every detail properly and consistently. Stoddert's position compares in many ways to Revere's, since both men had to deal with a mixture of strategic plans and practical logistics, and they thrived on the opportunity to set and implement bold objectives that exploited new opportunities. A deficiency of entrepreneurship or managerial prowess in either man would have led to disaster.

Because of a short-lived and fortuitous bipartisan emphasis upon national defense, governmental stinginess did not constrain Stoddert. Republicans lost support in the patriotic fervor following the XYZ controversy and Congress voted large appropriations to the Navy Department along with explicit permission to build another three frigates, purchase more than forty ships of all sizes, arm merchant ships, and establish a marine corps. Stoddert interpreted this mandate in the broadest possible terms and immediately developed a network of naval agents who served as important links in the navy's administrative chain, acting as purchasing agents, recruiters, supervisors, local technical experts, and general "handymen." One agent generally supervised each naval yard, although they also oversaw transactions between the government and the numerous merchants and private contractors who supplied the navy.[14]

By the early fall of 1798, Stoddert dispatched his small fleet to the West Indies to engage the French privateers who had been terrorizing American shipping. Although the young navy faced the daunting logistical challenge of waging America's first naval war at a vast distance from home ports, its vessels performed so well that they more than paid for themselves through insurance savings and increases in trade within eight months. By 1800, American ships had defeated two powerful French frigates, seized at least eighty-six French privateers, and liberated more than seventy captured American merchant ships. One of the most dramatic and celebrated moments in this conflict occurred when the *USS Constellation,* one of the super frigates, resoundingly defeated and captured *l'Insurgente,* reportedly the fastest ship in France's navy. The price of shipping insurance plummeted during this period and hostilities came to a halt after the Convention of 1800 peacefully suspended all political ties between America and France in order to allow America to remain neutral and thereby continue trading with France during its ongoing war with Great Britain.[15]

An ardent and utilitarian Federalist, Stoddert frequently had long-term national goals in mind when he took short-term actions. While managing the conduct of the Quasi-War he also implemented many procedures to make the navy more efficient, modern, and self-sufficient. He embodied the "loose construction" philosophy of constitutional authority by repeatedly and creatively reinterpreting his mandates to justify the expansion of governmental power. For example, when Congress authorized only the pursuit of "armed" ships on the high seas, Stoddert informed his captains that any ship could be considered armed if any of its crewmen carried a musket or pistol. And even though Congress voted against the acquisition of public naval sites in 1794, 1797, 1798, and 1799, Stoddert used $135,000 of a $700,000 shipbuilding appropriation to purchase 6 navy yards. He also interpreted a $200,000 timber appropriation as permission to purchase 2 wooded islands for $22,000. Later congressional investigations condemned both of these purchases, but the government nevertheless chose to retain all of this property, which proved valuable to the nineteenth-century American navy. Stoddert realized that it is often easier to ask forgiveness than permission and he knew how to minimize the fallout from his repeated stretching of his instructions. Because his actions clearly resulted from the spirited desire to augment America's naval power, and not from personal ambition, he retained a high standing in the opinion of fellow cabinet members and President Adams.[16]

The Tentative Growth of American Manufacturing

On December 29, 1798, Stoddert wrote a letter to House Committee of Naval Affairs chairman Josiah Parker, illustrating the full scope of his farsighted Federalist industrial policy. He began by observing that ships should be built locally instead of purchased internationally, because constructing them in America allowed government expenditures to enrich local industries and create essential infrastructure. For similar reasons he believed that ship construction should rely upon domestic raw materials and manufacturing establishments. America's long shipbuilding tradition facilitated this plan, and all commodities required for ship outfitting could already be manufactured in the United States with the exception of hemp, canvas, and copper. Americans could privately grow hemp and manufacture canvas if a steady demand could be assured via large government contracts, and thanks to Stoddert's encouragement the hemp and canvas industries soon flourished. But copper remained a trickier challenge until Revere entered the picture.[17]

Stoddert took a similar approach to the development of American foundries. In 1798 America faced a shortage of quality cannon of all sizes. Even though the War Department had contracted with government foundries and private founders such as Revere as early as 1794, the total output of American shops remained small. Few American foundries possessed sufficient technical skill to produce large guns and only half the finished cannon passed the difficult proving tests without bursting. Facing a dearth of weapons for the navy's steadily growing fleet, individual captains often frantically competed with each other to procure their own armaments. While Stoddert imported cannon to meet short-term needs, he also promoted a policy of strong support for American founders, even though many of their products did not meet quality standards. He realized that government support could expedite the learning process, and the development of a domestic industry would ensure America's ability to defend itself against blockades or foreign navies. By 1800, he could report to Congress that "The manufacture of cannon, and indeed of all kinds of arms, and military stores, is now so well established in the United States, that no want of them can be experienced in future, nor does it appear essential that large supplies of them should be laid up in store for the navy service."[18] Revere benefited from this armament process, as he sold twenty-eight cannon to state and federal governments in 1798, more than triple his sales for 1797 or 1796, as shown in the previous chapter.

Stoddert's belief that self-sufficient domestic industries lay in the nation's best interests placed him in the middle of a nationwide discussion over the benefits and dangers of manufacturing. America's continued rapid population growth increased the pace of the market economy's expansion, which fed the demand for manufactured goods. Britain developed new techniques for increasing its industrial output at this time, including mechanized textile production and new power-generating equipment such as steam engines.[19] Revere responded to this trend in his postwar activities, for example, by fabricating larger numbers of cheaper silver goods and building a furnace that could produce utilitarian iron objects. The vast majority of America's growing population lived on small farms in rural settings and procured nonagricultural necessities from imports or small owner-operated craft shops. The growth in the size and number of manufacturing establishments at the end of the eighteenth century jeopardized this centuries-old system and led to a lengthy political debate on the future of the nation.

Manufacturers in early America had to overcome a bias in favor of the agricultural pursuits that most citizens of this agrarian republic viewed as inherently virtuous. The debate about the impact of manufacturing on America's future began in earnest after the Revolution and continued well into the 1800s. This controversy originated in classical republican notions of the nature of society. This social theory posited an ongoing struggle between "power" and "liberty." Power enabled governments to decisively promote national interests, but power's aggressive tendencies threatened to limit or eliminate the vital liberty possessed by all free people. Constitutions, representative government, and other frameworks existed to limit power, but this limitation also depended upon the virtue of the people. Agriculture initially served as the ideal model for society because a geographically distributed nation of property-owning farmers would theoretically remain virtuous, thanks to their hard labor, egalitarian customs, and independent self-sufficiency. In the often-quoted words of Cicero, "Of all the occupations by which gain is secured, none is better than agriculture, none more profitable, none more delightful, none more becoming a freeman." In contrast, most observers equated manufacturing complexes with urban overcrowding and poverty, the oppression of the propertyless poor by factory owners and urban politicians, and excessive production of luxury goods that made the people self-indulgent and vile. The debate over America's future, pitting agriculture against commerce and manufacturing, had obvious ramifications for the artisan community. The decision to allow and enable unrestricted commercial and industrial development jeop-ardized the small-producer tradition, and favored mercantile interests hoping to maximize overall economic expansion, efficiency, and output.[20]

In addition to the need to address financial and logistical hurdles, early manufacturers often had to justify their work in an ideological sense. Thomas Jefferson and James Madison opposed the establishment of large-scale indus-try in America due to concerns over the corrupting influence of luxury goods, shifting of population from farms to cities, and creation of a large class of poor laborers who would be oppressed by a capitalist aristocracy. The debate over America's best path to prosperity intensified in the 1780s during the period of economic distress, leading Jefferson to famously comment in his 1785 book *Notes on Virginia,* "let our work-shops remain in Europe." Anti-industrialists did not oppose the increased use of machinery—Jefferson in particular loved the scientific aspects of mechanization—and often supported widespread manufacturing in households or smaller distributed shops because these ac-tions did not inherently threaten their republican vision like large manufac-tories did.[21]

Other writers, growing more numerous as the economy matured, expressed the need for a balance between agriculture, commerce, and manufacturing in order to maintain independence, observing how manufacturing augmented the work of agriculture by enabling the more efficient use of natural resources. The promise of prosperity made possible by manufacturing and other fac-tors allowed many Americans to overcome their earlier Puritan-inspired fear of luxury goods, often described as a corrupting influence. In the words of Benjamin Franklin: "Is not the Hope of one day being able to purchase and enjoy Luxuries a great Spur to Labor and Industry? May not Luxury, therefore, produce more than it consumes, if without such a Spur People would be, as they are naturally enough inclined to be, lazy and indolent?"[22] This shift in attitudes presaged the coming of industrial capitalism, a time when people did not perceive material benefits as good or evil as long as their recipients properly earned them. Similarly, Americans no longer exclusively identified manufactures with luxury goods after defense armaments and productivity-increasing tools grew in volume and importance.[23]

Late eighteenth-century discussions mark the beginning of industrializa-tion discourse in America, decades before the real shape of industry began to be understood or experienced. Physician and statesman Benjamin Rush started the ball rolling in a 1775 speech urging Americans to promote manufactures in order to reduce imports, produce changes that could aid agriculture, employ the poor, encourage immigration, and lessen dependence upon England.[24] In later years these comments became distilled into two primary pro-manufacturing talking points: manufacturing made America economically stronger and more independent; and manufacturing employed the idle and thereby increased efficiency while reducing sloth. Tench Coxe took over where Rush left off. In addition to holding appointed offices as assistant secretary of the treasury, commissioner of the revenue, purveyor of the United States, and other positions between 1790 and 1810, Coxe became America's greatest spokesperson for technology: "These wonderful machines, working as if they were animated beings, endowed with all the talents of their inventors, laboring with organs that never tire, and subject to no expense of food, or bed, or raiment, or dwelling, may be justly considered as equivalent to an immense body of manufacturing recruits, suddenly enlisted in the service of the country."[25] Coxe eloquently reiterated older arguments about manufacturing as a source of independent self-sufficiency, adding that new machines and power sources allowed even greater productivity with less labor, further underscoring the connection between technology and republican virtue.[26]

Many leading Federalists did not follow Coxe's lead, at least, not wholeheartedly. Part of the controversy surrounding manufactures resulted from the shifting and divided loyalties of merchants. Some merchants invested in industrial establishments (particularly during embargos when they could not invest as much in commerce) while others opposed all forms of domestic production that competed with their import trades. As late as 1810, Coxe specifically extolled the great mutual advantages received by merchant-manufacturer alliances in Britain and questioned the absence of similar alliances in America.[27] Due to a fear of damaging America's commercial prowess, Alexander Hamilton's 1791 *Report on Manufactures* refrained from proposing government sponsorship of manufacturing establishments. Like most Federalists, Hamilton primarily attempted to support commercial and financial institutions and saw manufacturing more as a means to this end rather than an end in itself. His report avoided any pro-manufacturing proposals that might frighten merchants and as a result he failed to address the proto-industrial segment's most pressing concerns, such as savage competition from cheap British imports, high labor costs, and lack of capital. Tench Coxe, who happened to be Hamilton's assistant at the time, tried to promote his pro-manufacturing viewpoint in his draft version of Hamilton's *Report*, which proposed protective tariffs, government loans and land grants to manufacturers, improved transportation infrastructure, and prohibitions on the import of certain foreign items. Hamilton rejected most of these proposals, undoubtedly avoiding a confrontation with strict constructionists, agricultural interests, and anti-manufacturing merchants. The *Report* emphasized the need to promote division of labor and manufacture armaments for national defense, and urged both a protective tariff and bounties to reward manufacturers, while reiterating Rush's points about employing the idle and reducing foreign dependence.[28] The federal government failed to take a prominent pro-manufacturing position at this early date, but the banner of future industry would soon be advanced via subtler methods.

Other groups attempted to fill the void left by federal inaction. State legislatures jumped into this effort by offering monopolistic patents and other forms of support to prominent manufacturers. For example, Massachusetts passed a protective tariff in 1785, lent 200 pounds to Robert and Alexander Burr to help them finish textile-related research in 1776, and gave lands to the Beverly Cotton Factory incorporators in 1779. Unfortunately, these modest efforts did not produce significant returns. Beginning in 1785, manufacturing societies crept up in different cities in an attempt to share knowledge and coordinate lobbying efforts for protective tariffs. Some organizations took an information-based approach, such as Benjamin Franklin's American Philosophical Society, which encouraged science and technology by corresponding with British experts such as the Royal Society and promulgating knowledge of inventions. Many pro-manufacturing groups followed conventional class assumptions and aimed their appeals to merchants who might invest in manufacturing firms, instead of artisans or other individuals familiar with the practical business of manufacturing. Other groups did include practicing artisans who banded together to address common issues faced by members of different trades. Manufacturing societies coincided with and in some cases gave rise to large-scale manufacturing experiments such as the Boston Sail Cloth Manufactory, Boston Glass House, Hartford Woolen Factory, New York Manufacturing Society, Beverly Cotton Manufactory, and the high-profile Society for Establishing Useful Manufactures. Other than Samuel Slater's groundbreaking Rhode Island spinning mill, which succeeded due to Slater's combination of technical and managerial experience, all other large-scale manufacturing experiments failed by the early 1790s. America's lack of competitiveness resulted from chronic shortages of skilled labor, capital, and entrepreneurial experience, which could have helped these firms navigate the turbulent political and economic climate.[29]

Benjamin Stoddert's successes in stimulating hemp, canvas, and armament manufacturing become all the more impressive in light of America's prevailing economic and ideological environment. Stoddert shared the views of Tench Coxe and other Federalists hoping to strengthen America's manufacturing infrastructure, but unlike Coxe, he had the freedom and authority to act on his views as long as he could justify their connection to America's naval program. In some cases he could do this easily, as when he promoted domestic hemp and canvas production by guaranteeing a market for these products. By the close of the century, Stoddert found himself one step away from realizing his dream of a powerful navy built entirely in America, by Americans, from American materials. He merely needed to solve the problem of rolling copper.

The Search for Sheathing

In his search for copper sheathing, Stoddert took on a technological challenge that had vexed naval experts for more than one thousand years. It had been known since Roman times that all oceangoing wooden ships, especially ones cruising in tropical or Mediterranean waters, faced hull problems as their service life grew long. Most wooden hulls acquired a "sea mat" of unwanted marine life within six to eight months, starting with clusters of barnacles but soon including vast masses of seaweed or other debris. This "mat" could be four inches thick and weigh more than one hundred tons, running the risk of rotting the hull, sinking the boat, decreasing its speed, or ruining its maneuverability. A second and far more serious threat resulted from the onslaught of shipworms such as the *Teredo navalis*. These creatures weakened hulls by boring into planks, turning ship bottoms into "leaky sieves." Ship hulls usually became dangerously weakened in one or two years and the expensive refitting process required months of hard work, during which time the ship occupied precious dry dock space.[30] A solution to this problem would not only save money and labor, but would increase the number of operating ships in the fleet at any time by decreasing the need for tedious dry dock repairs. The mid-eighteenth-century British navy tried using combinations of different types of wood, lead sheathing, tar, hair, and paper to prolong hull life, with no success. Copper sheeting finally provided a solution, greatly prolonging a ship's working life by deterring or killing sea creatures that attempted to adhere to it.

The first ship to receive an experimental layer of copper sheeting upon its hull was Britain's *HMS Alarm,* coppered in 1758. By 1763, the British navy recognized the effectiveness of that technique and by the end of 1781 almost the entire British fleet had coppered hulls. One historian commented that copper sheathing represented "the most important technical innovation to be implemented by the naval protagonists during the American War of Independence."[31] Contemporaries echoed this praise, and copper-sheathing technology received many accolades from naval officers: British Admiral Thomas Graves reported that wooden ships could not act in concert with copper-sheathed vessels because the wood hulls could not keep up, while other captains attributed some of their combat successes to maneuverability granted by copper sheathing, such as Admiral Lord George Rodney's capture of six Spanish ships of the line in 1780. After one engagement an angry French naval commander complained that British copper sheathing allowed the British captain "to manoeuvre as he thought proper, and take any advantage that wind or weather might give him to avoid action if he judged it advisable to do so." Responding to the economic impacts of this new technology, British Navy Comptroller Charles Middleton estimated that copper sheathing might double the number of ships at sea at any time. As a result, he gave coppering precedence over all other ship repairs, including fitting ships with improved guns such as carronades.[32]

Unfortunately, copper sheathing produced a serious problem: copper in the sheets reacted chemically with the iron spikes and bolts used as fasteners in the hull, rapidly rusting the iron. This reaction, called galvanic action, led to problems that outweighed the many benefits of the copper sheets. British naval engineers conducted many experiments in search of a method for protecting the iron fasteners, usually involving the use of paper, lead, or tar to separate the copper and iron. Some of these experiments attempted to substitute fasteners made of copper or copper alloys for the iron ones, but assessors deemed these unsatisfactory because the metalworkers had trouble making the copper hard and strong enough. In 1783 British metalworkers successfully mastered the art of producing tough bolts from a copper-zinc alloy via grooved rollers, and within several years these new bolts became standard issue throughout the navy, solving the rusting problem and allowing copper sheets to be used without risk.[33]

Copper-sheathing technology slowly migrated to America and also to other nations such as France.[34] Prior to 1795, few U.S. ships used copper sheeting.

That changed as merchants and later the navy noticed the increasing presence of copper on the hulls of visiting ships, and certain ships in the New York harbor added copper sheathing as early as 1792. On April 12, 1794, Joshua Humphreys advised Secretary of War Henry Knox of the radical changes being incorporated into new ships, including copper sheathing. In a report on December 23 of the same year he highlighted the vital importance of copper-sheathed ships at greater length: "Their bottoms always clean and ready for any expedition; if they were not coppered, their guns and stores of every kind, must be discharged, and they hove down every six months, (having no docks to dock them). This will not only be an expensive job, but strain the ships exceedingly, and injure the hull and rigging more than can easily be calculated, which ought to be avoided by all means; not only so, but an expedition may be lost by the delay."[35] Identifying the need for this technology was a big step, but America had a long way to go before producing it on its own. In the short term Americans had to rely on Britain for these copper items, which perpetuated the producer-consumer relationship dating back to colonial times.[36] Before America coppered its navy, its copperworking community consisted of itinerant smiths and small, poorly managed copper mines, all operating in the looming shadow of overwhelming British technology. As the early copper mines failed, foreign copper imports provided nearly all the items used in America, and local smiths increasingly focused on simple repair jobs. If not for a growing navy that demanded large amounts of reliable high-quality copper products, this mercantilist situation would have prevailed for much longer, with copper remaining a costly luxury item easily replaced by other metals. The navy saw copper in a completely different light, as an essential resource able to save money by prolonging ships' operating lives. America's lack of copper mines and expertise suddenly became an expensive and potentially disastrous liability.

The failure of the most prosperous British copper mine in 1798 added further urgency to the copper situation in America. Britain first restricted copper exports in 1798 by declaring copper a vital naval store, thereby depriving the United States of its primary supply. While the U.S. Navy purchased refined copper from many places—copper plates from Sweden and Turkey, pigs from South America, and bars from Russia—Britain remained the most trusted source of quality copper. Demand for copper bolts, spikes, and sheeting reached unprecedented levels during the merchant boom and naval armament period of the late 1790s, and British exports became highly unreliable due to the war. Britain's demand for copper ship sheathing caused the price of copper to rise from 25 cents per pound in 1791 to 50 cents per pound in 1798.[37] Copper procurement plagued the navy throughout the late 1790s: Stoddert and his naval agents sent many letters in search of supplies of copper and had to ship available resources from port to port to meet the demand wherever it was most urgent. At one point they contacted the superintendent of the mint to request clippings of cents or any other copper on hand.[38] By the end of the eighteenth century, America faced rising copper costs, questionable supplies of copper, inadequate technical know-how, and prospects of naval conflicts with various European powers—a litany of related concerns in need of an imaginative problem solver. Benjamin Stoddert had arrived in the right place at the right time.

Stoddert's interest in naval self-sufficiency had permanent ramifications for the future of copperworking in America. Instead of continuing the status quo, he hoped to use government resources to facilitate the birth of a domestic copper industry, because he realized that this technology lay far in advance of Americans' current capabilities. Copper sheet rolling was an even greater challenge than other forms of copperworking, primarily because the manufacture of copper sheeting remained a guarded technological secret in Britain and Europe. Coppersmiths made most copper objects from copper sheets, and used the "battering" process to make these sheets before the advent of water-driven rolling mills. This slow procedure involved the repeated hammering and measuring of copper bars until they became sufficiently thin, long, and uniform. The battering process sufficed during periods of low colonial demand, but any form of mass-produced sheet copper had to be rolled in a water-powered rolling mill. Such a machine probably did not exist in America before 1800.[39]

Under Stoddert's direction the navy made several attempts to foster American copper refining, manufacturing, and rolling, by dispensing loans—a controversial measure not covered in the Constitution—to defray research and startup costs to private firms, followed by promises of large contracts. Stoddert first attempted to create an American copper industry on May 16, 1799, when he granted a $10,000 naval loan to John Ross and Benjamin Henfrey, referred to as "Esquires" in the contract. Stoddert insisted they use the copper output of American mines to produce bolts, spikes, nails, and copper sheets "of quality equal to the rolled English, patent copper." He gave them seven months to produce bolts and spikes, and an additional year to learn to roll sheet copper.[40] Succeeding correspondence only indicates that Stoddert had abandoned this attempt within three months, long before the first deadline, although he quickly tried again.

On August 22, 1799, Stoddert wrote to South Carolina's Robert Goodloe Harper, the chairman of the House Committee of Ways and Means. "On the subject of coppering our ships in London," he began, "don't you think it high time we should be out of leading strings—I hope the whole Copper for the 74's will be the produce and manufacture of our own Country and have made arrangements to that end which I believe will be attended with success."[41] Although it might appear that the "arrangements" mentioned in this letter referred to the Ross and Henfrey endeavor, his next piece of correspondence indicated that Stoddert had already given up on it and shifted to a second plan. The very next day (August 23, 1799), Stoddert signed Articles of Agreement with Jacob Mark and Nicholas J. Roosevelt of New York City, whose professions are listed as "merchants."[42] This contract stated that, in exchange for a $30,000 government loan, Mark and Roosevelt would erect machinery for the production of copper bolts and sheeting equal to the best British goods, and exclusively use domestic ore supplies from the Soho mine near Newark, New Jersey, which Roosevelt owned. This contract was significantly more legalistic and defensive than the Ross and Henfrey contract, as it invoked several provisions to protect the government against default, probably a sign of the government having learned its lesson from the prior failure. The navy's principal stated object was "establishment of domestic manufactures of copper on a scale proportional to the probable demands of the United States for the Navy Department." The Roosevelt enterprise, Stoddert's most ambitious and costly effort to secure an American source of copper, intended to incorporate all aspects of the copper production process, from mining the ore to smelting, refining, and forging usable naval products. If this proposal succeeded, the United States would achieve self-sufficient copper production in one stroke. Stoddert gave Mark and Roosevelt one year to produce copper bolts and spikes, and an additional six to nine months for the copper sheeting.[43]

Stoddert revealed his doubts about the success of the Roosevelt venture in a letter to General John Swan, his Baltimore purchasing agent, on January 17, 1800. Despite the contract that should have supplied all of the navy's copper needs, Stoddert asked Swan to look into importing bolts and spikes. For the domestic-minded secretary, this request was an admission of defeat. Regardless of the funds Stoddert allocated, Americans seemed incapable of refining and rolling copper. On the following day he informed Roosevelt that Joshua

Humphreys had inspected the first batch of sample spikes and declared them unusable. Apparently, Roosevelt had hammered some of the spikes while still hot (rendering them too soft) and cast other spikes directly (rendering them too brittle). Stoddert's letter betrayed a sense of exasperation, and provided Roosevelt with a brief lesson in metallurgy:

> The process is to take a piece of copper, heat it, and draw it near the size of a spike then cool it in water, hammer point & head it cold. If hammered hot, the spike will be too soft, will not drive, but bend in driving & often so far in the ship as not to be got out without great difficulty. As to cast spikes, there never can be certainty that they are solid . . . You can judge whether you can make such spikes, & if you cannot, the best way will be to give up the thing at once.[44]

In the midst of Roosevelt's failure, Stoddert renewed his devotion to a self-sufficient American navy in a new letter to Josiah Parker. In particular, he focused on the vital importance of an American copper industry: "as for copper the manufacture of this article requires the expenditures of so much money, that without effectual Public assistence it will not soon be established in the United States."[45] This seemingly obvious statement marked Stoddert as something of a visionary. He continued to defy proponents of agriculture, strict constructionists, and champions of federal armories, even in the face of repeated failure, by taking the novel step of loaning money to a private manufactory in the hope of increasing national industry. Learning to sheathe a ship in copper required vast outlays of money and effort, but Stoddert knew that the process would more than repay this investment . . . assuming someone succeeded. The nation that mastered copper-rolling technology took a major step toward forming a powerful navy and merchant fleet, and the person who provided this sheeting was assured of profit and service to his country.

Stoddert concluded his letter to Parker with the observation that Britain no longer exported any copper items because its largest mines had failed. America was on the verge of being completely cut off from copper supplies, and even a controversial $30,000 loan did not enable Roosevelt to master the technical challenges of copperworking, to say nothing of the more complicated smelting and rolling processes. But hope was not lost. Revere had produced many tons of malleable copper bolts and spikes by this time. After a lifetime of anticipation, the stage was finally set for his entrance.

The Road to Rolling Copper

Paul Revere's copper-rolling ambitions arose haphazardly, in contrast to Stoddert's methodical strategies. Revere always sought to expand the scope of his operations and spent most of the 1790s adding new functionality to his operations while gradually increasing their scope and profits. However, he also desired something different, as illustrated by his local government offices, attempts to serve in a federally appointed position, ferocious defense of his Revolutionary War record in a long, drawn-out court-martial, and leadership roles in fraternal and labor organizations. In addition to financial prosperity Revere craved societal authority, respect, and the satisfaction of serving the public good. He first contacted Stoddert in an attempt to win a new copperworking contract and increase his income, but as time passed he came to realize that he might also satisfy his societal goals through higher-profile naval contracting.

By the end of 1798, Revere had more bolt and spike orders from merchants, shipbuilders, and naval contractors than he could handle, primarily because copper shortages constrained his output. Malleable copperworking brought him great profits, but he already had his eye on copper sheeting, the highest step of the copper ladder. In a January 27, 1800 letter to Robert Goodloe Harper, chairman of the House Committee of Ways and Means, Stoddert estimated a $170,000 copper cost for 6 proposed 74-gun ships, compared to $300,000 for the frames and $180,000 for other timber. In general, copper constituted around 15 percent of a ship's total construction cost, and sheeting made up 60 percent of the copper expense.[46] If Revere could master this new product he would have at least a short-term monopoly on a product desperately needed by his country. First he had to persuade Stoddert to give him a chance.

On the last day of December in 1798, Revere wrote the first of many letters to Stoddert, whose name he misspelled as "Stoddard." This letter is quoted at the beginning of this chapter. Although he eventually worked with Stoddert on copper-sheeting innovations, at this early time he primarily wanted to produce as many items as possible for the Navy Department. Revere listed the many copper products he could manufacture, casually offered to learn to produce sheet copper if Stoddert could find a reliable copper source, and concluded by asking for information about any potential domestic copper sources.[47] In spite of the fact that this letter said all the right things and arrived at the perfect time, Stoddert ignored it and instead made contracts with the two other would-be copper manufacturers before turning to Revere. Stoddert was a careful planner, unlikely to invest money in a man who was hardly known outside Boston. Revere's artisan and working-class roots shone through his letter, and as a moderate-scale manufacturer he employed a small team of workers and owned limited equipment and capital. In contrast, Stoddert addressed Ross and Henfrey as "Esquire," and related to Roosevelt and Marks as fellow merchants who happened to own a copper mine. Revere lacked a sufficiently persuasive advocate or the social standing needed to help him receive a contract. And in Stoddert's defense, many of Revere's claims would have struck most observers as exaggerated: his advertised skills, particularly the ability to anneal copper until it achieved both strength and hardness, lay beyond the reach of nearly every American practitioner at the time. As mentioned above, the knowledgeable Joshua Humphreys dismissed Revere's claim out of hand because it exceeded the skill set exhibited in every American shop he had ever visited. Humphreys and Stoddert found Revere's claims so farfetched that they did not even bother to reply to the letter.

Revere had a much stronger case by 1800. He wrote Stoddert a second letter on February 26, 1800, presented in Appendix 7. The letter begins by reminding Stoddert that Revere had already produced large quantities of malleable copper products and that his capability surpassed that of Philadelphia and New York metalworkers. He then offered to build a special furnace and learn to smelt domestic copper ore into malleable copper if the government provided him with copper ore and paid his expenses. Revere admitted, "I have never tried, but from the experiments I have made I have no doubt I can do it." To sweeten the deal he promised to teach his two sons his business, and referred Stoddert to Congressman Harrison G. Otis and President John Adams as character references. The mention of his sons served a critical role, because it demonstrated to Stoddert how this contract might produce a skilled shop that could outlive its current owner and offer vital products to America's navy for years to come. Best of all, Revere now enlisted Joshua Humphreys as an ally. Humphreys may have urged Stoddert to reject Revere's 1798 letter, but after visiting Revere's works, Humphreys asked Revere to contact Stoddert again and make use of his name. Clearly impressed, Humphreys wanted to fast-track Revere's application in order to get Revere's production process started as soon as possible.[48]

Revere's letter to Stoddert reveals a combination of experience and audacity. He dropped his earlier offer to roll copper into sheets and instead focused on procuring contracts to produce bolts and spikes and to smelt copper ore into usable metal. Although his self-confidence ran unchecked, he supported his claims with a convincing record of accomplishments. At a time when other Americans could not make copper malleable enough to form strong bolts, he had finished many tons of products for various clients. Still, his boldness is evident: without ever having handled a substantial quantity of raw copper ore, he knew he could smelt it. Some of this confidence possibly originated from his experience with refining iron (and possibly copper) in his furnace, although he surely realized the major differences between refining and smelting. However, a new awareness of his own mortality partially offset this confidence, moving him into a slightly defensive position. He was 65 years old, but offered his sons as a guarantee of future production. His old views of family continuity in apprenticeships and artisan shops still affected his thinking.

Revere followed up this letter on March 11, 1800, by writing his version of a "Short History of that valuable & necessary Metal in this Country" for Congressman Otis, the representative from Revere's district, quoted in the previous chapter. In this attempt to enlist the government, Revere discussed his long experience with bell and cannon manufacture and estimated that he had "made the greatest improvements in that metal of any man in this State, if not in the United States. I am confirmed in this opinion, from conversing with Col. Humphries of Philadelphia." Revere had acquired a national mindset by this time and started imagining himself as the founder of a new American industry, a valuable servant to his country.[49]

These letters soon bore fruit. Revere traveled to Philadelphia to meet with Stoddert and Humphreys in May 1800 and received a contract to produce bolts and spikes for two ships under construction in Boston and Portsmouth. The results of this meeting exceeded his expectations, because Stoddert finally trusted him with his own pet project and reminded Revere of his 1798 offer to learn to roll sheet copper. In addition to signing a contract for bolts and spikes, Stoddert asked Revere to smelt and refine some domestic ore and use it to roll sheet copper for testing and evaluation. On May 21, 1800, Stoddert told Stephen Higginson, his Boston naval agent, that it was more important to prove that Revere knew how to make sheeting than to have the sheeting meet any specific or immediate demand, testifying to Stoddert's prioritization of long-term goals over short-term needs. Revere impressed Stoddert, who told Higginson to hire him on "such terms as may be considered liberal."[50] Perhaps

Figure 6.1. Charles-Balthazar-Julien Fevret de Saint-Memin (1770–1852), drawing of Paul Revere, 1800 (photographic reproduction from Library of Congress Prints and Photographs Division, LC-USZ62–7407). This drawing represents an older, heavier, and more genteel Paul Revere who stands apart from Copley's portrayal of Revere the artisan. Now 65 years old, Revere wears fancy clothing, sports a visible double chin, and poses in profile, signifying his social and financial success. This painting roughly coincides with Revere's entry into the copper-rolling field, an important line of work that added to his prestige.

after the uninformed claims made by the two earlier groups of gentleman-would-be-manufacturers, Revere's straightforward hands-on attitude was exactly the right approach.

On June 28, 1800, Stoddert asked Higginson to judge whether Revere could manufacture sheet copper equal to that of the British. If so, he would soon be asked to manufacture a large amount. On October 31 of the same year, Stoddert reemphasized the importance of determining if Revere could manufacture sheeting, "for it is very uncertain whether we shall long be permitted to import this article from any foreign country—England will not now supply it." Stoddert then laid out the terms of this contract. Higginson should offer Revere a $12,000 government loan, later reduced to $10,000, to help him establish his copper mill.[51] After years of independent frustration, Revere and Stoddert had finally found each other. The success or failure of their partnership now lay entirely on Revere's shoulders.

Revere returned from Philadelphia in May 1800 with three objectives: smelt and refine a barrel of domestic copper ore provided by Stoddert, work some of the copper into spikes and bolts, and roll the rest into copper sheets. His first task, smelting the ore, offered the largest challenge because he had absolutely no exposure to mining or smelting outside theoretical texts he might have read. In addition, smelting had the most technical complexity of these tasks. Copper smelting is a physical and chemical process requiring great heat above 1084 degrees C and a chemically "reducing" atmosphere rich in carbon but lacking in oxygen. British smelters used a sixteenth-century German process in which they roasted the ore to drive out sulfur and then melted it to draw off slag. They repeated these steps dozens of times on each batch of ore over a period of weeks or months until the copper attained the desired consistency. Most miners switched to the simpler "Welsh process" after it was developed in 1750, because it used enormous quantities of coal to minimize the length of the operation.[52]

Revere already had extensive experience making malleable copper bolts and spikes, but sheet copper posed entirely different problems. He needed to construct and operate a rolling mill, using a waterwheel to transmit power via shafts and gears to two adjustable rolls that would flatten the copper to any desired thickness. As usual, he turned to established technical fields when approaching a new one, and in this case he relied on his own silver shop and observations of larger iron establishments. Revere had used his silver-rolling mill to produce major quantities of sheet silver since 1785. By 1800, he had become quite proficient in its use, but it was a simple device used to process extremely malleable silver without the benefit of waterpower. It taught him some of the principles of copper rolling, but not all. In contrast, iron-rolling mills had much in common with copper mills, but Revere had indirect exposure to them at best. Iron-rolling mills appeared in America with the first ironworks at Saugus, but were extremely uncommon. Colonial ironworkers passed red-hot bars through a set of water-powered rolls repeatedly, making the iron thinner each time. A cutting wheel or shears could eventually slice the sheet into strips for wagon wheel ties and barrel hoops, or into smaller strips for nail making. Low-quality remains from Saugus indicate the lack of skilled expertise on these tasks, and the existence of other iron-rolling equipment is uncertain.[53] Revere had to master the most relevant details independently: low-quality work sufficed for nail making but not for Revere's exacting clients, and copper differed from iron in nearly all respects.

Many details of Revere's copper-rolling preparations fell outside any of his written accounts at the time. He bought an ironworking mill in the town of Canton for $6,000 from "Messrs. Robbins, Leonard, and Kinsley" in either late 1800 or early 1801.[54] This site previously housed the gunpowder mill that Revere helped to establish in 1776 on behalf of the Massachusetts Provincial Congress. Leonard and Kinsley had operated an ironworks on this property for many years, and Revere's new purchase included the right to utilize the water in the Neponset River and a deed to several buildings. These buildings contained useful equipment, including a water-powered slitting mill that could cut iron or copper into workable bars or rods. By replacing the cutting wheel with parallel rollers (also called rolls), Revere could convert this machine into a rudimentary rolling mill, sparing him the problem of installing and adjusting the shafts and wheels associated with water power. Revere originally estimated that he could renovate the existing buildings and equipment by June of the same year. Undoubtedly, he saved time and money by renovating instead of buying or building new equipment. He derived the greatest benefits from the waterwheel equipment because he had never attempted to harness the power of running water before this point.[55]

One cryptic record in Revere's memoranda book indicates that he built a new furnace in September of 1800. The notation reads:

Memorandom Septem 1800

It took to build the new furnace   3000 Red &

200 White Brick

4 Cask lime

Sano came to live at our house Tuesday Octo 14 1800 @ 2½ doll p week

David Oliver came Octo 20 2½ p week[56]

This notation does not mention whether the new furnace was located in Canton or in Revere's Boston workshop, and does not include prices for the bricks, lime, or total labor. We can concoct a plausible explanation for either location:

Figure 6.2. Sketch of a rolling and slitting mill. From Edwin Tunis, *Colonial Craftsmen and the Beginnings of American Industry* (New York: Thomas Y. Crowell Company, 1965), p. 154. This sketch portrays a rolling and slitting mill, similar in some ways to the one Revere used in his Canton manufactory in the early 1800s. The waterwheel on the right provides immense power that turns the upper and lower iron rollers. A worker uses iron tongs to feed a bar of heated copper between the rollers, which compress and elongate the copper until it emerges from the other side. The worker would then decrease the separation between the rollers, reheat the sheet, and feed it through again until it attained the desired thickness.

his efforts to smelt and refine copper may have induced him to build a new furnace in Boston while he shopped for property in Canton, or this might have been one of his first modifications to his Canton property, anticipating all the furnace work he would perform there in the years to come.

On January 13, 1801, Revere asked a friend traveling in England to purchase a set of iron rolls, the large cylinders that would use waterpower to flatten the copper. "I can procure them here," he explained, "but not in such perfection as the English ones, neither are they so good." In addition to specifying dimensions for the rolls (twenty inches long by nine inches diameter), he asked his friend to try to observe some British copperworks and determine how they heated their copper. This is an understandable question, since copper sheets had to be heated before each pass through the rollers. If this could be made more efficient, the rolling mill's productivity would increase dramatically and the sheets might be less brittle. Heating practices also related to fuel usage, and improved efficiency could substantially reduce his expenses. Whether Revere received any benefits from his friend's espionage is unknown, but other references in this letter to Joseph Pope, "a good mechanic and an ingenious man," show that Revere had other sources of information upon which he could draw, even extending to Britain. A letter written in May 1802 reveals that by then Revere used rolls of his own making, but he still desired British rolls, this time from Bristol.[57] Poor-quality rolls probably accounted for many of his early production errors such as irregularly sized sheets or "pitting," the small indentations in copper sheets resulting from bumps or holes in the rolls. But thanks to his ingenuity in iron casting and turning, at least he *had* rolls.

The new year of 1801 brought good news for Revere, Stoddert, and America. By January 17, 1801, Revere sent Stoddert a piece of sheet produced from the barrel of domestic ore, proof of his mastery of the smelting and rolling processes. The creation of this one sheet was grueling: by the time he finished, Revere had converted approximately four hundred pounds of ore into thirty pounds of copper sheathing. Along with the sample sheet, Revere submitted a letter describing the many difficulties he faced: "My apparatus is not calculated for smellting, but for refineing only, by which means it gave me more trouble, & a greater expense . . . I had to refine that small quantity on a hearth, where I refine 1800 lb at one time, & I had to roll it by hand in a silver smiths plating mill."[58] Because of his lack of equipment, Revere had to pound the ore by hand, smelt and refine the copper in an inefficient, oversized furnace used for cannon casting, and manually roll it into sheets on his small silver-plating mill. He claimed he could double his yield with the proper equipment, and he probably did not exaggerate. However, he was one of the few Americans, if not the only one, to own and understand this strange combination of silver-working and founding equipment, and his success was no small matter. In a later letter to Joshua Humphreys, Revere apologized for the small width of his sample sheets, but pointed out that "the mill I rolled it in is the best there is in this Town."[59]

In the same letter that accompanied his sample sheet, Revere reported that he also had finished a large shipment of bolts and spikes (sixty thousand pounds) and was working on his copper-rolling mill. He estimated producing rolled copper sheets as early as June 1801. This letter also contained the first reference to his government loan. According to the final terms of the loan, after proving he could refine and roll copper he would receive a $10,000 interest-free advance that he could pay off in finished copper. This advance was intended to allow him to research and construct a copper-rolling mill, since these high startup costs blocked most entrepreneurs from entering the field. However, Revere had not waited for the advance. By the time he submitted his sample sheet he was already in debt from his purchases of the Canton property, equipment, and copper. To expedite the repayment of his debt he asked to receive payment for his bolts and spikes whenever he delivered each order, a reasonable request that would not be heeded.

Ironically, the tests Revere passed in order to secure his loan and contract from Stoddert were far more grueling than any of the copperwork he later performed. Compared to smelting, all other copper fabrication processes seemed simple, and the five hundred-pound sample from Maryland was probably the last unworked ore Revere ever handled. The equipment he used for the sample sheet also added layers of complexity to the process, since none of it was intended for the task at hand. When Revere finished rebuilding his Canton rolling mill, his technology greatly surpassed his original equipment. Revere's workers regularly refined copper by pounding it with water-driven triphammers and heating it in air furnaces, and they then drained the pure end product into flat bar-shaped molds. They heated these bars and passed them repeatedly through parallel rollers, decreasing the distance between the rollers each time until attaining the proper sheet thickness.[60]

After sending his first sample sheet to Stoddert, Revere recapped his success in triumphant letters to Harrison G. Otis and Joshua Humphreys. Revere informed Humphreys, by now a good friend, that he was modifying his recently purchased iron-slitting mill to roll sheet copper according to "the English method." He proudly told Otis to see for himself the sample sheet in Stoddert's possession, adding, "It is one Evidence that Copper can be got in our own Country & manufactured into Materials for Ship Building." Revere concluded his letter to Otis by remarking, "What a dreadful change in Politicks," perhaps the most prophetic statement of his life.[61] Election results had arrived, Thomas Jefferson would soon take office as America's third president, and Revere's rising star was about to fall.

Thomas Jefferson's election and the impending change in administration were worse than dreadful for Revere and his business. The Adams administration expired on March 3, 1801, and Jefferson's officers quickly replaced most of Adams's appointees, including the irreplaceable Benjamin Stoddert. Jefferson was the first truly unwelcome president in the eyes of New England Federalists such as Revere, who virtually treated the new administration as a hostile coup, looming with uncertainty and menace. The resulting political chaos that soon dominated the end of Revere's career was merely the latest skirmish in a rapidly escalating struggle between America's first two political parties.

Alexander Hamilton, George Washington's secretary of the treasury, was the prime mover of the Federalist Party practically from the start of Washington's first administration. A brilliant thinker, tireless worker, and persuasive writer, he envisioned a powerful centralized national government led by an almost kinglike chief executive in command of a powerful army and navy capable of dealing with both internal and external threats. Because America had not yet become a world power, Hamilton favored close ties with Britain to ensure continued trade revenues and support against other European powers. In order to realize this vision America needed a solid financial footing that promoted economic stability and commercial growth, explaining why Hamilton implemented fiscal strategies such as forming a new national bank, consolidating all state and federal debts, and issuing loan notes to serve as currency. Hamilton's proposal to offer government support of large-scale industry, alone amid his entire economic program, went unfulfilled due to powerful opposing interests. A dynamic and headstrong figure who always stirred controversy as he fearlessly imparted his ambitions and goals to the infant government, Hamilton inspired heated opposition from supporters of states' rights and agrarian interests who erroneously feared he might make America dependent upon Britain, turn the president into a monarch, encourage a military overthrow of the republic, or create an aristocratic class that wielded power without accountability. His most resolute enemies, Thomas Jefferson and James Madison, created the Democratic-Republican Party as a way of marshalling opposition to some of his policies, and his ongoing feud with John Adams, Washington's vice president and successor, split and weakened the Federalist Party.

John Adams and the Federalist administration lost the election of 1800 for many reasons, including the bitter internal rift between Adams and Hamilton, unpopular taxes to pay for national infrastructure and the Quasi-War with France, the repressive Alien and Sedition Acts, and the association of Adams's party with elitist pro-British policies. The Democratic-Republicans attempted to reverse these trends when Thomas Jefferson assumed office. As one of the most philosophical intellectuals of the founding fathers, Jefferson occupies an exalted place in America's history, although his complex and seemingly contradictory views have earned him the label of the "American Sphinx." Jefferson believed in democracy to a greater extent than many of his contemporaries, and viewed many of the Federalist policies as subversive attempts to entrench power in the hands of a small number of moneyed men. During his tenure as secretary of state and as John Adams's vice president, Jefferson promoted agrarian interests, strict interpretation of the Constitution in favor of states' sovereignty, reconciliation with the new French Republic, and proud popularization of democratic ideals. The Jeffersonians constituted a vocal minority under the Washington and Adams administrations but gained control of the legislative and executive branches in what Jefferson referred to as "The Revolution of 1800," with dire consequences for manufacturers such as Revere.

Jefferson's succession to the presidency immediately shattered Stoddert's long-term naval plans. Stoddert's term lasted until April 1, 1801, and his successor, Robert Smith, did not take office as Jefferson's secretary of the navy until July 27, although interim naval secretaries such as Secretary of War Henry Dearborn and General Samuel Smith filled the office for brief periods. The Democratic-Republicans' vocal opposition to a strong American navy deterred most potential candidates for the secretary of the navy position, most of whom were Federalists anyway, and different prospective candidates rejected Jefferson's offer until Robert Smith finally accepted.[62] This slow and clumsy changing of the guard contributed to administrative delay on all outstanding contracts, and Revere's correspondence includes inconsistent messages from each interim secretary, much to his confusion. During this bewildering period the Quasi-War drew to a close as word of a peace treaty from Paris gradually reached administrators and naval commanders.

Jefferson began his first term of office with many goals, chief among them quieting the partisan strife, minimizing the size of government, and repaying the huge federal debt. His major administrative measures often used fiscal policies to attain ideological goals, primarily by abolishing excise taxes (while maintaining import taxes, which he approved), shrinking the army and navy, and cutting budgets wherever possible in hopes of reducing government bureaucracy and repaying the debt. The navy presented a particularly appealing target for Jefferson's cost reduction for many reasons. He could realize huge overall savings by targeting the navy because its budget was so large. And ideologically speaking, Jefferson had no love for the navy: it represented a version of the "standing army" that earlier Patriots deemed a threat to liberty and it served the commercial interests of primarily Federalist merchants. Secretary of the Treasury Albert Gallatin, by far the most qualified financial expert among Democratic-Republicans, launched an economy drive in 1801 and repeatedly decreased all naval appropriations. Gallatin reduced the 1800 naval appropriation of approximately $3 million to less than $1 million by 1802, and it continued shrinking until 1805, when Jefferson increased funding to build numerous small coastal gunboats in lieu of expensive ocean warships. Navy Secretary Robert Smith had to deal with the angry naval officers protesting demotions or the dry-docking of their ships, but had no opportunity to mitigate this policy: in 1807 he told his brother-in-law "never was there a time when executive influence so completely governed the nation."[63] This was not an opportune time to ask for naval funding.

Revere's hopes and expectations were nearly smothered in early 1801. He wrote to Stoddert on March 5, attempting to collect the promised "advance" loan and use it to defray his mounting startup expenses. Revere had fulfilled his portion of the deal by preparing contracts and obtaining sufficient collateral to cover the loan. However, when he brought them to Higginson, the Boston naval agent, Higginson mentioned his new orders from "government," as well as personal misgivings about whether Revere really needed the loan after all. Revere's letter to Stoddert reiterates that Revere did, in fact, desperately need the loan:

> you, Sir, I am sure, are Sensible, that a business of this kind can not be carried on without money . . . I was in hopes that I should be able to git thro with out the loan not doubting that I should obtain the loan if I wanted it. I now find I cannot go on with out it. I beg Sir you be so kind as to give Mr. Higginson such orders, as that I may be able to go on with the business. I have no doubt by the Month of June next I shall be able to produce Sheet Copper equal to the Brittish.[64]

Stoddert's brief response arrived quickly, informing Revere that "About to retire from office, I must refer to my Successor your application for the advance of 10,000 dollars to enable you to complete your Machinery for rolling copper into sheets."[65] One small consolation appeared on the envelope and again at the bottom of the letter: Stoddert addressed Revere as "Esquire," perhaps in belated recognition of his abilities, competence, and national service.

Without the $10,000 advance from the navy, Revere suffered from a critical cash shortfall and could not repay the private loans he had incurred to purchase the mill property, equipment, and operating expenses. The payment delay affected all aspects of his vulnerable new business. As an illustrative example, in a letter to William Bartlett of Newbury Port, he regretfully turned down an offer to purchase some copper, a rare opportunity he would ordinarily have seized, given the perpetual copper shortage. He explained that the change in the administration (although he first, suggestively, wrote and crossed out the word "Government" instead of "Administration") had made him wary, and given his scarcity of funds he had to decline any new offers, especially until he knew the identity and plans of the new secretary of the navy.[66]

Throughout this trying period of administrative confusion and opposition, Revere sent a barrage of nearly identical letters to all relevant government officials (Stoddert, Samuel and Robert Smith, and Levi Lincoln), repeatedly restating the terms of the promised loan, his massive debt, and his pressing need for the government to pay him immediately. Robert Smith echoed the prevailing Jeffersonian belief in limited government power through strict constitutional interpretation when he told Revere, "I know of no law which authorizes this Dep't to lend money for the creation of copper works." Stunned that the federal government would even consider this sort of constitutional question so late in the process, Revere pointed out in a follow-up letter that "I had no doubt but the present Administ. would have fulfilled what the last had promised. It is exceeding [missing words] individual should suffer, when he is exerting himself for the good of the Government." Revere was not going to be drawn into a theoretical debate regarding loose or strict constitutional authority: he wanted the money he had been promised.[67]

Revere's persistence, string pulling, and letter-writing campaign eventually paid off. On June 6, 1801, Samuel Smith told Samuel Brown, the new Boston naval agent, "altho I do not entirely approve of such Contracts yet as the faith of the Government is in some measure implicated it has been determined to comply therewith."[68] Although this approval and funding arrived frighteningly late, it still represented a victory for Revere. The monetary advance kept his business solvent, and he could repay the loan in copper products instead of cash.

Revere's ledgers confirm the financial hardship mentioned in his letters. One account record titled "1801 Furnace" lists his income and expenses between January 1, 1801, and March 1802.[69] The debit and credit lists begin with the entry "to foot of account in Old Book," which carries more than $10,431 in expenses and $14,351 in credits from 1800, the year before he began rolling copper. These values can be taken as a rough approximation of Revere's costs and expenses before he began rolling copper, producing an estimate of approximately $3,900 in profits for the year 1800. In the 1801 ledger, Revere recorded a total of $28,608 in expenses and $30,364 in income for a 15-month period, yielding a profit of $1,756 on a far higher volume of transactions than in the prior year.[70] The fact that Revere showed any profit at all sets him apart from the great majority of entrepreneurs starting new manufacturing operations in early America. Revere's costs do not simply include operating expenses, but also one-time startup costs such as purchases of property and equipment, and yet he still balanced them with income streams. Table 6.1 presents a breakdown of Revere's income for 1801 and the first two months of 1802.

The $10,000 loan stands out as the single largest revenue source for this period, representing almost one-third of his total income, and without this loan he would have faced more than $8,000 in losses for this period. The fact that Revere included this loan in his tally of income for the year gives us a window into his view of accounting and his shop finances. We now recognize that a loan is not a source of income, and Revere certainly understood that he had to repay this money in future years. His inclusion of a loan alongside moneys received indicates his emphasis upon immediate liquidity, necessary when his mounting expenses threatened to bankrupt him. The loan kept him solvent when he needed it most, and represented an easily repayable obligation, basically an advance on future copper-sheeting contracts. Revere usually had to wait for months to receive the moneys owed him by different clients, and in the government's case months often turned into years. A loan reversed this trend, and not only gave him much-needed cash, but also tied him to the government by forcing the government to purchase enough copper sheathing to at least settle the account. In practical terms, the loan also enabled him to repay his private loans, which carried high interest rates.

Revere's records reveal other factors that contributed to his financial survival. His next largest sources of revenue were payments of $7,468 and $3,343 from two federal naval agents, Sam Brown and Stephen Higginson, who con-

Table 6.1.   Revere's "Furnace" Income, 1801

Client	Amount (dollars)	Total (percent)
U.S. Loan	10,000	32.9
Sam Brown (U.S. naval agent)	7,468	24.6
Stephen Higginson (U.S. naval agent)	3,343	11.0
Amasa Davis (state of Massachusetts)	2,916	9.6
Bell Sales (nine different clients)	1,973	6.5
Other	4,663	15.4
Total	30,364	100

tinually bought large quantities of bolts, spikes, and other copper items. Unfortunately, these payments did not equal Revere's sales during this period: naval agents usually bought on credit. Revere lobbied to receive his unpaid money while simultaneously pushing for his $10,000 loan, illustrating how he needed to keep these different accounts separate from one another: he would soon owe the government on a $10,000 loan that he could repay in rolled copper, while the government owed him thousands of dollars for copper bolts and spikes that he had already delivered. In a letter to U.S. Attorney General Levi Lincoln, Revere mentioned that he already delivered 84,718 pounds of his bolts and spikes to the navy.[71] Subtracting the cost of the copper, the government owed Revere 24 cents per pound for these bolts and spikes, totaling more than $20,000, and it had only paid him $14,500 by December 1801. Revere continued full-scale production throughout this trying period, amassing new government charges that had yet to be paid. The payments that he did receive kept him afloat, however late and incomplete they might have been. Even while he personally struggled to learn the copper-rolling process, Revere's workers produced a steady revenue stream by churning out bolts and spikes. Payments from the state of Massachusetts, sales of bells, and other transactions also helped him survive the transition period. Without a set of diverse and mature manufacturing operations under his belt he would not have had enough capital or equipment to give him the necessary time to master a difficult new practice.

Revere's ability to keep his shop financially solvent depended upon the careful management of his many expenses, listed in Table 6.2.

As always, raw materials represented Revere's largest expenditure, at almost 40 percent of his total. However, several charges for the purchase of his Canton property, stonework, and insurance for his rolling mill constitute the second

Table 6.2. Revere's "Furnace" Expenses, 1801

Expense	Amount (dollars)	Total (percent)
Raw Materials (copper and iron)	11,213	39.2
Property Purchase, Insurance, and Modification	6,589	23.0
Loans and Interest	6,258	21.9
Labor	4,267	14.9
Fuel (coal and wood)	281	1.0
Total	28,608	100

largest cost. Payments of loan principal and interest charges almost equaled the property costs, which makes sense when we consider that Revere's reason for taking out the loan was to enable him to rapidly purchase the Canton property. Labor fees, while less than the cost of metal, property, or loan payments, still represent an enormous sum, testifying to the size of Revere's skilled labor force at this time. These high fees collectively drove Revere to the verge of financial loss, but had he cut back on any of them he would have either had to curtail his present shop output or reduce his construction of a Canton manufactory to meet the needs of future contracts.

In addition to granting him a loan, the government also helped Revere procure copper stockpiles. A letter to Robert Smith on October 26, 1801, mentions that Revere had received "nearly 120 thousand weight of copper most of it refined ready to draw into bolts and spikes."[72] This immense quantity of copper (60 tons) was the property of the government, delivered to his shop in anticipation of the important products he would make. Revere's billing system enabled him to account for copper easily, because he charged clients 50 cents a pound when he provided the copper, or 24 cents a pound when the client provided the copper. No other client could lay their hands on this much metal, and it served Revere well through many future government contracts, allowing him to keep working without having to wait for materials. The large expense for copper and iron listed in Table 6.2 illustrates his ongoing need for additional metal on behalf of his other clients, and correspondence reveals that this procurement challenge continued to give him headaches.

By the time Revere received his government loan, the mill had cost him more than $15,000. Because he started setting up his works before receiving the loan his equipment was fully operational when the money arrived, and he rolled more copper while maintaining his production of malleable copper bolts and spikes. He expressed great pride in his achievement: "I have Erected my Works & have Rolled Sheat Copper which is approved of by the best judges here as being equal to the best Cold Rolled copper." The navy concurred, giving his copper a rating of "excellent" in November 1801, and by the following May Revere informed Robert Smith that he had rolled more than twenty thousand pounds of sheet copper. Adding the payment for this shipment to charges for other copper products requested by Smith, Revere fully repaid his $10,000 loan and completed his contract.[73]

Unfortunately, the government's administrative paralysis had only begun. Revere's steady stream of government correspondence did not abate with the receipt and completion of his loan terms. On July 1, 1803, he asked Smith for the return of the surety bond he originally provided as collateral for the government loan. Also, the government had not yet paid for further copper bolts and sheeting that Revere had supplied, and he needed the money. This situation reached the crisis point in November 1803, as described in a letter Revere sent to Robert Smith. He began by reminding Smith that he had delivered more than 64,000 pounds of worked copper to the government and had yet to be paid. Worse yet, in expectation of receiving his fees, Revere took out another $14,000 loan and used it to buy overseas copper, since the government was no longer providing enough metal to match his production. By the end of 1803, he had been paying interest on this loan for the past 13 months. He was owed $15,000 at this point, and Smith soon ordered and received an additional $10,000 of copper without paying for it. Revere's frustration was especially visible by the end of this letter, when he said, "We beg leave to mention, that it is more than two years since we have received one Shilling from Government 'tho we have been at work for them the whole time." He tried to appeal to Smith's sense of fairness, saying, "We are now Sir distresed for Money . . . you must be sensible that it requires a considerable capital to carry on a Business the Stock of which cannot be purchased but with cash."[74] This comment certainly shows the great distance Revere and America had traveled from the barter and credit relations of colonial Boston.

This letter illustrates the type of institution that Revere was dealing with, a government that bears little resemblance to the bureaucratic institution of today. He procured and paid for the copper used in the government contract, incurred debt and interest payments in doing so, and was reimbursed more than two years late. One receives the impression that without constant goading from Revere the government would have kept taking his copper without bothering to pay for it. This administrative lethargy was not a calculated attempt to save money or trick Revere into doing extra work. The government was, it seems, incapable of doing any better. Administrative inexperience or shorthandedness exposed all transactions to major postponements. Constitutional questions such as the navy's right to offer loans to new industries produced huge delays as novice officeholders sought precedents and considered different legal interpretations of their authority. Officials had an even harder time answering practical and technical questions because expertise on copperworking was hard to find anywhere in America, and was especially scarce in the nation's capital. Revere learned that innovation has its costs as well as its benefits: doing something for the first time might earn contracts and praise, but it also produced confusion and delay.

Without naval funding and guaranteed contracts, Revere could have never afforded to research copper rolling. Without Revere's experience and technical flexibility, the navy seemed incapable of founding a copper-rolling industry. In many ways, this was a match made in heaven. Or so it seemed.

The Revere-government partnership's effectiveness can only be measured according to how it enabled each party to achieve its goals. Although Stoddert had to deal with constitutional and financial limitations, he had clear ideas about the direction he wished to take the navy. Stoddert knew the risks involved in loaning money to Revere since he had already failed, twice, to jumpstart a copper industry with government funding. By the time he made his final investment in sponsoring the copper-rolling process, Stoddert had spent $76,000, a princely sum that produced results only after Revere entered the picture.[75] He undoubtedly considered his investment in Revere a major success because it finally gave him the solid domestic source of copper sheeting that completed America's ability to produce all naval commodities. His short tenure in office prevented him from realizing major practical benefits from Revere's operations, but his immediate successors received many tons of copper products from the small Canton mill, and in the years leading into the War of 1812 America did not have to worry about its ability to outfit its navy. Other indirect benefits of Stoddert's investment included the transfer of Revere's technology to other American rolling mills, which added to America's technological infrastructure and made future innovations possible. In this case as well Stoddert could not have been happier with Revere, a man who freely shared his experiences and knowledge with fellow manufacturers for the remainder of his life.

Revere's copper-rolling startup experiences also illustrate the ideology and practical limitations of the early government, and in particular, of Benjamin Stoddert and the Federalists in John Adams's administration. Stoddert's bold and loose interpretation of the federal government's powers to foster new industry reveals that the government sponsored private endeavors as early as the 1790s. At a time when even Alexander Hamilton could not arrange for the House of Representatives to vote on his *Report on Manufactures*, Stoddert quietly found ways to support American industry, fund private research, and develop a powerful, self-sufficient navy. But at the same time, the government's administrative and bureaucratic limitations are more evident in this period than in Revere's earlier transactions. No government official seemed capable of paying for goods at the time of purchase or meeting the government's contractual obligations.

Once again, Revere exploited his ability to apply one form of technical expertise to new endeavors, broadening the definition of technology transfer. In this case, he pioneered a new field and became the first American to roll sheet copper. Although this path-breaking achievement sounds dramatic, it was not a terribly new experience in his life since he had been jumping into new fields and mastering fresh technical challenges for years. The difference lay in the risk and scale of the rolling venture. Revere had no local practitioners to emulate or consult; British copper rollers were distant and secretive; and the financial risk was vastly larger, threatening the savings he had accrued over many decades of lucrative work. But although he could not rely upon equipment or personnel from Britain, he did draw upon the practices of other metallurgical fields, and after mastering the basic rolling process he gradually refined his techniques by comparing his products to British goods. The rolling mill, like his silversmith flatting mill, was a machine capable of producing fairly standardized output. Familiarity with machine use enabled him to participate in the technical experimentation and modification process, and then step back and teach his laborers to perform the bulk of the work after the major problems had been solved.

Revere probably did not understand the risks he was about to face because many of them fell outside his own experiences, or indeed, the experiences of any American. He surely appreciated the technical challenges he had to overcome, but confidently and correctly assumed he could master the rolling process. However, he never realized the government would consistently fail to pay him on time, or question the propriety of a loan to private industry after it had been offered. Someone who valued reputation and status as highly as he did probably assumed that an organization as powerful and important as the federal government would honor all of its commitments unquestionably. And in the long run, in spite of painful delays, this is exactly what happened.

Revere rolled enough copper to repay his loan by May 1802, and successive huge contracts with the navy continued to keep his business profitable for years, although the government's payments were late more often than not. If this was the price of a relationship with the early federal government, it was one Revere gladly continued to pay. The size of these government contracts dwarfed all his other work and also augmented his reputation. Without the promise of a loan and the assurance of a huge contract that let him repay the loan in goods instead of cash, would 65-year-old Paul Revere have risked bankruptcy to begin a new American industry? The answer is clear: without the U.S. navy, no copper would have been rolled at Canton.

CHAPTER SEVEN

The Onset of Industrial Capitalism

Managerial and Labor Adaptations (1802–1811)

Writing to Secretary of the Navy Robert Smith on May 24, 1802, Paul Revere mused about the manufacturing startup process: "You must be sensible, that every new manufacture that is attempted in our country will be attended with many dificultys . . . this has been the case with us, but we have succeeded better than we feared, considering that the business was as new as if it had never been done any where; for we could git no information respecting it . . . it can be no wonder if we made some mistakes."[1] As anyone would guess after reading this preamble, bad news was on the way. Revere and his workers had not yet mastered the nuances of the rolling process and many of the sheets in this first shipment were too short. Indeed, saying he faced "many dificultys" was an understatement: he had to confront deep debts, experiment with new equipment, search for skilled laborers, and compete with cheap imported goods, all while maintaining the steady output of bells, cannon, bolts, and spikes from his Boston foundry and Canton mill.

Revere elaborated upon his tribulations in a December 19, 1803 letter to Joshua Humphreys, a letter that starts warmly with "It gave me Pleasure to recognize my old Friend's hand writing." After discussing some errors made in a recent shipment, he offered a frank assessment that mixed the good and bad:

> We are dayly gaining experience. I have had & now continue to have the whole to feel out, for I have not been able to get any information from any person and should I live & be able to take care of the Business for seven years to come I should not get to the Zenith, we are obliged to pick up hands as we can & as they begin to know some thing of the Business they leave us.
>
> I cannot help acknowledging that I have done better than my expectations. Our sheets are as well finished and as soft & as free from scales & cannot be distinguished from English.[2]

By 1809, the situation had stabilized. In response to a government survey on manufacturing endeavors, Revere reported:

> The principle manufacture and that on which we most depend is Manufacturing Copper into Sheets, Bolts, Spikes, Nails, and every article used for fastening and Covering Ships Bottoms and Copper Smiths use. Ours is the only one in the United States . . . Benja Stoddert Naval Secretary as long as he remained in office fostered our manufactures. We manufacture Church and Ship bells of every size and weight cast every kind of Brass ordnance all our manufactures we warrant equal in goodness and manufacture to the English.[3]

Although he did not estimate his capital and profits, they were substantial by contemporary standards. He employed more than ten workers and owned a variety of new machines, including two rolling mills, two air furnaces, a trip hammer, a wire drawer, and massive equipment for turning and boring his own cannon. Revere had branched out into new markets for copper sheeting, including cooking utensils and boilers for Robert Fulton's steamboats while still practicing the casting processes he began in 1792. By this point, he thought in big terms, bigger than his own growing operations, and he strove to convince Congress to raise a protective tariff on imported copper sheets. And at last he permanently cemented his partnership with his son Joseph Warren Revere, who by this point played at least an equal role in all technical and managerial decisions. Joseph Warren shared his father's technological aptitude and inquisitive intelligence, and added a fresh new managerial perspective that eventually carried the company through exciting expansions and new product lines. In 1811, less than a year and a half after responding to the government survey, Revere retired from the business entirely and lived with his wife on their Canton property, confident that he had left his business in good hands.

Revere's journey from his neophyte rolling operations in 1801 to the success he achieved by his retirement in 1811 encapsulates the major issues faced by all American manufacturing concerns during the first forty years of the new republic. As the market economy quickly encompassed all aspects of American life, new business practices enabled the most entrepreneurial shop owners to function on a larger scale. The application of technology also increased throughout this period as manufactory owners tried to boost their productivity and the consistency of their output in scarce labor markets. At the same time, society's relationship with its natural environment evolved in courts, in manufacturing practices, and in the consciousness of key individuals. All of these changes, occurring in unison, heralded the dawn of the age of industry.

America's Transition to Industrial Capitalism

In 1800 Americans celebrated the arrival of a new century. The year held a mixture of promise and concern because the Revolutionary generation, most active nearly twenty-five years before, had begun to pass on the torch to a new group who had grown up in an independent American nation. Federalists George Washington and John Adams had left the stage, and Jefferson proclaimed the dawn of a new age with his "Revolution of 1800." These grand political changes echoed overseas events that promised turbulent foreign relations in the years to follow. The French Revolution, which started in 1789 with the storming of the Bastille and promulgation of the *Declaration of the Rights of Man and of the Citizen*, had finally run its course, culminating in the ascension of Napoleon Bonaparte to the dictatorial position of first consul in November 1799. Escalating warfare soon overtook Europe and produced major impacts in the New World as well, such as the sale of the Louisiana territory, economic embargos and blockades, and a renewed American war against the piratical activities of the Barbary powers. The new century offered countless local developments as well, threatening the upheaval of all aspects of daily life.

Dynamic revolutions always run the risk of eclipsing the traditional settings from which they spring. Through the first two decades of the 1800s, industrial establishments, cities, and power machinery represented anomalies on the American landscape. In 1810 90 percent of the country still followed agrarian work and lifestyle patterns, and clocks and watches remained rarities. Rural "work time," determined by natural and seasonal cycles, had a much greater influence than the carefully calibrated "clock time" starting to appear in factories and industrial establishments.[4] The individuals operating in this new culture, and particularly the risk takers who first took this leap, constituted a small but active minority that played a primary role in shaping the nation's future. Paul Revere stands out amid this unusual minority as a prominent participant in the older labor system who nevertheless pioneered the transition to the new industrial mindset.

America's government indirectly encouraged the growth of industry by building up nationwide communication, transportation, and marketing infrastructures conducive to new endeavors. The U.S. Postal Service played an essential, if unheralded, role by fostering a new communications network that bound citizens together in many ways. During colonial times, America's postal system provided a relatively high level of service, especially along the Atlantic seaboard and in the higher-populated northern colonies. Most mail carriers made three trips a week and the cities of New York and Charleston were connected to each other and to England by a naval packet service. The system declined during the war, but the Post Office Act of 1792 set the stage for future development by giving newspapers extremely favorable shipping rates, prohibiting mail surveillance, and facilitating the expansion of postal routes into new areas. As a result the U.S. Postal Service blossomed into a major instrument of local and interstate communication and connection. By the first quarter of the nineteenth century, the postal network had major impacts upon American life: it facilitated the growth of literacy and an independent press via its support of newspapers; expanded the market economy by disseminating financial information and transporting bills of exchange; and increased participation in government by publishing federal activities and aiding communication between representatives and their constituents. These changes, and particularly the unification of the market economy, made it easier for industrialists to procure funding, recruit a sufficiently skilled workforce, locate clients and suppliers, and carry out the strategic aspects of their business in an informed manner.[5]

Transportation networks also expanded dramatically throughout the nineteenth century. At the turn of the century most common roads "were hardly more than broad paths through the forest," generally poor, unpaved, and seldom maintained, often blocked by stumps, rocks, ruts, or unbridged streams and rivers. The limited repair work required to maintain passability on these roads—primarily bridge building, stump removal, and log laying—remained a community responsibility often performed by farmers after the harvest season due to a shortage of labor and government capital. Most towns had road systems in place to facilitate the transportation of goods to market, but routes between towns or cities were problematic at best. The economic prosperity of the 1790s and nationalist sentiment following the ratification of the Constitution inspired the creation of better roads for transporting people and cargo. State governments sponsored and supervised the construction of some roads but when their resources ran out they sought out private and public companies to build and maintain toll roads, effectively transferring the cost of road construction and maintenance to those who used them. The completion of the Lancaster turnpike in Pennsylvania in 1794 produced a boom in road financing that lasted until 1820 or so, eventually resulting in an interconnected network of quality roads densest in the Northeast and in areas of high population. Rivers offered an additional form of transportation, obviously most useful in the downstream direction. Up to the start of the nineteenth century, most farmers interested in shipping their produce lived near large cities or along rivers. While water transport made more sense for bulky and low-value items, wagons proved more reliable, quick, and efficient for higher-priced items, especially when transported over shorter distances. Manufactured goods often fell into the high-priced goods category, and the improvement of roads facilitated the development of new markets.[6]

Building upon these growing transportation and information networks, America's economy underwent a makeover in the late eighteenth and early nineteenth centuries, shifting from the colonial mercantile system to a market economy that promoted capitalist attitudes and industrialization.[7] The market economy affected nearly every household: by the late eighteenth century, every rural community had some connections to the greater Atlantic world through shopkeepers and general wholesalers. Even members of predominantly barter exchanges maintained informal debtor and creditor relations and established common prices for goods that eventually interacted with those in more distant markets. The major impact of this shift took place in urban centers. While America's overall population more than doubled from 3.9 million in 1790 to 8.6 million in 1820, the urban population of the East more than tripled, with the highest increases occurring during times of rapid economic growth and in seaport towns benefiting from thriving foreign trade.

By 1800, more than one-third of Massachusetts citizens lived in urban areas. Business changed dramatically throughout this period: merchants and stores grew more numerous and specialized, cash replaced barter exchanges, putting-out work became a common income source for many families, banks and banking increased, and hired labor accounted for a larger proportion of the workforce. Market growth favored industrial development by greatly increasing the demand for manufactured goods and the ability to exchange those goods over great distances. The expanding economy created a series of new problems, opportunities, and government policies that defined the nineteenth century as the age of capitalism.[8]

The growth of capitalist practices had both ideological and practical components. Capitalism received a major boost from post-Revolutionary rhetoric, which shifted public thinking away from classical republicanism's emphasis upon the virtuous pursuit of communal goals. While individuals such as Revere illustrated that communal service still held an important position in the American mindset, liberal republicanism's prioritization of the freedom to pursue private interests achieved a higher profile. Most Americans considered the enhancement of personal prosperity a major objective of independence, and society became more consumption-oriented and materialistic throughout this period. Capitalism flourished in this fertile environment, since materialistic consumption directly fostered the desire to realize profits and amass capital in order to facilitate future earnings. The early republic period was the era of capitalist transformation, and by the 1830s capitalism penetrated society and cast its shadow over the distribution of income, wealth, and land; legal precedents; technological changes; labor, work, and leisure patterns; political discourse; and many other aspects of American life.[9]

Fueled by population expansion, thriving markets, improving infrastructure, capitalist attitudes, and surging consumer demand for manufactured products, industrialization gathered speed in the early 1800s, particularly in New England, where regional circumstances magnified the benefits of machines while minimizing their harmful impacts. Massachusetts farming increased greatly in productivity around the turn of the century due to the better use of labor and the application of new strategies. Thriving agricultural productivity, abundant profits, and a rapidly increasing rural population fostered manufacturing by producing surplus laborers for workshops as well as surplus farm products for sale to urban laborers. Prosperous farmers then purchased urban manufactures in return. Farmers had already occupied most of New

England's arable land by the late 1700s and the population boom produced numerous younger sons who began migrating to the West. New Englanders often took a positive view of mechanization: machines could not "steal" labor from an agricultural sector that produced surplus migrating workers, but they could increase output and provide new employment opportunities. New England also contained fewer recent immigrants and slaves than other areas, further diminishing the fears of a corrupt and dependent concentrated labor force. By 1800, New England's manufacturing operations included boots and shoes in Lynn; glassware in Braintree; hats and caps in Danbury; paper in Norwich; wood clocks in New Haven; pottery in Roxbury and Concord; shovels and axes in North Easton; jewelry in Providence; and bronze, brass, and other metal trades in Bridgeport, Wallingford, New Britain, Waterbury, and Meriden.[10]

American society increasingly viewed manufacturing as an asset, and celebrated the hardworking laborer's efficient use of the nation's natural resources to protect his countrymen from the profiteering foreign merchant. The barriers to industrialization faded significantly by the War of 1812, and Americans started identifying national progress with technological advancement by the 1820s, a trend that accelerated after railroads appeared.[11] Of course, industrialization did not have a purely beneficial impact, nor did everyone hail the onset of new technological systems and production methods. Many new industries offered a web of positive and negative consequences, impossible to disentangle or fully understand until years after the fact. Women, immigrants, unemployed rural workers, and other groups often increased their income via newly created employment opportunities such as factory work, but worried about the growing authority of their powerful and exploitative employers. And while artisans might benefit from skilled or supervisory positions in new firms or factories, they had a larger chance of losing their shops and autonomy due to competition from industrial production. Industrialization raised standards of living while simultaneously increasing income gaps and exacerbating class differences. Even though American society had always been hierarchical, in earlier times the limited amount of capital and the importance of labor kept the relationships between classes cooperative and symbiotic even when they were unequal. The great stratification of wealth and the growth of a permanent worker class changed matters, creating one group of people who profited from the labor of a second group, and fewer opportunities to cross that divide.[12]

Many entrepreneurial Americans took advantage of the new economic climate with a flurry of new inventions, adaptations to existing processes, or managerial changes that produced even more economic growth. For example, Jacob Perkins's nail-making machine decreased the time and expense associated with home building and other construction. Eli Whitney's cotton gin transformed both the southern and national economies by vastly increasing the efficiency of cotton farming, and Joshua Humphreys' shipbuilding advances, as well as the improvements in steamboat technology started by John Fitch, James Rumsey, and Robert Fulton, improved the American navy while increasing the speed and decreasing the cost of American shipping. Other developments centered on increasing America's productive output: Oliver Evans's automated flour mill in 1787 inaugurated the age of the automated factory, saving large amounts of labor with a combination of spatial organization, mechanization, and synchronization of work processes, while other inventors, including textile pioneer Samuel Slater, clockmaker Eli Terry, and weapons manufacturer Thomas Blanchard, laid the final groundwork for standardized mass production. Many of these new techniques reflected the objectives and primary bottlenecks of earlier technological systems. Textile and papermaking factories needed to reduce labor and other operating costs to allow them to compete with cheap foreign goods, so they developed new machinery that raised productivity and efficiency without imposing higher skill requirements upon the largely unskilled labor force. Managers of armament workshops, in contrast, struggled to increase precision in order to finally solve the problem of interchangeability. These advances improved the prospects of individual firms while raising America's overall technological expertise.[13]

New England cities became centers of industry and fostered technological advancement via a strong technical network that disseminated knowledge through subcontracting, equipment comparisons, open shops, a pool of skilled labor, and technically competent entrepreneurs such as Paul Revere. New England firms grew more specialized, especially in metalworking trades, and different shops often subcontracted work to each other to maintain their areas of expertise. As a result, many shops became centers of learning that transferred knowledge to other manufacturers, even in different fields. Nathan Rosenberg coined the term *technological convergence* to describe this phenomenon: when a large number of firms work on similar challenges and trade knowledge and solutions, rapid technological advancement usually results. New England's industrial advantages become evident in a comparison between the Harpers Ferry and Springfield armories: the Harpers Ferry Armory in Virginia was hampered by a lack of discipline and constant resistance of skilled workers to change, while Massachusetts's Springfield Armory drew support from its community and had an easier time purchasing equipment and hiring technically proficient laborers.[14]

Following his successful copper sheet rolling trial and the ensuing confusion as the government clumsily transitioned from a Federalist to a Democratic-Republican administration, Paul Revere faced some of the most promising business conditions the nineteenth century had to offer, and he knew it. He was one of the most experienced and diversified members of a small manufacturing community, operating within a nation whose vast public and private demand for goods continued to increase. Revere's success in the final decade of his professional life benefited from the prevailing political and economic climate, but also resulted from his own skills, decisions, and background. By continuing his production of earlier items such as bells, cannon, bolts, and spikes, while improving the quality and quantity of his copper sheet output, Revere maintained a profitable line of products and increased his profits and his reputation in spite of occasional setbacks from embargoes, lawsuits, and technical challenges. In particular, Revere set up his growing enterprise in a way that allowed him to marshal the four fundamental resources required for proto-industrial manufacturing success: investment capital, labor, technology, and the natural environment. These four categories provide a helpful framework for a study of Revere's final working years, and frame the analysis of America's transition from crafts to industry.

Investment Capital, Managerial Practices, and the Role of Government

The technological and managerial aspects of early manufacturing firms evolved in parallel and became increasingly interdependent, particularly after 1790. As Revere's experiences with individuals, businesses, and governments have illustrated, many Americans struggled with both technological and managerial challenges. The colonial artisan, for example, had to master more than the tools and construction procedures of his craft, because he also made pricing, advertising, labor oversight, equipment purchasing, and many other decisions every day. We can divide these decisions into two categories. Most issues corresponded to short-term needs related to the smooth and profitable functioning of a shop, which we now recognize as the field of management. Other issues related to the longer-term raising and allocation of funds, which we describe as investment capital (or simply capital) issues. Managerial skill and investment capital remained in short supply throughout the colonial and early American periods, and successful artisans and manager-entrepreneurs had to overcome these deficits in order to thrive.[15]

The simultaneous realization of political independence and growth of a market economy inspired many artisans and entrepreneurs to abandon their familiar craft principles in order to invest in new techniques, equipment, and managerial practices aimed at producing more goods for broader markets. With the growing prevalence of machinery, owners brought larger numbers of workers together in factory settings, accompanied by specialized supervisors to coordinate and even synchronize their operations. The prohibitive capital cost of land, buildings, labor, and equipment also inspired new forms of corporate ownership and the growing separation of owners and managers. As a former artisan, Paul Revere might have resisted these changes, but instead he embraced them. His reasons for joining the vanguard of the industrial and managerial revolution included his lifelong eagerness to master new processes and his desire to enter the increasingly anachronistic gentry class. Since a mid-eighteenth-century gentleman would not sully his hands with manual labor, Revere quickly adopted new practices that enabled him to identify himself as an owner and manager instead of a laborer, even though he previously stood among the most skilled laborers in the nation.[16]

Investment capital, or more accurately the shortage of capital, played a key role in defining early American economic growth, particularly in the manufacturing sector. Beginning in colonial times but continuing throughout the nineteenth century, artisans struggled to raise capital when they wished to start a new business, relocate, expand, or modify their production. Craft shops often grew slowly, periodically directing some of their profits back into the business to cover fixed (land, buildings, machines) and working (wages, inventory) capital costs. Workshops and factories on the other hand required larger investments and spent more on land and buildings than machinery. Many of these bigger operations, especially ones selling to distant clients, often required more working capital because they had to pay wages and purchase supplies continually, while their own income from clients might be delayed for months.[17] When artisans or manufacturers found themselves in need of additional credit they usually attempted to secure a loan from merchants or wholesalers if possible, because banks rarely wanted to take a risk. For example, the *Evening Post* reported in 1804 that for most New York banks, "the application of the laborious mechanic is treated with contempt and rejected with disdain." Without an accommodating bank, artisans had to depend on individual lenders or patrons. In the first generation of metalworking production in many areas, growth proceeded at a crawl until the mid-nineteenth century due to capital scarcity. Even successful or high-profile inventor-entrepreneurs suffered from capital scarcity in the 1790s: steamboat pioneer John Fitch faced immense capital shortages (among other problems) that curtailed his operations, and Oliver Evans attempted to develop a series of inventions related to steam power, including steam vehicles, but the lack of capital doomed him every time.[18]

The federal government indirectly helped fill this investment capital void by building infrastructure and creating favorable trade conditions that facilitated the growth of national markets and credit sources to aid capital accumulation. After 1806 new policies started to encourage investment in manufacturing: bounties and premiums for technological improvements, fairs to display them, fundraising lotteries, and even support from Thomas Jefferson and James Madison in times of national crisis all helped investors funnel millions of dollars into new mills and factories, primarily for the purchase of buildings and land.[19] And in a small but increasing number of cases the government directly helped manufacturers with support or long-term contracts, as it did with Revere. In this sense the government paralleled the natural resources of the young nation: it was a powerful asset offering tremendous benefits to those who knew how to access it.

Paul Revere addressed his biggest need for capital when he started his copper-rolling mill. As with most manufacturers, land and buildings represented his major expenses, though he also invested in new equipment and additional workers. He made it to the end of 1801 thanks to a combination of his prior earnings, ongoing bolt and spike contracts, and government loans. From that point forward he did not face any major capital outlays but he did have to keep a careful eye on his balance sheet, putting the maximum possible amount of his profits into expanded operations while making sure he had enough available cash to cover expenses. This task became even harder when such a large percentage of his business depended on the government.

In 1802 Revere thought of his rolling mill in restricted terms. Funded by a hard-won loan from the Department of the Navy and faced with thousands of pounds of government copper waiting to be rolled into sheets, he saw this new enterprise primarily as a tool for the completion of federal contracts. Although his mill had only operated for a short time, he had no trouble imagining it on a large scale, as a pillar of America's defense program. This confidence and enthusiasm was an important prerequisite in Revere's attempt to create a major manufactory. He had abandoned much of his small craft shop mentality and was willing to think big, to lobby for major contracts and federal protection. Now he needed to convert big dreams into big successes.

Revere's early contracts paint a picture of great productivity and prove the degree to which the government counted on his ability. He received more than 193,000 pounds of copper from Boston naval agent Stephen Higginson in 1802, enough copper to satisfy all the bolt, spike, sheathing, and other copper needs of the two 74-gun ships under construction. By November he had already delivered almost 85,000 pounds of bolts and spikes to the navy yard, and another 40,000 pounds of bolts and spikes stood ready for shipment. He had refined and melted the more than 68,000 pounds of remaining copper and was ready to make it into whatever was needed, probably sheeting. The navy trusted him with all the copper manufacturing for two of its major vessels, and he was well on his way to finishing the entire contract.[20]

Unfortunately, the federal government and Revere had many frustrating interactions in the years to come. Federal payments were more often late than on time, and most government representatives wasted his time with repetitive or confused requests. His requests for tariff protection, federal appointments, or new contracts almost never bore fruit. He dropped a clue about his opinion of the relative importance of his federal and state contracts in a November 6, 1802 letter to Robert Smith. Revere could not account for all of the sheets he rolled a few months earlier, because "about that time I had an application for thin sheet copper to cover the Dome of the new State House, which was an order of so considerable consequence to us, that we undertook it."[21] Unlike the federal government contracts that always involved late payments, misunderstandings, and voluminous correspondence, this Boston job for a local government agency able to pay promptly and with cash did not require a single letter or headache. Revere later provided sheathing for the dome of New York City Hall, and the transaction unfolded just as smoothly. These state and local governments had more experience than the new federal government, and operated on more manageable and tangible scales. Revere would have loved to deal with them more, but a city or state could only provide so much work, while the federal government's needs seemed to grow without bounds.

Paul Revere put forth his best efforts in dealing with inexperienced government bureaucrats and managed to receive more funding than most manufacturers, but the deck was stacked against him. Revere's Federalist sensibilities did not endear him to the Democratic-Republican administration, and he returned the favor. His ideological dissatisfaction became evident in an 1804 letter: "I very much doubt my influence with the present Administration. My sentiments differ very widely from theirs in politicks—My friend, you know I was allways a warm Republican; I always deprecated Democracy as much as I did Aristocracy; Our Government is now completely Democratic, they turn every person out of office who are not nor will be of their way of thinking & acting."[22] In this statement Revere declares himself a republican, a foe of both democracy and aristocracy. He refused to accept the leadership of a hereditary few, but also feared the consequences of rule by an unqualified mass. Instead, he confirmed his lifelong and even die-hard adherence to the principle of a republican meritocracy: he wanted a minority to represent the interests of the people, chosen because of their talents and experience instead of their birthright. Unsurprisingly, Revere considered his lifetime of hard work, technical accomplishments, public service, and leadership more than ample qualifications for societal leadership, even though the Jeffersonians did not. In more pragmatic terms, Revere also disagreed with the Jeffersonian administration's plan to limit the size of the federal government, restrict naval funding, and retire the large oceangoing warships that needed his copper sheathing. After numerous letters failed to produce new contracts or any form of encouragement, Revere was ready to write off his elected leaders. At least he had plenty of work orders from merchant ships and Massachusetts contracts.

Jefferson's non-importation and embargo policies eventually made matters worse. The period from 1793 to 1806 was, with minor exceptions, one of vast American prosperity. A large proportion of this economic boom resulted from wars between Britain and France that increased demand for American goods while removing overseas competition. Starting in 1805, however, Britain increased its attacks upon American shipping in order to minimize American trade with France. Other practices such as impressment of American sailors continued unchecked. Jefferson's administration responded to attacks on American shipping with the Non-importation Act of 1806, a prohibition on certain British imports intended to economically coerce Britain into amending its aggressions without dragging America into a war. When Britain continued its hostile activities into June 1807, Congress passed the Embargo Act to end all American shipping to all countries, hoping that this economic punishment would force them to negotiate better terms. Between December 1807 and March 1809, Jefferson's administration maintained this policy against violent protest from merchants, particularly those from New England. Incidentally, this policy had little, if any, effect upon the economies of Britain or France.[23]

Although disastrous for merchants, the embargo actually produced several substantial benefits for manufacturers. The cessation of overseas trade immediately ended any competition from foreign imports, opening the entire American market to domestic producers. In addition, many merchants and bankers found themselves in an unfamiliar position, holding a surplus of capital that they could no longer apply to shipping ventures. These moneyed interests endorsed a variety of new endeavors and invested heavily in manufacturing operations that might make up some of the shortfall of imported goods. Unfortunately, Revere deviated from typical manufacturers in many ways and the embargo hurt him more than most. He did not need or want investment capital from outsiders at this late date, having already purchased his property and established his manufactory. With the prohibition of all trading, the merchant vessels that provided a large percentage of his clientele no longer operated. In addition, his raw materials usually arrived from abroad, so he found himself cut off from nearly all metal supplies. Fortunately he found a silver lining, described in a March 6, 1809 letter to Joseph Carson in Philadelphia: "The miserable conduct of our Rullers in laying that Cursed Embargo has nearly deprived us of selling copper for ships, but as good sometimes comes out of evil and there being no chance of gitting copper for stills from England we have turned our attention that way. We are supply some Gentlemen in New York with upwards of 16,000 lb of sheet 3 feet wide by 5 feet long to make two boilers for two Steam Boats."[24] The New York gentlemen in question were none other than Robert Fulton and Robert Livingston, whose relationship with Revere will be discussed later. In spite of his complaining, even Revere had to admit he benefited from the lack of British competition. His market for naval sheathing temporarily evaporated, but other markets appeared. And while he had a harder time finding sources of raw copper, he and his purchasing agents managed to obtain enough, often by increasing purchases of recycled copper, to keep his output from drying up.

Even after the Democratic-Republicans lifted the embargo in 1809, Revere's relationship with the government, James Madison's administration by this time, did not improve. In September 1810 Revere offered Philadelphia merchant Joseph Carson a commission if he could iron out any contracts with the federal government to produce bolts, spikes, or sheeting. In Revere's own words, "We are not favorites of Washington." In spite of political differences, he desperately wanted to renew his ties with the Democratic-Republican government to help him receive more contracts of exceptional size. Revere realized by this point that large volumes of business offered the best solution to capital shortages, and no client could match the purchasing power of the federal government.

Revere amassed plenty of managerial experience from his prior business endeavors, but it could only take him so far. In the first decade of the nineteenth century, amid great political and economic volatility, he struggled to perfect his new copper-rolling technology while juggling his other operations. Strategies for his company's organization, labor relations, raw material procurement, accounting, and operational practices included a mix of familiar traditions and new techniques. Revere also had to adopt new external policies, as the increasing scale of his operations brought him into contact or competition with other organizations. In particular, he had more conflicts and litigation with other manufacturers in his area and he spent more effort lobbying the federal government for direct and indirect aid.

Revere left clues about his managerial priorities in his surviving record books. Although he did not radically alter his accounting methods during his copper mill career, he continued evolving the procedures he had begun forty years earlier. For most of his life, Revere led his countrymen in terms of his advanced bookkeeping methods. William Mitchell published the first American bookkeeping text in Philadelphia in 1796, explaining double entry accounting, periodic balancing, and profit and loss calculation.[25] Revere had been using these methods since the 1780s. However, by the time he included the Canton mill in his list of operations, some of his older strategies led to confusion and omissions.

Revere's records reflect the priority he assigned to the different elements of his business, as he spent the most effort documenting issues he deemed valuable or pressing. His most rigidly recorded transactions, understandably, were the sales he made to customers and the payments or barter he received in exchange. From an early period he used the double entry system to keep track of these changing totals, and he continued the system throughout his life. Wastebooks allowed him to record sales and payments as they took place, and ledgers allowed him to compute the amount a client owed him (or vice versa) at any point. General operating expenses, in contrast, occupied a much lower priority in his mind. If an expense such as fuel could not be directly attributed to a specific client, Revere's system lacked a consistent place to record it. Occasional receipts or records of contracts describe specific payments or contractual agreements, but the very existence of these tidbits further illustrates the lack of a methodology to track such practices. Furthermore, Revere either did not understand or chose not to estimate intangible costs such as depreciation allowances upon his equipment or buildings. Most businessmen remained unaware of these abstract costs, and even someone with the experience of Eli Whitney only began estimating and billing intangible expenses for his Mill Rock armory in 1812, shortly after Revere's retirement.[26]

In general, Revere took a pragmatic view of his business and recorded any information he believed would help him in the future. Sales and cash receipts had an immediate practical value because they helped him collect debts, but lists of his overhead expenses did not serve an obvious purpose. Revere preferred to deal with employees and suppliers on a personal basis, using his best judgment to determine how much of a certain good to order or what salary he should pay. The unpredictability of the nation's economy and international trade conditions might have contributed to Revere's lack of long-term planning: because prices, raw material quantity, and consumer demand changed radically from year to year, predictive recordkeeping might have seemed pointless. One result of this policy was the fairly common raw material shortages he experienced, although external conditions certainly had a greater impact than any recordkeeping oversight on his part.

Revere continued changing the credit and payment practices he had used since the start of his foundry. As of 1805 he offered sixty days' credit on all sales, although when he produced material for a ship he expected the payment by the time the ship launched. He also adopted a new policy to account for the interchangeability of many of his products: Revere and Son would buy back any bolts or spikes not used. This served as both a warrantee on the quality and standardized nature of his output as well as a testament to the volume of sales he experienced: unused bolts would soon find a new buyer in the Boston market, and bolts or spikes made for one ship could serve the needs of another.

He offered a more explicit warrantee on all his composition copper and tin, offering to provide a full refund if requested.[27]

In 1808 Revere implemented a new payment policy: customers now had four months of credit before their bill came due, but received a 5 percent discount for paying immediately. By this time he was willing to offer significant discounts that awarded his customers a fee in exchange for faster payments. This reflected the new business climate, in which quick access to cash produced great benefits. In addition, Revere hoped to avoid the hassle of chasing down late payments, a task he now subcontracted to a professional. He hired Attorney Thomas Selfridge in February 1806 to collect some of his outstanding debts. Selfridge received a commission of $6.40 per debt collected, and immediately brought in 4 debts ranging from $23 to $501.[28]

Many of Revere's new concerns and policies arose from the increasing scale of operations at Canton. Until this point he had conducted most business interactions personally, soliciting new orders in person or via letters to friends or contacts. He also relied upon intermediaries to spread the word of his product line or to tip him off to a potential new client or contract, but these men were usually friends such as Joshua Humphreys. This system sufficed when he ran a smaller workshop but did not adequately connect his larger business to the growing markets of the early nineteenth century. Revere recognized his problem and sought the help of an expert located in America's largest manufacturing center. Joseph Carson was a Philadelphia factor, a merchant intermediary who received commissions for arranging transactions between distant suppliers and producers. He first wrote Revere on June 6, 1809, with good news: the Philadelphia braziers he met with "all seem disposed to purchase your copper instead of English and seemed to regret not knowing of the establishment before." Carson's letter seethed with enthusiasm, as he described all of his contacts among the industrial centers in the Philadelphia area.[29] A prolific correspondence followed, and many contracts and raw material purchases resulted from Carson's work.

Revere forged another advantageous partnership with New York merchant Harmon Hendricks relating to raw material procurement. Hendricks was one of many suppliers who sold copper to Revere, but beginning in 1803 the two men became close associates who visited each other's works. Their constant stream of letters included news about local events such as the fever raging through New York and pleasantries such as "I pray my best respects to the Ladies." Hendricks procured many large shipments of copper, tin, and other materials for Revere, and clearly appreciated Revere's value as a continuing customer, writing in 1803, "I am induced to offer you the very longest Credit with the very lowest price . . . I have sold last week for 2 ships considerably higher price & shorter Credit." Hendricks usually gave Revere between four and six months of credit before bills came due, and offered other customers only two to four months. In 1805 Revere and Hendricks agreed to jointly purchase copper in the Boston and New York ports and to equalize the quantity and price between them. By 1806, the relationship between Revere and Hendricks had evolved further, as they occasionally exchanged Hendricks's imported copper for Revere's finished copper at the going market rates—a throwback to barter practices. Hendricks also sent Revere information about the prices of different finished copper products in New York and Philadelphia, information that undoubtedly helped Revere set his own prices at competitive and profitable levels. Revere reciprocated with information about his own business, and in 1813 Hendricks and his brother-in-law purchased and revitalized the Soho copper-rolling works in Belleville New Jersey, adding their own rolling mill.[30] Long accustomed to recruiting technical specialists to augment his own expertise, Revere had finally made the step toward subcontracting some administrative aspects of his work, such as debt collection, purchasing, and marketing, to experts such as Selfridge, Carson, and Hendricks.

Revere's managerial innovations and success in raising sufficient quantities of operating capital helped him maintain profitability, which inspired him to continue improving his practices. Revere's sheet copper sales, expenses, and profits over two one-year periods, from June 1804 to June 1805 and again from June 1806 to June 1807 (see Table 7.1), offer a representative sample of "typical" production quantities for Revere's mill.[31]

During the two years of analysis listed in Table 7.1, Revere's $10,642 in sheet copper sales constituted only 17 percent of his $62,600 sales of all goods. This illustrates the diversity of his operations: in addition to sheet rolling he continued casting cannon and bells while producing enormous quantities of bolts, spikes, nails, and other items. When one product faltered he depended on continued income from others. Because Revere did not present any details about how he allocated labor and other resources between the different aspects of his business, these figures assume that 17 percent of Revere's total costs apply to sheet copper production, and the estimates ignore all costs other

Table 7.1. Revere's Sheet Copper Profits, 1804–1805 and 1806–1807

Transaction	Cost	Total Costs (percent)	Total Sales (percent)
Costs			
Labor	$424	6	4
Transportation	$300	4	3
Energy	$300	4	3
Copper*	$6,208	86	58
Total Costs	$7,232		
Total Sales	$10,642		
Profit	$3,410		32

* This copper cost includes an estimated 10 percent "wastage," which was the figure used by Revere. He was contractually responsible for absorbing the cost of all wasted metal.

than labor, transportation, fuel, and copper purchases. Although many hidden costs such as advertising were negligible in comparison to the listed ones, others—like equipment purchase, repair, and insurance—were larger. The profit listed in Table 7.1 is most likely a minor overestimation, but even if we adjust it downward we still see the lucrative nature of Revere's rolling business, as he pocketed around 32 cents of every dollar of sales that he made. The categories and relative sizes of these operating costs mirror Revere's earlier experiences: metal purchases resoundingly dominate the balance sheet, followed by labor and fuel costs.

An October 21, 1810 letter to army paymaster Josiah Snelling catalogued Revere's output three years later, near the end of his involvement in the business: "We have manufactured in one year 122976 lbs of copper Bolts & Spikes at 50 cents—27659 pounds [of sheeting] at 50 cents 12976 pounds of brass cannon at 55 cents—10845 pounds bells and 16547 pounds composition casting at 42 cents amount including stock $93959.16 the works are now capable of extending the mark to three times the amount that has ever been done."[32] According to these figures, he sold almost $94,000 in merchandise in one year, approximately 15 percent of which came from sheeting—consistent with the earlier period when 17 percent of his sales came from copper sheets. Revere's total sheeting sales of $13,830 for this year dwarf his earlier sales volume. This increase resulted from many factors, including the removal of all embargo and non-importation restrictions, renewed naval armament, increasing prosperity, and, of course, Revere's increasing skill at producing and marketing copper.

Revere's growing business and large federal contracts made him more aware of the ways America's commercial and foreign policies affected him. Tariff law had its history in the earliest years of the new government, when Britain's immediate imposition of tariff duties against American vessels clobbered the new nation's fledgling economy. Under the new Constitution America equalized matters somewhat by imposing duties on goods from British ships. These duties usually hovered around 5 percent, enough of a tariff to raise revenue and soothe American merchants hurt by Britain's tariff without angering the many consumers who depended upon cheap foreign imports.

As with most policy issues facing the earliest administrations, the question of protectionist tariffs, free trade, and federal revenue sparked intense ideological debates between merchants and manufacturers, northerners and southerners, Federalists and Democratic-Republicans. Tariffs polarized society. Merchant importers hated them because they raised the price of their cargo while protecting domestic products. Southerners hated them because they helped the North's commercial interests while raising the prices of much-needed merchandise. Artisans came close to hating them because they increased the price of tools and imported items, revealing their growing consumer mentality. But manufacturers and their investors strongly supported tariffs as a weapon to free America from the shackles of economic colonialism.[33]

Revere cared about tariffs for more than ideological reasons, since his business depended upon his ability to successfully compete with high-tech foreign metal products. Revere had good company on this issue: countless other groups, including general mechanic societies and specific trade associations, lobbied state and the federal government for protective measures. Congress initially resisted such requests because a high tariff hurt agricultural interests by raising the price of goods they wished to purchase while making it harder for them to sell their produce to other countries. Congress relented in June 1794, raising the tariff on many imported manufactured items, including copper and brass products, from 5 to 15 percent, and then raising it again in 1804 to 17.5 percent.[34] Unfortunately, a small loophole led to large problems.

In April 1806 Revere wrote a short letter to Treasury Secretary Albert Gallatin by way of his congressman Josiah Quincy, pointing out two related problems. First, the tariff duty on "old" copper raised his raw material costs considerably. And second, the lack of a tariff on imported sheet copper failed to protect him from low-cost foreign competition. According to the existing law, imported manufactured items such as copper sheets and bolts should receive a 17.5 percent tariff to protect American manufacturers by making imports more expensive. Raw materials such as imported copper plates, bars, and pigs remained duty free. This law was misinterpreted at most customs stations because inspectors often classified raw material copper as "sheets" and added the import duty, while classifying imported copper sheeting, a manufactured item, as duty-free "plates."[35] At the heart of Revere's request lay the misunderstanding of terminology: when words like *plates* and *sheets* could alternately relate to raw or finished products, customs inspectors understandably had a difficult time making the proper call. This misunderstanding injured his operations by raising the cost of his raw materials while allowing imported British sheets to be sold at extremely low prices.

Unfortunately, powerful interests stood to benefit from the continuation of the status quo. Merchants purchased more copper sheets than any other client, and did not care about the competitive disadvantages faced by Revere's rolling mill. They wanted sheeting as cheaply as they could get it, and for the most part they preferred to stick with high-quality British products wherever possible. Many other copper workers purchased sheet copper for the manufacture of utensils or stills and also hoped for low tariffs and cheap prices. The combined actions of a subtle moneyed interest and a blunt popular one proved formidable.

Several drafts of a letter to Josiah Quincy reveal Revere's serious concern about tariffs. Quincy was Revere's congressional representative, and Revere asked him to deliver a petition on his behalf to officially clarify American policy. He explained the confusion of terminology in great detail, but then tied his argument to his own efforts: "for Instance the duty on Bolts, Spikes, +c, makes them come so high that no person presumes to import them, and we sell Cheaper (and of equal quality) . . . in Boston no person for several years have imported Sheet Copper for sail but Inglish merchants. We sell for the same prices they do; if they had the duty to pay it would give us a sufficient profit."[36] Revere clearly imagined his business to be large and important and directly tied his company's welfare to that of the nation. In particular, he implied that he could provide all sheeting needs for the country if he had sufficient federal encouragement, beginning with tariff protection.

This was exactly the wrong approach to take. On January 21, 1808, the

Committee of Commerce and Manufactures submitted their report. They agreed with Revere that no duty should be collected on "old" copper, because raw materials should be duty free. So far so good. The committee then dismissed his request for a protective tariff on sheeting, citing contrary petitions from "Sundry Copper Smiths" from Philadelphia and "Sundry Manufacturers of Copper" from New York City. Both of these groups as well as the Congressional Committee took offense at Revere's contentions. According to the final report:

> To induce the national legislature to impose the aforementioned duty on copper in sheets, the petitioners state that they have at considerable expense erected works which will enable them to supply copper in sheets commensurate to the demand of the United States . . . [should Congress take this action] and the quantity of copper in sheets furnished by them prove to be insufficient for the demand, the copper-smiths and braziers, a respectable class of manufacturers, would sustain injuries . . . The committee have had no satisfactory evidence offered to them that copper in sheets, in quantity sufficient for the use and consumption of the United States, can be supplied for the petitioners, or that the quality of the same is equal to that which is imported.[37]

Throughout this petition the language reveals the committee's Jeffersonian viewpoint: they sought to avoid "hasty legislative acts" and "legislative interference" that might "fasten on the community oppressive monopolies" as has been done "too often" in the past. In contrast, the committee praised the confident assertions in the petition from large groups of "respectable" manufacturers. The committee made one factual error in describing "the flourishing state of manufactures, which have supplanted foreign articles of the same kind." Without a doubt, imported copper sheets and other fasteners were more plentiful than domestic ones.

In a long response to Josiah Quincy, Revere fumed about the committee's verdict. He leveled most of his ire at the petitions of the coppersmiths and braziers, which he considered a completely unfounded betrayal. He defended his petition as "just" and claimed it would benefit other copperworkers as much as himself: "If we had whished for an exclusive right we should have applied for it when we erected our works, had we then done it we have no doubt the government would have granted it. It was as much a new invention as any thing for which patents have been granted by our government." This statement proves that Revere knew about patents and their monopolistic consequences, but still chose not to apply for one when he first perfected the rolling process. In the remainder of the letter he repeated the great expenses and trials he incurred while mastering copper rolling, and proudly claimed, "we were willing that any Americans who could attain to so usefull and necessary a manufacture should be on a footing with our selves." Concluding his rebuttal, Revere angrily denied the petitioners' claim that

> a sufficient quantity for supplying the United States can not be afforded by our manufactory or that we do not manufacture enough for the Town of Boston! The fact is that we have supplied different Merchants in Quebec, Newhamshier, Connecticut, Rhode Island, N York, Philadelphia . . . & that the sheets with which the Constitution's Bottom was covered before she went to the Mediteranean was manufactured by us . . . We again assert that if we were properly encouraged we have not the least doubt that we could Manufacture a sufficient quantity for the US. We have now 20,000 lbs by us and have but one pair of rolls whereas we can make use of six pair.[38]

By changing the petition from a clarification of terminology to a request that his business receive preferential treatment in exchange for its functioning as a national industry, Revere destroyed any chance he had for its passage.

Years later, Revere shared all his information on the subject with James Prince. Apparently, some copper importers realized Revere's petition would increase their material costs, and rallied the Philadelphia and New York copperworkers and braziers against Revere, insinuating that he wanted to raise a duty on raw copper. If true, this rumor explained much: a bloc of manufacturers carried more persuasive power against Revere than a group of wealthy merchants. Revere also explained how he discussed the goals of his petition with the Boston copperworkers, who then refused to join the coppersmiths of other cities against him.[39] Of course, many of these Boston copperworkers were probably his friends.

Revere's quest for tariff protection never left his thoughts. In August 1809 outgoing Secretary of War Henry Dearborn (now a collector for the port of Boston) asked him to participate in the survey of the Report on Manufactures. Seeing at last an attentive government audience, he brought Dearborn up to speed on the tariff situation and presented his request for the encouragement of manufactures, and sheet copper in particular, via a 17.5 percent protective tariff of domestic made goods. He also wanted the federal government to hire him for all of its copper needs since many judges rated his work equal or superior to British products. He reasoned that large contracts would help him improve even more. Unfortunately, Dearborn could not help him.[40]

By 1810, Revere was ready to try again, using a new approach. An August 20, 1810 letter to Treasury Comptroller Gabriel Duval reiterated all his old concerns about ambiguous terminology and the unfairness of protecting all imports except sheet copper. He was quite frustrated by this time, and pleaded for protection that he felt was both just and vital for his economic survival. Duval responded on September 4, simply stating that the courts had decided that bolts and sheets would also not receive a protective tariff, and only Congress could protect him at this point with new legislation. To make matters worse, Revere learned in December that Secretary of the Treasury Gallatin submitted a circular letter to all customs collectors requesting that copper bolts also be imported duty free. This measure prevented Revere's two major products from receiving any tariff protection against cheap foreign imports. Revere once again planned to petition the government to collect the same tariff on bolts and sheets as on other items, and in 1813 Paul Revere and Son joined forces with fellow copper manufacturers Levi Hollingsworth and Harmon Hendricks on yet another request for tariff protection. None of these efforts bore fruit, and no duty was collected on copper sheeting until the Civil War.[41]

The tariff controversy embodies many of Revere's complaints about the Democratic-Republican government's hostility toward him. It is strangely ironic that the man who always wanted to become a merchant ended up losing his fight against a merchant lobby in a Jeffersonian administration. However strongly Revere might plead his case, he would never win as long as pride in his achievements caused him to stress the uniqueness of his business. His impassioned arguments offer his conception of both justice and efficiency: he felt he had earned the right to receive federal protection through tariffs as well as continued direct support through large government contracts. In addition to rewarding him, large contracts would provide new funds he could use to expand his manufactory and better serve his country. Revere's political arguments favoring a merit-based republican system instead of excessive democracy came to fruition in this dream scenario: the government should reward and support the most qualified manufacturer and enable him to improve his craft. The U.S. government, especially under the Jeffersonians, would never levy a tariff that protected only one copper mill at the expense of merchants and copperworkers.

In spite of this defeat, Revere's persistence, energy, confidence, and pride are evident throughout this lobbying period. Yet the same characteristics that helped him receive his first government loan and enabled him to overcome innumerable technical challenges could not win over the new administration, a testament to both the changing times and a former artisan's constancy.

As a silversmith, Revere never gave a moment's thought to the type of business he would run. The question was virtually meaningless. A skilled artisan who owned his own tools had enough independence to work for himself, making his operation a sole proprietorship by default. Other than paying the proper tax on his property, he didn't have any papers to file. Similarly, his business had no independent existence or a name: anyone wanting silver would hire Paul Revere even if an apprentice or journeyman completed the work. As Revere expanded his business into new product lines throughout the end of the eighteenth century he had no reason to alter this status quo by incorporating or taking on partners. The primary reasons for changing one's business organization were the need to raise capital and to pool expertise. Revere made out quite well in these areas even though he was "only" a sole proprietor.

In the midst of ever-increasing prosperity a few years after building his rolling mill, Revere initiated a momentous change: "I have spent for the last three years most of my time in the Country where I have Mills for Rolling Sheets & Bolts, making Spikes, and every kind of copper fastening for ships. It has got to be a tolerable advantage business. I have one of my sons in partnership with me; he takes the care of the business in Boston, I take the care at Canton about 16 miles from Boston."[42] On June 7, 1804, Revere signed Articles of Agreement to form a partnership with his second oldest son, Joseph Warren Revere. Paul listed "Gentleman" as his title and occupation, while Joseph Warren appeared on these forms as a "Bell & Cannon Founder," a title he later changed to "Merchant" in spite of the fact that he spent nearly all his time working at the Canton mill. Neither Revere nor his son wished to emphasize his manufacturing-oriented livelihood when they could augment their reputation and self-image by calling themselves gentlemen and merchants. The signing of these papers created a new entity, "Paul Revere and Son," technically considered a partnership that combined elements of old and new business arrangements. The partners agreed to practice "the art & mystery" of bell and cannon founding for three years, to work for the company's good and not their private interests, to pay all taxes and expenses equally, to divide profit equally ("share & share alike"), and to have equal access to all company books and records. In the case of the death of a partner or other reason for dissolution of the partnership, company assets would be divided according to the ownership ratio at the time of the contract's signing. Since Paul contributed $32,400 of stock and tools while Joseph Warren contributed $16,200—both figures representing rough approximations, as demonstrated by the lack of inventory records—Paul became the "senior partner" and claimed two-thirds of all company assets.[43]

Revere's new partnership arrangement explicitly acknowledged the relationship that had evolved over the past fourteen or so years. Joseph Warren helped his father operate the foundry from the beginning and served as his father's right-hand man in nearly all matters. When Paul devoted the lion's share of his efforts to managing the Canton mill and property, Joseph Warren provided an invaluable service by supervising the Boston shop. This arrangement persisted until a storm destroyed the Boston foundry's roof on October 9, 1804, inspiring the Reveres to consolidate all operations in Canton, although they still maintained a shop for retail sales in Boston.

When this partnership expired on June 7, 1807, father and son renewed it until March 10, 1810. The renewal specified a fixed date for the termination of Paul Revere's career: if he was still alive in 1810, he would allow Joseph Warren to run the business alone after that point and would sell or lease his share of all stock and property at a fair price, with the exception of his house.[44] Apart from delaying his retirement until the beginning of 1811, Revere followed this script exactly.

The partnership form of business organization shared many features with the sole proprietorship. Both business types imposed unlimited liability upon the owners, who could be sued or imprisoned because of the firm's debts or practices. Interestingly, the personal liability aspect of partnerships often helped them to prosper: clients and lenders usually engaged with a firm for personal reasons such as the trustworthy reputation of the owner. A firm that existed separately from its owner lost this personal advantage. Partnerships also had short durations, often even shorter than those of sole proprietorships. A partnership expired when either partner died, or if either partner chose to terminate the agreement, or at the natural end of the agreement, usually several years. While partnerships offered the possibility of raising additional capital or pooling twice the expertise of a sole proprietorship, in reality many partnerships featured one member with far more experience and assets than the other. Partnerships avoided the stigma of servility by elevating both members to the position of owner, a vital asset in a society that placed such a value on independence. Family firms, a popular form of business organization representing a modification of typical partnerships, offered unique challenges and rarely lasted more than one or two generations. Paul Revere and Son was an exception. One of the biggest barriers to family business success involved arguments over succession or management of the firm, particularly in larger families with more than one potential heir. Where corporations included formal legal structures that governed shares of ownership and voting procedures, family partnerships depended on the common sense and cool-headedness of relatives, never a safe bet. But families offered other benefits such as loyalty and goal sharing that could help small firms focus on long-term issues vital to their survival. Especially in a patriarchal society, the founding family member had the opportunity to impart a defining vision to the enterprise that could last for the life of the company.[45] Joseph Warren Revere may have contributed less than his father, but his contributions held unquestionable value. Surviving correspondence illustrates a warm and trusting alliance between father and son, with Joseph Warren deferring to his father's leadership and vision while Paul credits his son with a shrewder business sense and outstanding technical intuition. Common goals and methods allowed their partnership to thrive, and Joseph Warren attempted to involve other family members after his father's retirement.

Revere had another option in addition to forming this partnership. Some manufactories overcame their inability to raise large amounts of capital by forming corporations. If they could arrange to receive a corporate charter from the state legislature, the manufactory owners could then sell shares of stock and raise enormous quantities of funds. Although limited liability did not appear as a corporate benefit until 1829, corporations often received corporate privileges such as monopolies, tax relief, or land grants from local or state governments. However, incorporation remained a rare organizational strategy at first. State legislatures only granted corporate charters in the 1790s and early 1800s to endeavors that served the public good, and did this sparingly to avoid creating unnecessary monopolies. Most early corporations such as colleges and libraries did directly serve the public welfare, although profit-seeking endeavors such as banks and turnpike companies that still provided public services became increasingly prevalent. Massachusetts granted only 3 corporate charters to manufacturers between 1789 and 1796, but granted 15 between 1800 and 1809 and another 133 in the next decade. The legislature rarely rejected these petitions and even allowed the scope of these charters to include peripheral activities unrelated to the primary mission. The growing popularity of corporations illustrates the overlap between public service and private enrichment, as this business structure seemed to accomplish both ends at once, although the public service aspect of corporations diminished in the following decades.[46]

Revere definitely knew about the existence of corporations, as he was the first president of the Charitable Mechanics Association, which actively lobbied for a corporate charter in the 1790s and received it in the early 1800s. He probably gave little thought to incorporation since he overcame the worst symptoms of his capital shortage in 1802 after receiving his government loan. Perhaps he did not want to share the ownership of his shop with strangers, knowing that investors, a board of directors, and perhaps even the state legislature would then have a say in his company's decisions. He almost certainly preferred to involve the increasingly annoying government as little as possible in his business. The fundraising capabilities of corporations remained modest in the early nineteenth century when merchants and banks, the only sources of significant investment capital, typically expected personal securities as a guarantee for loans rather than using property as collateral. By keeping things personal and private, Revere remained in step with most of his fellow manufacturers. In spite of entrepreneurs' tendency to expand their operations and adopt more modern managerial methods, the vast majority of shops throughout the country remained traditional small sole proprietorships or partnerships well past 1850. Even Alexis de Tocqueville commented, "What astonishes me, is not so much the marvelous grandeur of some undertakings, as the innumerable magnitude of small ones."[47]

Revere and Son eventually incorporated in 1828 under the able, forward-looking leadership of Joseph Warren Revere. The organizational forms chosen by father and son exactly mirror their different approaches toward management and capital. Paul Revere selected a partnership, an intermediate stage between sole proprietorships and corporations that allowed some pooling of capital and expertise within a comfortable family structure. Joseph Warren took the next step toward industrial capitalism, modernizing into a corporate framework to maximize his ability to grow the business.

The Changing Face of Labor

Paul Revere was one participant, though older than most, in a widespread movement to replace skilled craftsmen with machinery and machine-operators in response to new opportunities presented after the Revolution. These trendsetters contrasted with a larger proportion of the population who retained older methods, resulting in divergence among the producing classes and strikingly disparate employment outlooks for different artisans. Small, struggling shop owners and journeymen unable to raise capital lost economic and social status and usually became wage laborers. The most successful artisans, such as Paul Revere, became businessmen by raising sufficient capital to employ others and prosper in the new world of mass output. And the majority, living between these two extremes, entered the "middling" class, continuing their craftwork in any way possible, often supervising the labor of others in small shops, working as skilled laborers for larger employers, or becoming supervisors or machine tenders in factories. The term *mechanic* eventually identified this growing community of middling men unified by their belief in industry, frugality, and the "dignity and utility of manual labor."[48]

Paul Revere's background helped him dodge many of the pitfalls afflicting other manufacturers' labor relations. As a skilled worker with impeccable artisan credentials he understood where his laborers came from and what they wanted. His shop brought in enough income to allow him to pay reasonable wages without placing undue demands on the local labor market. But even Revere had his frustrations and setbacks when it came to employee management, because America had not definitively decided how its nonagricultural workforce should be hired, supervised, compensated, or treated.

Labor relations are vital components of all operations, particularly manufacturing shops, as they relate to the deployment of capital, the mastery of technology, and the efficiency of resource use. In the nineteenth century America wrestled with two particular labor challenges: labor shortages in some regions or trades, and friction between employers and employees. The choices made by the owners, managers, and workers in each manufacturing operation often spelled the difference between profitable growth and dissolution. Paul Revere's ability to blend old and new managerial styles and his impressive networking abilities allowed him to maintain a group of productive skilled workers even in the face of high turnover.

America suffered from a chronic labor shortage throughout its colonial years, a situation that persisted well into the nineteenth century. With local exceptions in major port cities, regions of high immigration, or congested areas, most Americans took advantage of cheap land prices and a welcoming frontier to purchase and operate their own farms, which often promised financial independence and prosperity. Many historians have used the "labor scarcity" thesis to explain America's rapid industrial progress, willingness to employ machinery, consumer behavior, and other trends.[49] The perceived consequences of the labor shortage even appear in Revere's correspondence. In 1803 Caleb Gibbs of Providence requested Revere's copper sheet and nail prices on behalf of several merchants, since "it will be better for them to purchase here, than to send to Europe." Gibbs then added, "but they ought not to think that matters of this kind can be had in our Country yet as low as in Europe when we have to give such price for labour." Clearly, American manufacturers often viewed labor as a limiting factor, but reality was a bit more complicated. While America certainly lacked the ample quantities of laborers found in England and other nations, America's labor supply improved by the turn of the century and continued growing in the decades that followed, particularly in urban areas and in the Northeast. New sources of nineteenth-century labor, including apprentices escaping their contracts in order to earn wages, immigrants, and surplus population leaving farms, competed with journeymen for jobs and aided the shift to more modern business practices. Although many employers could locate laborers seeking employment, the need for dependable workers possessing the proper skills remained pressing.[50]

The relationship between management and labor changed in the early nineteenth century. American artisans primarily stuck to earlier practices throughout most of the eighteenth century, including bespoke work, apprenticeships, and social obligations connecting apprentice and master. The growth of the market economy and industrial production eroded craft labor traditions and fostered new practices such as division of labor, wage labor, and standardized output, which Paul Revere and other pioneers had implemented many years earlier. Different permutations of old and new managerial and technical practices promoted a spectrum of nineteenth-century manufacturing operations ranging from traditional artisan shops to capitalist factories; from tool-wielding skilled laborers to powered machinery. In the middle lay entrepreneurial ventures that started as craft shops but expanded into larger-scale manufactories. As they grew, many of these businesses reorganized labor by dividing tasks into simple jobs requiring more stamina and less training; standardizing products; relying upon larger numbers of apprentices or semiskilled laborers, including women and children; increasing machine use; paying wages for time worked instead of tasks completed; and rigorously supervising the work process to maximize productivity. Managers often attempted to make their employees work in a more regular and efficient manner instead of the skilled workers' traditional "alternative bouts of intense labor and idleness" described by E. P. Thompson in his classic work *The Making of the English Working Class*. Even the definition of the "labor problem" changed over time to reflect an increasing identification with the managerial mindset. In the eighteenth century, "labor problem" referred to a widely perceived shortage of skilled artisans, but by the mid-nineteenth century this term shifted to identify a perceived shortage of hardworking laborers, and eventually, to a shortage of discipline or drive among workers whenever laborers failed to perform as efficiently as the managers hoped. Many artisan traditions such as liquor breaks or flexible work hours eroded when masters attempted to increase productivity. Machine operators in particular needed to work regular hours in a diligent and steady manner in order to maintain the overall product flow throughout the workplace. The pace of work often depended upon the limits of the power system, such as a river's flow, more than upon the human need for rest or breaks. In addition, the shift from tool-based production to self-acting machinery shifted control over the pace and style of work from the laborer to the owner or supervisor. All of these workplace changes caused many artisans and journeymen to protest the loss of skilled laborers' hard-earned privileges and independence, and they came to resent what they perceived as inflexible and emasculating regulations.[51]

Employee relations gradually shifted from a mutual exchange of obligations between masters and apprentices to short, specialized training periods that produced employees able to perform standardized tasks for a fixed wage. The apprentice system faltered after 1800, since apprentices rarely needed six or seven years to learn the trade and often preferred to leave early and start earning wages. Masters also deemphasized their educational responsibilities and focused on deriving benefits from apprentice labor. For example, Harpers Ferry Armory initially offered apprentices a fixed monthly wage of $12 that largely paid for their room and board, buttressed by a piecework wage for all their output above a certain minimum level. Their apprenticeship contracts bound them until they were 21, and promised to teach them the "art and mystery" of gunsmithing. They also received a new suit of clothing when they graduated. In 1809, managers changed this system to the "non-contractual training program" that treated apprentices as wage laborers. Harpers Ferry offered young boys a salary and nothing else, and their education suffered as a result. However, since many of the master armorers employed their own sons, many craft traditions and secrets propagated to the new generation through less formal means, keeping the spirit of apprenticeship alive albeit in a diminished manner.[52]

Changing traditions led to friction and protest. Journeymen and master craftsmen enjoyed harmonious relations throughout the eighteenth century, with only occasional small strikes breaking out less than once per year. Journeymen's roles changed as their advancement prospects worsened. In the eighteenth century most journeymen were young and hopeful, earning money on the way to setting up their own shop. But even by the late 1790s, Boston had three to six times as many journeymen as masters, indicating their diminishing prospects of running their own business. By 1815, most journeymen were older and less likely to advance to shop ownership, often supporting families on their wages. Urban journeymen complained about wage reductions, longer working hours, the fading of apprenticeship, deskilling, and the increased use of day laborers, and in New York City more than two dozen substantial strikes took place between 1795 and 1825. Masters, journeymen, and apprentices flocked to Jefferson's political party to oppose these workplace changes, which they attributed to elitist Federalists. In short, they protested the erosion of the craft system, demanded to be treated as skilled laborers, and expected to enjoy all the privileges of the prior generation of artisans. These hopes were doomed to fail, as the workplace, and indeed, the larger economy, had already embraced most of the values of industrial capitalism.[53]

Paul Revere managed to sidestep the prevailing labor conflicts and did not share the "typical" artisan's frustrations. While operating his brass and copper endeavors throughout the 1790s he evolved a system of wage labor employment that adopted some trappings of the craft system, and he applied these practices to the new copper mill. The results were largely, though not entirely, successful, and he maintained a continual pool of at least five to twelve mill workers between 1802 and 1810. The number of employees on his Canton mill's payroll and the total salary they received is displayed in Table 7.2.[54]

These figures reveal a general trend of growth, broken only in 1808 when Jefferson's non-importation and embargo policies curtailed the sales and forced

Table 7.2.   Size of Workforce and Labor Costs in Canton Mill

Year	Number of Workers	Total Labor Costs (dollars)
1802	5	1,276
1803	5	1,108
1804	5	1,002
1805	7	1,573
1806	7	1,831
1807	10	2,292
1808	6	1,490
1809	12	3,420
1810	12	3,641
Total		17,633

Revere to reduce his workforce. Revere relocated all his Boston foundry opera-tions and equipment to Canton after October 1804, which might account for some of the payroll increase because these figures apply to Canton and do not include Boston-based foundry employees.

Revere's ledgers and correspondence do not indicate any chronic reduction of output resulting from labor scarcity, unlike the frequent slowdowns caused by raw material shortages. Revere did face short-term labor problems at times, as did all establishments at the time. For example, in September 1809 all of his workmen could not work for ten days because of sickness. Whenever workers and managers argued over the right to set schedules and the pace of work, many employees in other manufactories resorted to strategies such as a "slow-down" or feigned sickness to increase their leisure time. Revere's records do not indicate whether his crippling 1809 work stoppage had feigned or natural causes, but the former seems unlikely. Throughout all his employee records, frequent reports of absenteeism indicate that any worker could begin a leave of absence whenever he wanted as long as he was willing to forgo his salary. Thanks to Revere's lenient policy, workers did not need to act sick to earn free time. A second worker conflict occurred in 1810, when two of his principal workmen left in the middle of a job.[55] This abrupt and damaging departure might reflect unhappiness with Revere's working conditions, although it could also indicate a sudden business opportunity that lured them away. These were the only two instances throughout all of Revere's records when he explicitly attributed a production delay to labor problems.

Revere's labor relations relate to early nineteenth-century legal trends. Many disputes between laborers and managers reached the courts and in nearly all the early cases the courts favored the employer, a throwback to earlier court support for master craftsmen against their apprentices. Nineteenth-century labor relations became more explicit and quantitative than the verbal contracts under the earlier system, and often relied upon written contracts arbitrated in courtrooms. Most judges viewed labor contracts as "entire" agreements that became null if violated in any way. In several of the most egregious cases, employees signed yearlong contracts, missed a few days of work, and therefore had to forfeit all wages as their penalty for breaking the contract. Revere never faced this problem with his laborers, partly because he relied upon informal verbal contracts and related to them as fellow skilled workers whom he treated with respect as long as they earned his trust. Revere showed his age by favoring the older style of flexible personalized contracts rather than the standardized professional contract terms that gained in popularity from the 1780s to the mid-1800s. Joseph Warren Revere, more modern and formal than his father, recorded four explicit written contracts in his "Canton Ledger," a notebook often used for miscellaneous recordkeeping tasks. These four notes are the only ones of their kind, possibly signaling an early experiment that was not continued. On November 15, 1805, he agreed to hire Enos Withington for one year (beginning the prior September) for $18 a month, with a raise to $19 a month for the last five months. On November 18 he enlisted Thomas Pattersole until April 1 at the rate of 3 shillings, 6 pence per working day. On November 25 he contracted with Isaac Bosworth for $1 per day, and on February 18, 1806, he hired Timothy Allen for $1 a day. The different agreement types are interesting: Withington's contract reflected months of work that had already taken place at a monthly rate; Pattersole's rate used shillings and pence to pay a day wage; and Bosworth and Allen received $1 a day as agreed before work began. Clearly, father and son lacked a consistent method for contracting salaries and labor.[56]

The only surviving description of any worker's job title applied to the foreman position. Revere apparently delegated some authority to Willaby Dexter, succeeded by his brother Jeremiah Dexter after Willaby left. Revere clearly trusted and respected Willaby, who received by far the highest salary in the plant, three months' wages at $45 per month and five months at $49 per month. Willaby provided room and board for many of Revere's employees, as described below, and also gave evidence for the court proceedings in Revere's lawsuit against Leonard and Kinsley in 1804. Jeremiah Dexter received the most raises of any employee in Revere's service, and became the highest-paid worker after his brother left, as well as the first one to receive a daily wage. Revere paid his best employees accordingly, knowing from his own experience that this would lessen the likelihood of turnover while increasing their motivation and morale.

Managerial influence often extended past the walls of the shop, as many early nineteenth-century factory owners considered themselves responsible for the moral and spiritual well-being of their workers. Examples of managerial proselytizers include Eli Whitney, who styled himself a "steward of the Lord," and the Waltham and Lowell textile mills, which attracted young women from rural communities with their paternalistic emphasis upon regular church attendance and proper behavior. The desire to promote or mandate virtue among employees arose in response to the widespread fear of industrial decadence and oppression, as different factory owners attempted to demonstrate that manufacturing establishments could educate and enlighten their workers. These efforts often created employee dissatisfaction, as the owner's attempts at rigid supervision diminished or eliminated laborers' prior privileges such as liquor, swearing on the job, and workers' freedom to spend the Sabbath as they saw fit.[57] Revere never considered or implemented any policies along these lines. Despite all his attempts to become a gentleman, former craftsman Revere understood the needs of skilled labor and seemed capable of dealing with his workers amicably.

Revere did not unduly suffer from the scarcity of labor in America, but at the same time he certainly wished for improvements in his workforce. He expressed some frustrations in the 1803 letter to Joshua Humphreys quoted earlier. When discussing errors in an early shipment of bolts, he laid some of the blame on labor and management practices:

> The Bolts we sent them were drawn from Pigs of our own casting that weighed 250 lbs. The men whom we imployed were not carefull they did them by the job & did not take the pains they ought to in cutting the pigs they made finns & those finns made the scales. All these bolts were manufactured before we got our works agoing. Now they are all done under my & my son's inspection . . . we are obliged to pick up hands as we can & as they begin to know some thing of the Business they leave us.[58]

This letter alludes to two common labor problems: lax on-the-job practices and high turnover. Revere implies a solution to the first problem: now that his works were properly "agoing," he and his son could keep a closer eye on the shop to make sure the laborers followed rigorous procedures. The erroneous shipment took place when his Canton shop had first opened its doors and as a result he might have subcontracted the work, relied upon unfamiliar workers, or allowed the workers more than their usual independence while he focused on more pressing matters. The second problem, workers leaving after learning some skills in the shop, had plagued managers and master craftsmen since the days of colonial apprenticeships. Revere was no stranger to worker turnover, nor could he address it definitively. His only recourse was to offer the best possible working conditions and endeavor to recruit capable new workers when older hands chose to leave.

Revere continually looked for new skilled employees and occasionally wrote letters in search of recommendations or leads. One such exchange provides our only glimpse of a negotiation process. Revere received such a favorable report of Isaac Bosworth, a Plymouth copperworker, that he contacted him by letter to arrange a trial in his shop. Revere waited for his son to return from Europe before opening the negotiations, a sign of his faith in his partner. On October 19, 1805, Revere said he would not meet Bosworth's request of 9 shillings a day, approximately $1.30 to $1.50, because this high rate would upset the current wage hierarchy. At the time he paid one man $18 a month and had several job applicants request $20, so he offered Bosworth $26 a month, which translated to $1 a day for 26 working days (implying a 6-day workweek). Revere believed his wage system appealed to his laborers, saying, "you will judge wether that wages paid punctually in Cash is not better that the way you have received your wages." If Bosworth came for a trial month and both parties wished to continue the relationship, Revere then offered to engage him for a year. This is exactly what happened, and Revere later sent Mrs. Bosworth $40 in December at his request, informing her that Mr. Bosworth planned to work at Canton over the winter.[59]

Revere charted his workers' salaries and employment periods in a ledger prior to 1807.[60] Revere switched from a monthly wage to a daily wage between 1805 and 1806. This change had little practical impact upon Revere's treatment of issues such as absenteeism since he already converted monthly wages into daily rates whenever he deducted money from workers' accounts for missed days. Revere might have moved his unit of employment time to a smaller increment to reflect increasing worker turnover. Most workers stayed for only one or two years, rarely longer than four. New England presented many opportunities for skilled laborers, as well as the ever-present lure of running one's own farm after amassing enough money. Revere's manufactory represented a good opportunity for workers to learn useful technical skills while earning an excellent salary. This transient behavior occasionally became a problem, as noted above, but Revere's total number of workers steadily increased. As long as he could replace his losses and train new workers relatively quickly, Revere's labor situation remained beneficial.

As with so many of his manufacturing practices, Revere mixed old and new labor methods, even reaching into his own artisan experiences for occasional inspiration. Revere paid his foreman, Willaby Dexter, for the room and board of many employees, whose names he recorded as Jon Battler, Pettier, Vase, Story, Withington, Mr. May, Jacob Perkins, Asa Smith, Joel Fales, and finally, "Nath Morton & Apprentices." The "Nath Morton" notation implies that the apprentices may have worked for longtime Revere employee Nathaniel Morton, and not for Revere himself. The complexity of this transaction adds proof to the interconnectedness of personal and business relationships in small artisan-style shops like Revere's. Dexter owned a house near the Canton rolling mill and offered Revere a convenient way to provide his employees with nearby accommodations that might even build worker camaraderie . . . for a price. Dexter's board charges totaled $182, a sizeable sum. Some of these workers, such as Battler, May, Perkins, and Morton and his apprentices, do not even appear in Revere's ledgers, raising the question of why Revere paid their board. He might have employed them in non-mill operations such as carpentry or masonry, or they might have worked in the mill without written contracts, a common practice. After Willaby Dexter left his job, Revere paid his employees directly for their lodging at a rate of $2 a week, presumably to help them make their own living arrangements. Subsidized lodging hearkened back to artisan obligations, and Revere probably discontinued this practice in 1805 when he stopped recording lodging subsidies. In 1808 he revealed that he hoped to build "a large dwelling House" for his workmen on an unused part of his property, a plan he never brought to fruition.[61] Even the mention of this idea shows how Revere thought about his relationship with his workers, hoping to provide them with a place to live on his property and food to eat as well as a competitive salary. He continued to see them as a community of skilled laborers, fellow craftsmen whose hard work and judgment profoundly contributed to the success of his manufactory. In this manner he remained their leader and manager without engendering the backlash experienced in many other firms attempting to modernize. Perhaps Revere succeeded because he subconsciously applied one of the great lessons from the other revolution that had transformed America several decades earlier: he who loses the good-will of the people loses all.

CHAPTER EIGHT

Becoming Industrial

Technological Innovations and Environmental Implications (1802–1811)

Following a few early years of trials and errors, Revere's copper sheets earned universal praise, often from lofty sources. In May 1803 well-respected Captain Edward Preble of the *USS Constitution* informed the secretary of the navy that the *Constitution*'s copper sheathing was worn out and needed replacing, and "that made by Mr. Revere, is good, and of proper thickness." This endorsement echoed Revere's strong reputation among the merchants and shipbuilders of Boston. After he received a contract to replace the *Constitution*'s sheet copper, a June 18, 1803 article in the *Columbian Centinel* stated that Revere's sheeting "is the first ever manufactured in the United States, and will not suffer by comparison with the best sheets imported." And on June 25, 1803, the *USS Constitution*'s logbook recorded, "The carpenters gave nine cheers, which were answered by the seamen and caulkers because they had in fourteen days completed coppering the ship with copper made in the States."[1] Apparently Revere was not the only person who believed his successful mastery of copper rolling constituted an important national service, and public praise, among other factors, motivated his aggressive efforts to improve and expand his business.

The Canton mill's transformation in the early nineteenth century led to great strides toward standardization, an expanded product line, greater overall output, and an increased reliance upon natural resources such as raw materials, fossil fuels, and waterpower. Revere's operations had become a fairly advanced technological system by this point. Recall that a technological system represents a combination of physical artifacts such as machinery or buildings, administrative components such as company policies or lawyers, intellectual property and proprietary knowledge, legislative resources such as favorable tariffs or laws, and accessible natural resources. System-building inventors and entrepreneurs often begin by inventing a single revolutionary technology but develop that technology into a larger system incorporating more of the above components in an attempt to improve its productivity and better achieve its goals.[2] This theory sheds light on the new challenges Revere faced in the final years of his career when his small manufactory's success raised new technical and organizational challenges. Without technological improvements, productivity increases usually result from the application of additional resources into a system, which typically offers benefits in proportion to the added expenditures. For example, the easiest way to produce more copper sheets is to hire more laborers, buy more copper, use more waterpower, and build more of the same type of rolling mill. People operating within the confines of a system generally view the future of their system in these terms, extrapolating future increases from present circumstances. Technological changes, however, have the potential to rewrite the rules of the system and enable potentially exponential gains in productivity and efficiency. The ability to learn from others or innovate new methods becomes essential for continuing growth.[3]

Even though his mill already produced quality sheeting, Revere continued learning new techniques throughout the final years of his career. He improved his equipment, processes, and products in order to expand the reach, efficiency, and standardization of his technological system, but technical changes were merely the tip of the iceberg. Inevitably, his growing manufactory increased its impact upon the local and remote natural resources that sustained it, inspiring an array of technical, managerial, and even regulatory responses. Revere eventually found himself embroiled once again with the federal government, as well as with the local courts, and his struggles in both arenas foreshadowed larger changes facing the increasingly industrial republic.

Technical Practices and Improvements

Of all the cutting-edge technologies in 1800, metalworking had the greatest impact on technological development, for example, by giving rise to machine construction and precision manufacturing.[4] Even among the highly skilled practitioners of this advanced field, Revere was one of the most accomplished metallurgists in America. His ever-increasing workload, glowing reviews, and knowledgeable advice-laden letters to clients and colleagues attest to his combination of scientific awareness and practical know-how. Although some of his valuable background in different technical fields soon grew obsolete in the dramatic industrial expansion of the nineteenth century, his experimental spirit fueled his drive for improvement and further benefited from his son's fresh new perspective. By the time of his Canton endeavors, he had put aside his earlier career-jumping attempts at social climbing to wholeheartedly focus upon manufacturing, and as soon as he moved all of his different mechanical operations under one roof he began altering earlier production methods and incorporating new pieces of equipment. Revere remained the master of technology transfer, and used all available methods to continue his drive toward standardization. Many factors aid technological development, including assistance from governments in the form of patents or bounties; transfer of tools, blueprints, and skilled workers from more advanced societies; and innovations emerging from the inventive activity of native practitioners. In turn-of-the-century America the actions of the small and inexperienced government proved less significant than each manufacturer's ability to exploit the knowledge of available skilled workers while learning as much as possible from the example set by Britain.

Many turn-of-the-century Americans shared Revere's emphasis upon technological improvement. Government leaders such as Tench Coxe and Alexander Hamilton wrote at length about the need to foster inventive activity and help America close the technological gap with Britain. Widespread belief in "Yankee ingenuity" and American enterprise, combined with general outrage at Britain's well-known efforts to retain a monopoly on all of its advanced technological know-how, fueled the call for public support of practical experimentation and the dissemination of relevant knowledge. After the failure to adopt many of the suggestions in Hamilton's *Report on Manufactures*, the government relied upon the age-old practice of patents, even though this idea had its own controversies.

Rudimentary patents first appeared in the late medieval period, and took on a more modern form by the nineteenth century. According to patent theory, inventors need rewards and support to most efficiently advance their nations' technology, and states can inexpensively offer this support by giving inventors monopolistic control over the right to realize profits from their inventions. But patent theory raised a conflict between rewarding an inventor and disseminating the invention in a way that benefited society as a whole. Some groups, such as the London Society of Arts, opposed any form of secret withholding of knowledge, and required their members to refuse patents for their inventions. Patents also angered anyone who opposed monopolies.[5]

U.S. patent law rewards inventors by granting them a short-term monopoly that they can convert into profit either by being the only manufacturer of their patented device or by selling the patent rights to others. In this way a patent creates intellectual property, by converting an idea into something that can be owned or sold. The patent office also publicizes the patent after approving it, allowing others to make legal use of it when the monopoly period expires. After its passage in 1790, the patent system attempted to fulfill republican ideology by rewarding both the inventors and their society. Unfortunately, patents did not work as originally intended, particularly in the early days, when a complex examination process bogged down the system. Under the initial Patent Act of 1790 the secretary of state, secretary of war, or attorney general had to examine each patent claim, an extremely time-consuming and untenable process. The revised act of 1793 went too far in the other direction by merely registering each claim and allowing the courts to untangle the truth and merit of each invention. Courts had a difficult time weighing competing technical claims against each other, and some patent applications complicated the process by making overly broad assertions about relatively minor improvements. Even worse, for a long time inventors received minimal protection after receiving a patent. For example, Eli Whitney's patented cotton gin was so easy to copy that duplicates appeared all over the South and Whitney spent more money in lawsuits than he ever made in profits. And early steamboat inventors John Fitch, James Rumsey, and John Stevens all received patents for similar steamboat inventions and did not receive any protection or support from them. The patent system also oversimplified the process of technological innovation by identifying a single inventor and ignoring all of the prior work, collaborators, laborers, or other parts of the network that aided its success. A third patent act in 1836 finally resolved these difficulties by establishing a patent office. At any rate the patent system did nothing to address the scarcity of capital believed to be limiting inventions in the first place.[6]

Revere knew of the existence of patents, as demonstrated in his letter to Josiah Quincy during the tariff controversy described earlier: "If we had whished for an exclusive right we should have applied for it when we erected our works, had we then done it we have no doubt the government would have granted it. It was as much a new invention as any thing for which patents have been granted by our government."[7] In forgoing his right to receive patent protection, Revere again found a way to combine ideology and practicality. His lifetime of technological knowledge sharing did not endear him to a system that made this information proprietary. More pragmatically, a patent did little good at a time when other Americans lacked the ability to implement the copper-rolling process, as illustrated by Benjamin Stoddert's costly and unsuccessful funding attempts.

America's support of entrepreneurship took off after the Revolution when state governments and eventually the federal government promoted laws and policies strongly supporting practical improvements and inventions, unlike many British laws favoring tradition and property rights. America's manufacturing innovations multiplied in the decades after 1790 thanks to favorable factors such as communication and transportation infrastructure, technical knowledge accumulation, stronger technical communities willing to exchange information, readily available land for manufacturing establishments, and a growing labor force comfortable with machine operation. The concentration of skilled practitioners in urban centers favored new endeavors, as entrepreneurs could discuss financing options with local bankers, consult relevant books, read newspapers from any city in the nation, or brainstorm technical options with experienced workers.[8]

The biggest changes to technical systems result from the countless incremental improvements to equipment and procedures made by large numbers of practitioners over a long period of time. This ongoing inventive adaptation requires great creativity and effort, and can produce tremendous results. Americans proved adept at technological improvement thanks in part to workers' ability to perceive possibilities and opportunities, which depends upon their preexisting knowledge and experiences. Early Americans lived in a turbulent society, characterized by continuing migration, fluid communities, and ample opportunities to enter a new field or start a new life in a different region. Many Americans became jacks of all trades and made connections across different disciplines. Managerial styles also affected worker innovation. Many managers maintained casual shops that encouraged conversation and the free exchange of techniques even at a cost to short-term productivity, while tightly managed shops might feature great regularity of production at the cost of conversation and experimentation. British consul Phineas Bond commented on the limits and promise of American textile manufactories: although he realized that America was "essentially deficient in those main sinews of advancement, money, artificers, and fit utensils," he shrewdly recognized the importance of incremental improvements and correctly divined the way of things to come, concluding that "still their exertions are made with great zeal and the improvements tho' small are progressive."[9]

Revere and his workers altered many aspects of their production processes over the years to improve product quality and respond to changing demand. The vast majority of these alterations were subtle and unrecorded, only made visible through a study of the ways copper sheets and other products changed over the years, for example, by taking on different dimensions or receiving fewer customer complaints. Trial and error, aided by on-the-job experience, undoubtedly enabled Revere to make these improvements, but he also did his best to accelerate the learning process by going straight to the source. Along with many of his colleagues, dating back to colonial days, Revere tried his best to obtain technological knowledge and materials from England.

In spite of strict British laws prohibiting technology transfer, numerous American individuals and firms, encouraged by local and federal governments, relied on English expertise to learn new processes or improve existing ones. Particularly in the textile field a large number of machines and skilled laborers were smuggled out of the country and put to work in America. The emigration of skilled labor soon became the stereotypical representation of technology transfer, even though it did not apply to all industries.[10] As far as their origins can be determined with any certainty, none of Revere's employees came from Britain. This is not from want of trying. In 1802 he wrote to Mr. Bennoch, a former Bostonian who now lived in England. This long, detailed, mouse-eaten letter paints a vivid picture of Revere's limitations and concerns. He started with a summary of his product line, and mentioned that he had failed to find any information about the "English Method of Roling" in books or other information sources. And even though he had already rolled more than two thousand sheets that the "Inspectors appointed by our Government" declared "equal in quality to the English," Revere confessed that he still could not "finish it in the high stile that they do." Revere hypothesized that his difficulties might be attributable to the iron rolls he cast and turned himself, which he considered too soft for the task at hand. Revere asked Bennoch to procure rolls from Bristol, Liverpool, or anywhere he believed he could find quality workmanship. He also announced "I should be glad to give good wages to a Man aquainted with the business." Furthermore, Revere specifically sought more information about the British copper-rolling process, such as "the size & thickness of the pieces when first put into the Roles; what kind of Furnace or Oven they Aneal their Copper in; wether they role it single, or double; to what length they role it hot; & when they role it cold, wether they role it in water, and particularly how they clean it for finishing."[11] These questions illustrate Revere's progress in his industry: he possessed a basic knowledge of the equipment and procedures but had many questions about specific details. Since Bennoch never responded to this letter, most of these questions had to wait until Joseph Warren returned from his European travels three years later.

Revere's letter to Bennoch alluded to one critical component of his early rolling success. He not only had to learn the rolling process without any assistance, but also had to procure the necessary equipment. Americans understood waterwheel technology, but iron rollers presented a particular challenge since they had to be hard, durable, perfectly round, and free from blemishes. Instead of having to rely upon the limited and illegal availability of British rollers, Revere could cast and turn his own when the British supply dried up. His rollers were not perfect, but they functioned. In the period before machine-making shops made specialized equipment widely available, familiarity with ironworking techniques greatly aided Revere's copper-rolling success.[12] Over the years, Revere ordered rollers from firms in Plymouth, Massachusetts, Liverpool, England, and other locations whenever he needed to produce wider sheets or replace worn or broken parts. His instructions to each of these firms emphasized the need to match his exact specifications and use the highest-quality iron. In some cases he probably had greater ironworking expertise than the firm he contacted.

In the early 1800s Revere's technology transfer from England emphasized several indirect activities: occasional purchases of iron rollers from Liverpool, a letter to Bennoch that failed to produce a response, and the ability to reverse-engineer pieces of British sheet copper for use as comparisons against his own output. Indirectness ended in 1804, when Revere decided to initiate the boldest possible form of technology transfer. Four months after becoming the junior partner of Revere and Son, 27-year-old Joseph Warren Revere sailed to Britain to begin a yearlong overseas trip. Although he posed as a tourist and made a fair number of pleasure trips, he and his father had an ulterior motive. Joseph Warren was an industrial spy, perhaps the first in America's history, and planned to visit the major copperworks in Britain and northern Europe to learn their methods and study their equipment. By the end of his travels, he had studied plants in England, France, Holland, Denmark, and Sweden.

According to the technology transfer model discussed in Chapter Four, technological dissemination takes place through demonstration of potential, establishment of a pilot manufactory, diffusion of new technology, and modification to suit local conditions.[13] Industrial espionage represents a variation on the "modification" stage of technology transfer, in which managers and workers alter the transferred technology to better suit local conditions or the changing marketplace. Revere's letters illuminate his painstaking attempts to gather relevant information from books, experts, and the observations of travelers. Although he received the most help from American craftsmen and metalworkers, he looked to Britain for improvements and error correction after his mill opened. Industrial espionage was a highly efficient form of technological transfer because the search for information targeted the highest-priority questions. Joseph Warren had years of experience with Revere's operations and could orient his search at the most complicated aspects or biggest bottlenecks in the copper-rolling process. The fruits of this labor were guaranteed to fall on fertile ground. This espionage must be understood in the context of the times, as Revere and his son viewed it. Instead of apologizing for his son's spying, Revere cited it as a major asset of his operations for the remainder of his life. He was a borrower, trying to duplicate the best available technical processes, and industrial espionage represented the most direct and helpful manner of becoming a "master of the business" and helping his country close the technological gap with its primary competitor and former adversary.

Before Joseph Warren embarked, father and son diligently prepared him for different contingencies. Joseph Warren obtained a Commonwealth of Massachusetts passport signed by the governor, while his father obtained letters of introduction from the attorney general, influential friends, and himself. These documents enabled Joseph Warren to enter Britain, and once there to receive a passport from foreign minister James Monroe that allowed him to travel to other countries. In the age before photography, these passports contained a physical description, broken down into categories such as his height (five foot ten), eyes (light), mouth (small), nose (common), forehead (common), chin (large), complexion (rather light), hair and eyebrows (dark brown), and face (full). During his travels he received other official documents from American consuls in Hamburg, Rotterdam, and Kiobenhavn containing copies of his passport description. Joseph Warren preserved these documents for the rest of his life, as they reminded him of what might have been his greatest and most productive adventure.[14]

Throughout his trip Joseph Warren maintained a journal and sent many letters to his father detailing his discoveries. These letters contain a wealth of observations on European society from an American perspective. For example, he noted that Sweden had "by far the best roads I have ever seen" and marveled at repeated evidence of ongoing political turmoil in Napoleonic Europe: the French captured him on the road to Rotterdam, he was detained en route to Sweden, and he had to circumvent a blockade of the Elbe River. He also took the time to send several personal messages, informing his father of the death of an old friend and urging his younger brother John to stop smoking cigars because he felt the practice was harmful.[15] However, the majority of his observations concerned manufacturing matters.

Joseph Warren's notes reveal as much about his own operations as the ones he observed. For example, he often seemed frustrated by the lack of quantitative precision of the workers. While studying a London bell foundry on December 12, 1804, he complained, "They did not weigh the tin which they added but took out some metal & broke it & then added as they thought it wanted." This provides an interesting juxtaposition of working styles: the young American had already adopted more of a quantitative and exacting mindset, while the Old World artisans remained craftsmen to the core. He also felt less than impressed by their new furnace, declaring that it resembled a baker's oven "inside a one story building which I should have supposed would have burnt the first time they used it." But even amid disappointment he noted that their bell molds and other tools surpassed his own.

The trip enabled Joseph Warren to learn more about state-of-the-art waterwheels. On December 19 he visited an Eastbridge rolling and slitting mill built for copperworking but improved for iron, basically a high-technology version of his shop back home. He noted that the mill drew power from a "good" stream with only a four-foot fall. The rolls fit into a "cast iron frame similar to a goldsmiths plating mill" using only one shaft and cog to move the bottom roller, which itself moved the top roller via some gears. He included an excellent diagram of this machine, and a later diagram illustrated how the two rollers could be powered independently in a more complicated setup. By 1809, the Revere mill used separate waterwheels to drive the top and bottom rollers. British firms still used the horizontally inclined breastwheels, as opposed to the more efficient enclosed turbines that appeared soon after.

Joseph Warren also identified the importance of managerial coordination and efficiency. On December 22 he visited a gigantic, comprehensive copperworks in Harefield, which cast, rolled, and cut copper in different buildings. The Canton property eventually adopted a similar layout. Joseph Warren's concluding paragraph best expressed his amazement:

> This works all the time imploy 150 workmen night & day, before I saw these I thought the works for Iron extremely well fitted but this is beyond description. altho so many wheels & all of iron there is no noise no chattering but all goes like clock work & you can talk & be heard the same as in open air these works have been all newly built within fourteen years & should suppose from the immense stock not less than a million sterling capital.

One can imagine the noise and "chattering" that must have taken place at his own mill in contrast to this one. Joseph Warren seemed most impressed by the organization and continuity of operations, as workers continually added large quantities of copper and fuel to the casting furnace, which produced a steady output: "as soon as the pig on tile was black another was poured on the top of it & so on untill full." Although he noted the rolling mill's six-foot width and impressive set of waterwheels and cogs "as true as clock work altho of this imence weight," the degree of mechanization generally failed to impress him: workers even used manual cutting shears to slice the copper. Many of these efficient procedures eventually helped him improve his own operations. He answered his father's longstanding copper-cleaning problem by noting that one boy mopped the copper sheets with urine or chamber lye before heating them, making them "sinfully clean & not liable to tarnish." Also, workers rolled copper sheets "square then corner ways then square untill got to the width then length ways until thin enough to shear." These simple, easily copied processes immediately found a home in the Revere mill.

His other travels drew a complex picture of old and new methods. He could not believe that Birmingham watch, toy, and pin manufactories used epic quantities of manual labor instead of switching to simple and efficient machines. In contrast, brassworks in Bristol impressed him with their use of

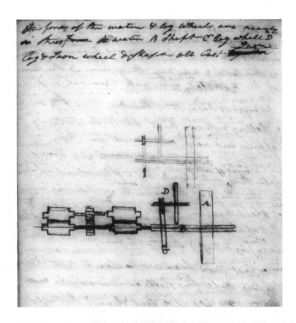

Figure 8.1. Joseph Warren Revere's sketch of British rolling technology, 1804, from Joseph Eayrs Financial Journal, volume 56, page 10. From Revere Family Papers, 1746–1964, microfilm edition, 15 reels (Boston: Massachusetts Historical Society, 1979), reel 15. Joseph Warren Revere made this diagram of a rolling mill in Uxbridge, England, during his 1804–1805 visit to England and Europe. His thinly disguised industrial espionage mission uncovered great quantities of usable information about state-of-the-art metalworking practices. This image illustrates how to use a single waterwheel to drive two horizontal rollers as well as a vertical blade for slitting iron or copper. Courtesy of the Massachusetts Historical Society.

steam power and ancillary equipment (a crane and a stretch of wooden rollers) to simplify the maneuvering and rolling of tremendous sheets. Best of all, a cog enabled the mill's rollers to turn in either direction, so the workers could simply move a sheet forward and backward until it was finished.

In light of Britain's strict policies regarding technological secrecy, Joseph Warren had a surprisingly easy time gaining entrance to a range of establishments. He occasionally mentioned that an owner refused to share proprietary information or turned him away, but most of his journal pages contain detailed observations and drawings. His journal from other countries on his itinerary is not included in the Revere Family Papers, but correspondence and other writings allude to interesting visits to mines, foundries, and copperworks.

Revere never codified the specific consequences of his son's espionage, but many of his technological improvements over the next few years drew upon these observations and sketches. As important as these improvements were, Joseph Warren's most valuable discoveries related to organizational procedures. He witnessed gigantic concentrations of labor and capital, dwarfing anything in America at that time, all laid out "like clock work" to efficiently transform raw materials into finished products. In the years to come, Joseph Warren applied these overseas lessons by altering the Revere mill's processes and machinery, bringing it a major step closer to the modern factory ideal.

Paul Revere practiced what he preached. As a lifelong beneficiary of technology transfer, someone who easily left behind any adherence to the artisan's protective and proprietary attitude toward technical knowledge, he repeatedly shared his own experiences with others. At some point before September 1808 he received a letter, now lost, from Philadelphia merchant-manufacturer Levi Hollingsworth. In 1804 Hollingsworth constructed his Gunpowder Copper Works along Gunpowder Creek outside Baltimore to provide copper products for Baltimore's thriving port, but by the time of this letter he had run into numerous production problems and requested assistance from the far more experienced Revere. Revere did his best to clarify many details of the copper bolt-manufacturing process. Revere suggested a high-tech method to use special rollers, described in a following section, to roll copper bolts far more efficiently than the previous tool-based process. He even included a labeled diagram. Revere could have ignored the letter or offered limited advice that avoided disclosing his most advanced techniques. Instead he went out of his way to share everything he knew that might help a fellow manufacturer. Revere's generous support eventually provided him, or more accurately, his firm and descendents, with a form of karmic payback, because the corporate successor to Hollingsworth's Gunpowder Copper Works later merged with Revere's company. In a shorter time frame, Hollingsworth remained friends with Revere and collaborated with the Revere and Son firm on a new petition for tariff protection in 1815, again ignored by Congress.[16]

Revere also offered assistance in an 1810 correspondence with H. M. Salomon, a Philadelphia plating mill owner. Salomon initially purchased copper from Revere in February, optimistically expecting to need a large quantity. A May letter to Revere explained how disaster had struck:

An accident has happened in our mill which will unfortunately prevent our Roling out any more plated metal until it gets remedied. One of the large cast steel rolers broke last week directly thru the middle in two pieces. In this City and Philadelphia we cannot find a person capable of making another. Mr H Hendricks has just now informed me that you make rolers in your foundery. If so please to let me know how soon you will be able to furnish us with a good Cast Steel roler of the size we require . . . The Iron casters here have thought that good Iron ones would answer. I have made 2 pair for us but found they would not do. If no one here can do it, I shall send to England for one.[17]

This letter portrays Salomon's desperation but fails to offer many specifics, as Revere indicated in his rapid response.

the break of the Roler is more from want of attention than accident, they must be constantly greased. Mr Hendricks information is not quite correct we have cast Roles but only for our own use. We now have them cast but turn them ourselves, everything depends on their being turned true, we cannot give our opinions as we do not know the dimensions of your roles, we use English & American Roles they are Cast Iron we esteem them equal to Cast Steel. You do not mention the cause why the Roles you got last in New York will not do. Ours are cast from the best of Iron.[18]

Salomon responded with a grateful two-page letter that illustrated some of the differences between himself and Revere. Salomon knew nothing of the business apart from what his workers and his superintendent told him, revealing himself to be an owner lacking in technical experience, a prevalent category by the later nineteenth century. Salomon's workers repeatedly attempted to make their own iron rolls but could not remove the small holes from their surfaces, and therefore the rollers produced pitted, pockmarked sheeting. Based solely upon Revere's first letter, Salomon trusted him enough to ask him to procure a pair of rolls "such as you think in regard to quality." In spite of Revere's help, Salomon closed his shop doors in November and asked Revere to buy back the unused copper.[19]

Revere's willingness to help strangers and potential rivals such as Levi Hollingsworth and H. M. Salomon illustrates the camaraderie shared by manufacturers at this time. Many businessmen and manufacturers cooperated in order to regulate economic activity, exchange tools and personnel, or share information: profit maximization and collaboration often coincided. When firms were small in comparison to the market, cooperation usually helped everyone because information sharing spread risks and shared rewards.[20] In addition to his confidence in his mill's technical sophistication, Revere did not feel threatened by producers in other cities. Manufactories on his side of the Atlantic had never posed a threat to his sales, and could become allies in his battle for tariff protection against the true enemy, cheap imports from British facilities. Kindred spirits and potential sources of advice or support, these allied manufacturing pioneers worked together to advance the technical skill of American workshops, and America.

Standardization and a Tour of Revere's Product Lines

By the first decade of the 1800s, Revere's growing firm had to begin considering the challenges of transfer, growth, consolidation, and competition that face most mature technological systems. While his son eventually completed this journey after Revere's death by fully breaking with artisan traditions and beginning new organizational and production processes, Revere and his workers still altered their production line in order to make the manufactory larger and more efficient. Along with his fellow forward-looking manufacturers he focused his efforts on innovating, transferring technology, standardizing his output, and developing new products.

Revere's altered technical methods highlight contemporary trends of machine use and early mass production. The earliest American craft establishments used power machinery only for the initial processing of raw materials, for example, to saw lumber, grind grain, or smelt iron. Skilled workers typically took over at that point and converted the processed materials into usable goods with tools rather than machines. These practices changed after the Revolution, when entrepreneurs applied powered machinery to the production of finished goods such as textiles and copper sheets. Many American machine adopters did so because of the exceptional fluidity of labor: a machine could be relied upon over the long term whereas skilled laborers came and went. As craft and proto-industrial workers shifted from multipurpose hand-held tools to less versatile and externally powered machines, the duties and nature of the workforce changed. Creativity, versatility, and independence became less important than machine-based mechanical skills, discipline, and consistency. At the same time, population growth, increased consumerism, and circulating currency—in short, the expansion of markets and capitalism—inspired manufacturers to move toward the "American system," a predecessor to mass production.[21] Historian Eugene Ferguson defined the American system as "the sequential series of operations carried out on successive special purpose machines that produce interchangeable parts." One of the major goals of the American system (and later, of mass production) was the cheap, large-scale production of standardized output. Four technological components paved the way for large-scale standardized production: efficient machinery, high-quality raw materials, the intensified application of energy, and the use of gauges to measure the quality of the output.[22]

Revere's operations certainly met most of these criteria by 1802, when advances in other fields enabled him to purchase reliable machine parts and fairly steady supplies of quality raw materials. He never specifically mentioned any use of gauges but he did depend on accurate and consistent measurement of his products: for example, he classified his sheet output into different grades usually measured in ounces per square foot. The majority of his earlier production depended upon skilled laborers and produced carefully crafted individual items such as teapots or church bells. But Revere started incorporating machinery and experimenting with standardized output for his silver shop operations as early as the 1780s, and by the 1800s his Canton mill employed devices in a variety of operations. After relocating all his operations to Canton and exploiting its waterpower and concentrated workforce he made even greater strides toward the methods and goals of the American system.

By his retirement in 1811, Revere had set up a copperworking complex housed over three buildings. Several letters describe the layout and the titles of his facilities. The rolling mill building contained a rolling mill powered by two twenty-one-foot diameter waterwheels; two furnaces for heating the copper bars and sheets; "Machiniary for Boring from the solid"; and machinery for turning cannon, drawing wire, and heading spikes. The hammering mill building contained one heating furnace, two forges, and two trip hammers, powered by an eighteen-foot waterwheel. And the foundry building contained "two large furnaces for melting with the necessary apparatus for moulding."[23] Revere most likely designed this setup with raw material flow in mind. He initially processed all raw materials in the foundry building, where he might cast his own copper pigs or pour bronze into bell or cannon molds. Workers then brought the copper pigs to the hammering building for reshaping, annealing, and possibly some refining before moving to the rolling mill for drawing into spikes or rolling into bolts or sheets. Similarly, workers carried solid cannon to the rolling mill building for boring and finishing.

These processes had evolved considerably since Revere's earlier days. Forty

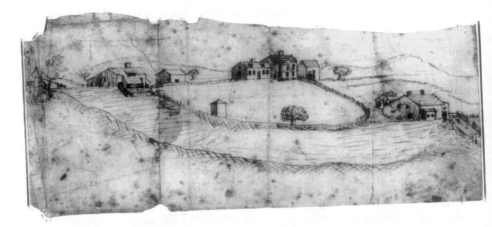

Figure 8.2. Paul Revere's sketch of his Canton Property, from pen and ink sketch of Revere Property on Neponset River in Canton. From Revere Family Papers, 1746–1964, microfilm edition, 15 reels (Boston: Massachusetts Historical Society, 1979), reel 15. This undated and untitled sketch was almost certainly drawn by Paul Revere himself, depicting his house in the center of the image and two of his shop buildings alongside a curving river. The sketch integrates Revere's home and work life, and juxtaposes natural elements such as the river and trees with fences and manufacturing buildings. Courtesy of the Massachusetts Historical Society.

years earlier, John Singleton Copley's portrait placed a younger Revere amid his tools and product, illustrating quite literally how the maker's hand is reflected in his work. As he neared the conclusion of a prosperous career, Revere's finest legacy remained his crafted items, and we best understand the magnitude of his accomplishments by studying his many product lines, the changes he made to his production routines, and the ways each deliverable moved closer to the ideals of standardized production.

Revere maintained his bronze bell and cannon production in the midst of sizeable copper bolt and sheet contracts. His output increased throughout his career, and his son continued both product lines for many years after his death. Revere's bell and cannon casting initially took place in his Boston foundry under the supervision of Joseph Warren, while the larger standardized operations related to bolt, spike, and sheet production took place in Canton under Paul's watchful eye. After the 1804 gale destroyed the Boston shop, Revere and Son moved all their operations to Canton and achieved greater efficiency for all their operations and access to waterpower for large cannon-turning equipment. Consolidation also helped Joseph Warren take a larger role in all aspects of the company.

Revere's pricing indicates that he consistently reserved his cheapest rates for bell manufacture and his highest rates for cannon casting. For example, from 1809 to 1810 he charged 42 cents per pound for bells, 50 cents for bolts and spikes, and 55 cents for cannon. At first glance this pricing might appear inconsistent with the complexity of the bell-making process, but Revere managed to simplify and perhaps even standardize his labor in two ways. First, greater experience allowed his workers to apply consistent procedures, such as an optimized ratio of metals and tried and true pouring processes, thus removing earlier uncertainty while avoiding time consuming "do overs." Second, he used a limited number of molds to produce a smaller range of bell sizes. After 1799 he restricted his output to bells weighing five hundred pounds or more (with two exceptions), usually used in churches as opposed to schools or ships. This restriction allowed him to focus on a more uniform product line. Since these large bells required an enormous weight of metal the amount of labor per pound was far less than for individually prepared spikes and sheets, which required annealing, hammering, and shaping. Even though bells remained his least standardized product his comfort with the process allowed him to accelerate his production by following an established process known to produce the desired acoustics for a bell of a specific weight and size. Revere made between two and twelve bells a year throughout the 1800s until his retirement.

Regarding the quality of his bells, Revere's correspondence reveals a single dissatisfied client whose complaints greatly affronted his sense of honor. After receiving their bell on May 31, 1802, Messrs Heywood, Flagg, and Stowell reported that "the general voice of the People seems to be that they are very much disappointed in their expectations that your Bell cannot be heard at a proper distance." Revere responded in July, briefly explaining that large bells have a more "Majestick" sound than smaller ones, but not necessarily a louder one: had the clients asked him earlier he would have told them as much. His somewhat pompous response failed to win them over, and the same clients responded with a far angrier letter, accusing Revere of a "want of honor and integrity." Predictably, Revere responded with utter outrage.

> You are the first persons who ever charged me with deceiving them or with the want of Honour, or Integrity, in business, and I despise the man who would charge me with either . . . If you are "mortified with a miserable thing" don't

blame me, after saying in your letter that "the sound is agreeable" and after hearing the Bell sounded in the Furnace forty times & approving of it, before you took it away; Then was the time for you to have disputed the sound, before you took it from the Furnace. Your Bell is the 52nd Church Bell which we have cast. We have never had occasion to write one letter to, or have had one word of dispute with either of the Committees; yours is the first.[24]

Throughout this exchange Revere professed his unshakeable faith in both the workmanship of the bell as well as his own integrity, and even though he wanted to make his customers happy he could not understand their dissatisfaction. His quality assurance policies do back up his position, since he let the clients listen to forty peals prior to their purchase. Revere eventually received the payment for this bell, and his other records indicate that these harsh complaints were in fact a rare occurrence. The records do not contain any other criticism of his bells, and one customer spoke for the majority when he told Revere, "The Bell proves a very good one—it gives universal satisfaction."[25]

Although cannon casting might appear similar to bell casting, it involved two expensive and unpleasant processes that explained the higher price. The cannon-casting process became fairly routine once a metal recipe and mold preparation procedure survived the test of time. After that, however, Revere had to bore the cannon, which he accomplished at Canton as early as 1804. In September 1810 he told factor Joseph Carson that his works could only bore one gun at a time "but should they [more cannon] be wanted at short notice we could extend them."[26] Boring was time consuming and difficult work: any misalignment between the cannon and drill created an asymmetric chamber that decreased the cannon's accuracy and greatly increased the chance of a deadly explosion. Following the boring, all cannon had to be proved under highly adverse conditions. Proving not only necessitated additional transportation fees but also ran the risk of bursting the cannon and starting the entire process from the beginning at Revere's cost.

Even in 1802 Revere rightfully considered himself a cannon-casting expert. Amasa Davis of the Massachusetts militia remained the primary customer for ordnance, buying pieces on many occasions. Revere also lobbied the federal government for contracts. In 1802 he informed Attorney General Levi Lincoln that he made "very great Improvements in the Foundry line, especially [words missing] Casting Cannon. I think I can say without vanity that I cast them as well as they are cast in Europ . . . I believe I have cast more brass cannon than [words missing] people in America. I have cast more than one hundred for this & some of the other states."[27] Revere received a small number of federal orders and occasional requests from other states. In 1810 he even received a secret request from the Spanish government to furnish twenty pieces of cannon, a tremendous order at the time that was probably never fulfilled.[28]

Neither Revere nor anyone else would ever mistake his bells and cannon for standardized products. Even the weights of different pieces made from the same mold varied. These processes still depended too heavily upon human judgment and skill, and the demand for bells and cannon never grew high enough to justify additional research into cost cutting or process improvement. Standardization simply did not make sense for these product lines, as the existing procedures satisfied his clients while affording a tidy profit. Revere continued these operations without any major procedural changes, and they provided him with a steady income and an expanding network of positive referrals.

Revere started producing bolts and spikes in his Canton mill shortly after buying the Canton property. The high opinion in which Revere valued his bolts and spikes, as well as his notion of honor, is evident in his response to a rare grievance from Mess. Beck and Harvey in Philadelphia, who complained that Revere's bolts and spikes were too brittle and out of proportion. His response speaks for itself:

> we were never so much astonished at any thing of the kind as at the contents of your letter . . . That "the bolts" should be "extremely brittle" is <u>impossible</u>, for they are drawn hot, under the forge hammer . . . the <u>spikes</u> are pronounced by judges here as good as ever was made, these sent you are exactly the same . . . We have manufactured the Bolts and Spikes for more than twenty Merchant vessels and you are the first persons who have found fault with our bolts or spikes . . . We have but one rule to do our business by, the strictest rules of Honour; we have sent you Bolts & Spikes such as our Carpenters approve of here, had you sent us patterns we should litteraly have followed them. We are Gentlemen.[29]

As with the bell grievance cited above, Revere saw no separation between his business and his personal reputation: a technical complaint about his product struck him as a personal attack upon his honor and challenged the "gentleman" status he awarded himself. In addition, he refused to accept the validity of this criticism in light of his own substantial experience, finding it "impossible" that these bolts and spikes might be deficient. Revere's next letter to these merchants was far less outraged, and indeed, almost humble. He discussed terms for reshipping the bolts and spikes back to Boston and clarified his defense of his bolts and spikes: "that they are as smooth as English we do not <u>contend</u> but that they are as good we do. Our Manufactory is yet in its infancy. I think I told you when in Phila. that we finished ours with a hammer which is the reason they are rough, the English is Rolled . . . we are improving & expect to finish our Bolts the same way."[30] Revere concluded by referring these men to Joshua Humphreys for his opinion of bolt and spike quality. Whether or not Beck and Harvey were correct about the quality of their bolts and spikes, Revere did not exaggerate his experience in this field. Throughout all his Canton operations, bolt and spike work proceeded with ever-increasing volume, eclipsing all other products in total sales. Any improvements he could make to this process would have a major effect on his balance sheet.

By 1805, Revere implemented the improvement mentioned at the end of his letter to Beck and Harvey. He either purchased or cast a special pair of rollers for the mill, for use in producing bolts rather than copper sheets. He explained this system to Levi Hollingsworth in 1808 and implied that in this instance he made his own rollers, which contained grooved indentations. If aligned properly, these rollers produced copper cylinders out of flat copper passed through them. The rollers squeezed the heated sheet and forced the copper into the rollers' indentations, producing what Revere called "strings" of copper—long, bolt-sized rods between four and ten feet long. Revere's reply to Hollingsworth included a cross-sectional diagram of iron rolls highlighting the diagonal indentations for bolt production.[31]

His workers still made other fastening products such as spikes at the forge, working them hot and finishing them cold to ensure that the points remained hard but not brittle. Revere alleged in one letter that his bolts and spikes were nearly as hard as iron and claimed his braces and pintles were the best in the state, because "we can give the metal a better heat and we cast the whole for one ship at one melting."[32]

Revere produced enormous quantities of bolts and spikes throughout his operations. The applicability of the term *standardized output* to products of this sort is a matter of debate. "Standardized" goods often imply complicated products such as mechanisms constructed from interchangeable parts: for example, rifles or bicycles. However, simpler items can be standardized in the same way that they can be heterogeneous. Revere's use of his rolling mill to produce copper bolts represents a major milestone in his manufactory because he shifted a previously handmade item into the realm of machine production. He made a similar though less dramatic improvement to his spike production process when he added machines that headed spikes, a process previously performed with a hammer and special tongs. As he offered products in several fixed sizes and weights, and as he offered to repurchase unused quantities, he increasingly treated them as uniform commodities. For example, a memorandum in his 1806–1812 bankbook lists the weights of 1-foot-long copper bolts of different diameters: 1.5-inch diameter bolts weigh 4 pounds, 1-inch diameter bolts weigh 3 pounds, and so on.[33] This calculation clearly assumes constancy among his output. In any early industrial endeavor the potential for irregularities still existed, but Revere's manufacturing and managerial policies indicated he wished to make them infrequent exceptions to a standardized rule.

In December 1803 Revere wrote to Joshua Humphreys to discuss his sheet-rolling process and how he learned by doing:[34]

> I agree with you that the manufacturing of copper in this country is a new thing & that every allowance ought to be made & every encouragement given to [so] usefull a Branch for tis by Experience that the Manufacturer becomes perfect & experience will not be gained without encouragement is given. We are dayly gaining experience. I have had & now continue to have the whole to feel out, for I have not been able to get any information from any person. Should I live & be able to take care of the Business for seven years to come, I should not get to the Zenith . . . I cannot help acknowledging that I have done better than my expectations. Our sheets are as well finished and as soft & as free from scales & cannot be distinguished from English.[35]

His final claim was not completely accurate. Revere continued his learning process on the job, and made his fair share of mistakes. As mentioned earlier in this chapter, he informed Robert Smith of some consistent errors in his first shipment of sheet copper in May 1802. Instead of producing sheets measuring 14 inches by 4 feet, a "small part" of the shipment measured 13.25, 13.5, or 13.75 inches wide by 4 feet long. Revere also mentioned that he consulted with some shipbuilders, and was informed that many of the sheets on a ship are cut down to a narrower size anyway. In other words, he believed his error would not have any harmful impact.[36]

This problem with the roller alignment was only one of many mistakes he and his workers must have made in their early years. His success in this field testifies to his workshop's ability to adapt and continue perfecting their processes. This in large part resulted from his experience with silver, iron, brass, and copper, which gave him a great instinctive awareness of metallurgical processes. He demonstrated his confident expertise in various letters to the Navy Department or other questioners, on subjects such as annealing, work hardening, and different finishing processes. Although his contract asked him to roll his sheets cold, he asked for permission to roll them hot, which would make them less brittle, allow nails to pass through them easily, and form a tighter seal with the ship hull. He knew this through practical experience and not from a study of the chemical and physical theories behind each process. Captain Edward Preble agreed with Revere's advice, and Navy Secretary Smith asked for more information. Revere responded: "The nature of copper is such that when it is in its pure State it is nearly as soft as Lead. It cannot be wrought till it is in that state, hammering it, pressing it, or Roleing it makes it hard & stiff; then heating it red hot, or Annealing it, brings it to its natural State again." He then mentioned that bolts and spikes were often finished while cold, to harden them enough to drive into wood, while British sheet copper was "finished as soft as the Annealing makes it." Revere then described his own method: "I think we have made an improvement, for after it is annealed and cleaned, we pass it once thro the Roles, which finish each sheet flat, smooths it, and adds a little to their stiffness."[37] Revere spoke with absolute authority on this matter and illustrated how he combined hot- and cold-working processes to produce both flexibility and strength. Smith adopted all his suggestions.

Just as Revere eventually treated bolts in a standardized way, he also came to view all copper sheets as interchangeable. As early as 1802, he mentioned that his sheets weighed "34 ounces to the superficial foot," or 34 ounces per square foot. The weight of a fixed area of copper depended on its thickness since the density of the raw copper did not vary much. Revere told Robert Smith that copper sheet inspectors recommended one-fourth of his sheets should weigh 30 ounces per square foot, a thinner sheet that he could easily provide if asked. Not only was he more in touch with the needs of the actual shipbuilders than the navy secretary, but he also began to consider the utility of producing different grades of sheeting for different uses. His 1806–1812 bankbook reflected this division of sheet thicknesses in a memorandum listing the weights of different thicknesses of sheet copper.[38] This method allowed

Revere to verify the uniform quality of his sheets with a well-understood and easily used measuring device, his scale. He did not mention whether he also used gauges to measure the width of his sheets, but this seems highly likely considering the increasing mention of sheet width in his correspondence with clients such as Robert Fulton. In spite of Revere's move toward standardization, his output occasionally suffered from imperfectly machined rollers. He was still a bit ahead of his time, and only when other industries perfected machining processes to create his own equipment could he truly establish standardized production.

Throughout his life, Revere sought new applications for his equipment and workforce. He had a remarkable ability to simultaneously perform older processes, perfect new ones, and research further opportunities. Revere's facilities produced a diverse product line by 1808, but he still wanted more. By this point his reputation had spread throughout the New England merchant community, connecting him to a lucrative opportunity related to the newborn field of steamboat construction.

New York State merchants trading along the Hudson River showed an avid interest in steamboats for many years, and members of the legislature offered corporate charters and monopolistic privileges to encourage entrepreneurs to perfect steam engine technology and operate franchises. Inventor John Fitch received a fourteen-year corporate steamboat monopoly in 1787 but failed to complete a functional vessel by 1798. At this point the legislature revoked his grant and gave influential merchant Robert Livingston a twenty-year charter, provided that he could build a working boat within one year. He and his partner, inventor Robert Fulton, managed to maintain their charter privileges without the existence of a working vessel and finally launched a successful steamboat in 1807. Fulton and Livingston's earliest attempts failed in part because they depended upon faulty steam engine equipment from Nicholas Roosevelt's Soho plant in nearby New Jersey. They succeeded after ordering designs and parts from Britain, including a reliable Bolton and Watt steam engine. As usual, Americans had to take great pains to remove an advanced technology such as a steam engine from England. Fulton and Livingston required a cheaper source of parts for their later steamboats, particularly the giant copper sheets used in their boilers. Eventually they discovered Revere.[39]

On October 8, 1808, Robert Livingston's brother John asked Revere for a price quote and size range for thick copper sheets weighing twelve pounds per square foot. Revere immediately responded that he previously had rolled only sheathing copper of fourteen-inch width but he could roll sheets as wide as his rollers, about three feet, and any length desired. He offered this copper at 55 cents per pound, and estimated he could roll a ton in approximately 10 or 15 days. Livingston soon accepted, and sent a draft diagram that carefully listed the dimensions of the different sheets he required.[40]

Boiler sheets differed from everything Revere had previously produced. Unlike copper sheathing, which merely had to remain on the outer hull of a ship, boilers withstood great pressure on a daily basis and required far more strength. All three boilerplate dimensions dwarfed naval sheathing, which only measured 14 inches wide by 4 feet long by 1/16 of an inch thick and weighed no more than 34 ounces per square foot. Boilerplate measured 3 feet wide by at least 5 feet long, and between 1/6 and 1/4 of an inch thick. In addition, several of the pieces required curved cuts along the sides to facilitate the construction of round portions of the boilers. Despite the novelty of this order, Revere completed more than 16,000 pounds of boilerplate early in 1809. Fulton placed a second order for 5 more tons of boilerplates in 1810, which Revere delivered in 1811. Fulton continued ordering more sheets from Revere well into 1814, after Joseph Warren had taken over the business. By this point the sheets sold for 70 cents a pound and were used in vessels intended to help the war effort.[41]

The steamboat contract had advantages and disadvantages. The boilerplate dimensions required altered production methods: bigger copper pigs, more men to lift and feed them through the rollers, and a different annealing process to compensate for the thicker sheets' longer heating and cooling times. These challenges led to errors, and Fulton complained in 1811 about excessive delays as well as brittle plates that cracked frequently. In spite of these protests he continued placing orders with Revere, realizing he could not do any better. On the positive side, the Fulton contract provided Revere with sizeable copper orders requiring less processing per pound in comparison with thin naval sheathing: Revere's workers had fewer sheets to process for an equal weight of copper. Because he charged by the pound, this contract promised major profits. Revere worked hard to encourage further sales, and by December 1810, he informed Robert Fulton that he had learned to make six-foot long sheets of any thickness, although he could not circumvent the three-foot width of his rollers.[42]

The Fulton contract prepared Revere for a host of creative new copper-sheeting applications that foreshadowed the business growth Joseph Warren experienced after he took over the company. Many of these new opportunities first appeared in 1809. Revere supplied local coppersmiths with heavy plate copper for stills, which also required sheets three feet wide by four or five feet long. The diligent efforts of Joseph Carson, his factor, turned up a long list of copper requests for still, sheet, "raised bottom," and "flat crown" copper from Philadelphia braziers. And the proprietors of a Charlestown bridge asked Revere to comment on the possibility of sheathing gigantic piles in copper to prevent them from corroding and accumulating barnacles, a proposal he wholeheartedly endorsed.[43]

Revere's business continued growing thanks to the new products and applications he and his son continued to locate. Each new process began in an experimental manner, characterized by trial and error as well as relatively large amounts of labor for each unit of output. But as he and his workers steadily gained expertise, Revere sought ways to standardize each new product, make it more profitable, and begin the cycle once more.

Revere and the Environment: Raw Material Shortages and Procurement Strategies

As a manufacturer, Revere realized that the success of his business depended upon his ability to procure the necessary quantities of fuel, metal, and water-power at reasonable prices. But neither Revere nor any other citizen of early nineteenth-century America connected raw material usage to the diminishment of nature, in part because the modern conception of the natural environment did not even exist at that time. While Americans and members of all nations valued nature, its bounty appeared infinite and unthreatened by humanity's ability to exploit it. Revere expressed resource constraints strictly in economic terms and focused upon the pricing habits of merchants, suppliers, and the government. Instead of connecting the rising price of wood with regional deforestation, for example, he looked for different suppliers or for ways to cut transportation costs. The market economy—far more tangible for him than abstract conceptions of the environment—mediated between nature and his industrial operations.

Of course, Revere and his fellow manufacturers did affect their environment in many ways. Beginning in his silverworking years, Revere played a small and indirect role in the larger American demand for metal and fuel resources. As the urban and manufacturing populations grew, the impact of their consumption increased. Revere needed more natural resources during his copper-rolling years although his fuel use remained small enough to be handled by the local market. However, his facility required so much copper by this time that he could no longer rely upon casual interactions with merchants, and instead had to increase his reliance on recycled metal while working with the government to secure a more dependable long-term supply. Revere's relationship with his natural environment became far more direct when he realized the finite nature of his water supply, provided by the Neponset River. In this case Revere's mill had a major impact upon upstream and downstream river users and the resulting conflicts foreshadowed major legal disputes of the later nineteenth century.

Ever since his earliest bell-rolling days in the 1790s, Revere knew all too well that copper was much scarcer in America than iron, and posed a far more dangerous limitation than labor shortages. The situation only worsened as his shop demanded more copper. Without a domestic supply he either reused "old" copper or relied upon overseas copper shipments, a choice between two untrustworthy options. "Old" copper often contained pieces of iron or other metals that ruined the finished product. In a 1796 letter to Nathaniel Gorham, Revere complained that "I was obliged to purchase it [copper] in small parcels which were utensils that had been burned in the West Indies and their was frequently among it Iron that I could not found." Imported copper usually had a higher quality but its quantity and price depended on ever-changing international trade conditions. The U.S. Navy purchased refined copper from many places, including copper plates from Sweden and Turkey, pigs from South America, and bars from Russia, but Britain remained the most consistent source of high-quality copper. Britain's restriction on all copper exports in 1798 deprived the United States of its primary supply and made the young country's precarious position impossible to ignore.[44]

Given the essential role copper played in his operations, Revere felt the need to secure a more dependable source and he asked the government to help him, since his manufactory sold much of its output to the navy. Revere summarized his copper procurement problems in a June 10, 1803 letter to Naval Secretary Robert Smith: "We find it extremely difficult to procure Stock, either old Sheeting or other Copper, sufficient for our small Manufactory, the merchants not being in the habits of importing it." He then offered two insightful suggestions. First, he asked Smith to reserve old copper sheathing stripped from government ships for the re-coppering process. This request extended his use of recycled copper, with the key advantage that recycled copper sheathing did not contain any iron and therefore had a higher quality. Second, Revere suggested that all naval ships returning from the Mediterranean sea should purchase cheap, quality copper at the port of Smyrna (now called Izmir, in modern-day Turkey) to use as ballast for the trip home. Revere added that Britain imported much of its copper from Smyrna, and voiced the Stoddert-like sentiment that "Should any difference arise between that government & ours would it not be difficult to procure a sufficient quantity from any other power for publick purposes?" Once again, Revere seemed remarkably well informed about international politics, and anticipated the War of 1812 about nine years in advance. Revere reiterated this advice in March 1805 to Naval Officer Jacob Crowninshield, hoping he might alter Smith's policy. Revere had conducted even more extensive research by this point, and determined that copper in Smyrna could be purchased for as little as 20 cents a pound, while it cost 27 cents in Leghorn, the next cheapest source. This information partially resulted from Joseph Warren's fact-finding mission in Europe.[45]

In addition to his attempts to alter naval policy, Revere did whatever he could to arrange a steady copper supply. He told Robert Smith, "We are endeavoring to git in the way of importing it but that will take time" as early as 1803, but lacked the contacts and capital to deal directly with overseas suppliers. He had better results from many small partnerships with different factors, or purchasing agents such as Joseph Carson, and with larger importers such as New York City merchant and copper importer Harmon Hendricks, as described above. In October 1809 Revere tried to arrange a partnership with a New Jersey mine that could have solved many of his problems, but the plan fell through. Since he couldn't do without copper he had no choice but to alter his work cycle and adjust his prices to respond to the changing international supply. Although he occasionally complained to friends or clients about temporary copper shortages, his records suggest that his output was never impacted for long.[46]

Revere had a much easier time procuring wood and charcoal. Although the land around Boston grew steadily deforested throughout the colonial and post-Revolutionary periods, numerous vendors sold fuel at reasonable prices. Throughout the year he repeatedly purchased two to three cords of wood at a time. He preferred buying oak and chestnut, long-burning hardwoods, for $10 to $16 a cord, and he purchased pine for under $7 a cord as a last resort.

He listed his suppliers in his records under their individual names, which indicates that he probably had personal relationships with woodlot owners rather than merchants. These transactions, combined with the availability of wood around Boston, kept prices fairly low and stable. He had to pay transportation as well as raw material costs but they represented a small fraction of his operating expenses. Factory-wide fuel costs were approximately $987 in 1806, and transportation costs for fuel and copper shipments were around $900. In contrast, Revere paid more than $18,000 for copper.[47]

Cheap fuel did not continue forever. In his 1810 report on Manufactures, Tench Coxe already noted the increasing number and severity of regional fuel shortages across the nation. He did not develop a comprehensive or long-term solution other than to recommend that industries such as wire drawing, cannon boring, and cutting establish themselves in low-fuel (coastal) regions, while the more fuel-intensive industries should relocate to forested regions.[48] Most existing manufacturers, including Revere, did not have the luxury of relocating. The most successful manufacturers found ways to build robust networks of potential suppliers and make accurate estimates of resource needs to allow enough time to meet all shop needs. Revere kept accurate records and amassed long lists of contacts since his earliest silverworking days and did not suffer from any major copper or fuel shortages. Waterpower, the most recent addition to his manufacturing system, proved more challenging.

Waterpower did not represent a revolutionary technology in the nineteenth century. The ancient Greeks made liberal use of waterpower, gear trains, and pumps, and civilizations since then used waterwheels to power grist mills, saw mills, and numerous other devices that processed raw materials. Until the 1790s, American manufacturing typically relied upon hand tools and hand power, and the first large-scale manufacturing attempts collected many workers in one location so they could produce items with traditional tools and methods. Waterpower entered the manufacturing world at the end of the eighteenth century, greatly expanding the potential for production and machine use while introducing a host of unanticipated managerial and control problems.[49]

The need for waterpower prompted Revere's purchase of the Canton property and never left his thoughts. Most machines require three types of components: a power source; transmission mechanisms such as gears, shafts, cams, and pulleys; and a tool to be manipulated by the machine. Revere's rolling mill consisted of a waterwheel driven by the river, gears and shafts turned by the wheel, and two large rollers turned by the shafts. The waterwheel and transmission components of this process involved technology that he had never encountered before, and he benefited greatly from preexisting equipment. The Canton property contained several buildings from an earlier ironworks whose waterwheels used the Neponset River's power to cut iron bars into small strips. Early waterpower almost exclusively relied on vertical waterwheels, usually undershot, with wooden crown and lantern gearing helping to direct the power in the proper direction. Revere modified this equipment to enable it to roll copper into sheets, thereby saving the major expense of building his own rolling mill or importing equipment from Britain. Revere used two waterwheels to drive his two rollers and another to power a triphammer via powered cams. Waterwheels probably drove the large bellows in his foundry and turned the boring device for cannon as well.[50]

Manufacturers had to select mill sites based on available waterpower, but this decision carried serious risks. A small river or stream, such as Canton's Neponset, seriously limited maximum output. However, a large river demanded far higher capital outlays to fully realize its potential. Assuming that basic hydrological conditions remained roughly the same between 1800 and 1909, an early twentieth-century study provides an estimate of Revere's available waterpower. In 1909 the Revere Copper Company offered its Canton land, buildings, and waterpower privileges for sale at public auction. The property description in the auction booklet describes this water privilege as an "average daily water power of 100 h.p. 12 hours a day." Waterpower depends upon the quantity of water in the watershed as well as the velocity of that water, which relates to the average distance it drops. The Neponset River drained over a 27.5-square-mile region (most of which was not owned by Revere's company), having a net fall of 16 feet. The 1909 property included a small millpond as well as the right to tap some or all of the water in the Kinsley, Massapoag, and Reservoir ponds upstream. The engineer observed that these additional water sources allowed twelve hours of flow even during the dry months. Table 8.1 presents his findings of the waterpower throughout the year according to two scenarios: open the dam and allow water to flow naturally over a twenty-four-hour period, or use the dam to store and release water over a more concentrated twelve-hour period.[51]

Revere's values probably fell between these columns: he often used his dam to store water and release it during daylight hours when his workers manned the shop, but he did not have access to all the storage capacity utilized by the

Table 8.1.   Neponset River Flow in 1909

Month	Twenty-four-hour Horsepower	Horsepower Concentrated over Twelve Hours
January	68.4	114.9
February	83.4	129.1
March	183.2	228.9
April	145.4	191.9
May	71.0	117.1
June	51.2	97.3
July	29.5	59
August	29.5	59
September	26.2	52.3
October	34	68
November	47.1	94.2
December	88.6	137.5
Average*	58.8	100.3

* The average excludes the rainfall in April and March that exceeded the rainfall in December. This was probably done to compensate for wasted water during periods of high flow.

company in 1909. The river flow pattern described in Table 8.1 reveals a dramatic seasonal cycle, with powerful flows in March and April, low flows from July to October, and intermediate flows in other months. This cycle explains the frequency of entries in his wastebook as well as his labor-hiring practices, since each year's major activity began in November and ended in June. This contrasts with Revere's work patterns when he operated from Boston, because he then adjusted his schedule to correspond to the buying patterns of merchants or his customers, as opposed to the rhythms of nature.

Waterpower increased the need for managerial control over manufacturing operations because shop operations and the deployment of water flow had to be coordinated and marshaled. Many early entrepreneurs had major difficulties efficiently using and adapting to the tempo of water flow, and often found it easier to shift to steam power later in the nineteenth century.[52] The Neponset River's flow left much to be desired. Although it usually met Revere's needs in spring and late autumn, the river froze for some of the winter, and summer dry spells forced him to cease all copper-rolling operations for weeks at a time. New York City merchant Harmon Hendricks mentioned this water shortage in an 1807 letter to Congressman Gordon Mumford. Hendricks told Mumford that Revere could not roll enough copper for all of America's needs, arguing, "I do not hesitate to say it is out of his power to supply Boston market solely. His miniature manufactory & works have not one third of the year the advantage of water, I have been at Canton where are established those works and I write from my own observations."[53] One possible solution to Revere's water shortage would have been to produce and stockpile large quantities of copper sheets when the river cooperated, and perform other tasks when it dried or froze. Although he did try to build a copper reserve, he often had to set aside his scarce supplies of copper for specific jobs. In general, demand for copper products remained fickle throughout Revere's tenure and stockpiling invited risk. In future years, when Revere and other manufacturers sufficiently standardized their output to minimize the uniqueness of different orders, companies had more confidence in sales estimates and practiced continual production. Revere held on to some aspects of his artisan mentality and preferred to receive an order before making a product.

Revere built and operated a dam early in his copper-rolling endeavors in order to overcome the river's variability and regulate its flow as much as possible. The dam allowed Revere to regain some control over nature by letting him store water overnight, on holidays, or when the river level was low, and release it in short, powerful torrents when he wished to run his mill. Although this solution might have seemed an extremely efficient way to make the best of a bad situation, other river users did not always agree.

Revere was far from the only businessman operating dams along the connected Neponset and Charles rivers. By 1800, the Charles River contained at least eight dams and twenty mills, the majority of which had been built since 1775. As the river became more congested, different constituents struggled to implement their competing views of the river. Peaceful negotiations gave way to private conflicts that eventually found their way into the courts, forcing judges to interpret and alter early riparian legal philosophies to correspond with changing societal demands.[54]

Natural resource legislation provides an excellent window into a community's environmental attitudes and property definitions. In colonial times all restrictive legislation emanated from Britain since England's deforestation alerted Parliament to the consequences of resource depletion. Colonists greatly resented these impositions and did not accept that the needs of a distant empire justified their lost economic opportunities. This attitude influenced the post-Revolutionary government, which produced as few restrictions as possible. Circumstances eventually ran ahead of legislation when numerous lawsuits, particularly concerning competition for waterpower, illustrated the finite extent of natural resources and provoked the formation of ideology and policy to equitably distribute these common goods.

The earliest water-use policies treated mills leniently because of their role in community development. Water-driven mills, typically saw and grist mills, often augmented the economic strength of frontier developments while also enriching their owners. These mills blurred the boundary between private and public institutions, and as a result, enjoyed significant legal protection under various "mill acts." Courts often invoked this protection in response to dam construction, which produced the majority of river-use conflicts. Dams and holding ponds allowed mill operators to efficiently regulate the flow of water, but also obstructed the annual migration of fish (evidenced in plunging populations of previously abundant alewife, shad, and salmon following industrialization), impeded human navigation, flooded the land upstream of the dam, deprived downstream farmers of a steady supply of water, and potentially increased erosion. Because of the obvious utility of early "public" mills, most judges and legislatures, particularly in New England, considered industrial river uses second in importance only to navigation and fishing. Under these acts mill owners could flood their neighbors' lands without permission as long as they paid a yearly sum to cover any losses incurred, treating this conflict in a quantifiable, purely economic manner. The mill act represented a major departure from the prior common law that allowed punitive as well as compensatory damages and also granted the injured party the right of "self-help"—destruction of the offending dam—to correct the problem.[55]

The New England mill acts did not create undue controversy in the early years of the eighteenth century for several reasons. The large quantity of usable land allowed towns to minimize the impact of flooding, effectively sacrificing some upstream property to enable mills to serve the economic needs of the community. Flooding remained relatively minor in most cases due to the small water demands of most early saw and grist mills, which therefore had smaller impacts upon river flow than the larger and more numerous manufacturing mills of the early nineteenth century. And the parties on both sides of these early disagreements typically shared a common ideology: mill owners, farmers, and fishermen all operated in an interdependent community economy that included elements of subsistence as well as markets. Even when they had different concepts of the best way to employ a river, most users viewed it as a communal resource theoretically able to serve everyone, thereby strengthening the community as a whole. It was even a safe bet that the local mill owner farmed some land as well, further cementing the interests of the different members of this relatively undifferentiated early economy.[56]

The relationship between mill owners and other river users deteriorated throughout the eighteenth century, particularly after 1750, when the number of blast furnaces and other large manufacturing operations increased. Blast furnaces did not fit into the values and work patterns of rural life, and instead required enormous volumes of water for sustained periods, devoted to the production of goods that would enrich the mill owner without visibly aiding the rest of the community. The increasing use of waterpower for personal profit instead of the public good necessitated a consistent quantification of the rights of different river users, and protests against the vast leeway granted to mill owners forced a reevaluation of the mill laws. These protests intensified against mill operators such as Samuel Slater, whose English heritage and large-scale technological operations marked him as an outsider and made him a lightning rod for local discontent.[57] Many farmers raised petitions against mill owners throughout the late colonial and early republican periods, requesting, for example, that mill owners construct fish ladders, open dam gates to aid migratory fish, or lower dams to minimize the threat of flooded farmland. Public petitions along these lines fared well prior to the Revolution but began to lose their impetus in the face of a strengthening market economy. Favoritism benefited certain private industries such as ironworks in New England, primarily because many legislators and judges had not fully distinguished between public and private enterprises. Many citizens and communities still believed all forms of industrial development lay in the public's best interests.[58]

Preferential treatment for industrial water users constituted a significant departure from British common law, which defined rivers and streams as transient, ancient entities not subject to exclusive ownership by anyone. British common law considered any alteration to a river's "natural flow" illegal. This doctrine never limited drinking, bathing, or the servicing of livestock—traditional activities that themselves appeared "natural"—but instead applied to larger alterations of the river's flow such as irrigation or industry. A series of eighteenth- and early nineteenth-century legal decisions, particularly in Connecticut and Massachusetts, began altering the "natural flow" precedent after it prevented the establishment of new industries. Early laws proposed the "prior use" or "prescriptive" doctrine, which implied that mills had the right to use rivers for industrial purposes as long as they did not "wholly" obstruct their flow. Once any industry or individual established a pattern of water appropriation, it could sue latecomers who interfered with it. The law eventually dropped the "prior use" doctrine in favor of the "reasonable use" theory, which contended that land ownership included the right to use and alter the river water for business purposes, as long as such use did not unfairly injure other present or future users. This concept allowed each judge to apply his own definition of fairness, modified by the details of each case. In Massachusetts the results usually favored large business interests.[59]

This situation grew more complicated when the majority of lawsuits pitted industries against other industries. By the nineteenth century, many rivers had become congested with industrial developments and dams regularly interfered with each other. The rapid growth of mills after 1815, primarily for textile production, forced courts to acknowledge that dams could interfere with both upstream and downstream users. At the same time, judges realized that the "prior use" doctrine favored monopolistic control of waterpower: the first user could claim that all future users at any point along the same river interfered with his established rights. As a result, the "reasonable use" criterion grew in popularity, particularly after 1825, until it became the established policy before mid-century.[60]

Three types of water use controversies took place throughout this period. First, downstream users sued upstream mills for altering the river's natural flow by constructing dams that either choked off their supply or released water so quickly that it overflowed downstream dams. Second, upstream users sued downstream neighbors who built their own dams so high that they produced "backwater," or elevated water levels in the river (perhaps leading to flooded riverbanks) that stopped their waterwheels. And third, neighboring landowners sued mills for flooding their land.[61] Although Revere settled some of his controversies out of court, he had the dubious distinction of being involved in all three types of dispute.

In October 1804 Revere sued Jonathan Leonard and Adam Kinsley for erecting and operating a dam that he claimed "obstructed and diverted the water" before it reached the waterwheel of his copper-rolling mill. Leonard and Kinsley owned an upstream ironworks (and, not coincidentally, are also the men who sold Revere his land four years earlier) and needed waterpower for their forge, saw mill, and grist mill. Revere's dealings with Leonard and Kinsley were not always antagonistic: after buying the Canton property he carried out many transactions with them through 1803, including rum, corn, cast iron, lime, and other purchases. He also shared surveying expenses with them to establish the boundaries of each firm's property in October 1802. This territorial action might also have foreshadowed the boundary disputes about to take place. In the early years, Leonard and Kingsley helped Revere, for example, by opening the gates of their dam for him on the rare occasions when he needed more water. In 1802 Leonard and Kinsley made their reservoir larger and deeper and controlled it with a more efficient mechanism, whereas the older one leaked. Revere found their dam use suspiciously hostile after that point: "It is in their power in a dry time to keep the water from us . . . if their saw and grist mills do not go, the water cannot come down the natural stream by reason of a mud sill which they have placed in the river . . . the water which passes their common works will not supply us one quarter of the time. They have frequently kept the water from us all day & lett it down in the evening after we had left work."[62] As a fellow manufacturer he knew Leonard and Kinsley had no logical reason to hoard water by day and release it at night. His description of his water shortage reveals his frustration, and he described at some length the actions he took to continue his production at a reasonable level. He reported that he "has been obliged to make a new wheel, a foot wider than the old one, to make a new floome so as to lengthen the gate a foot, or the mill would be of little service to him."[63]

Revere commissioned a study into the history of his land in preparation for an 1804 lawsuit that produced eleven pages of notes.[64] The notes were probably submitted to the judge or paraphrased by Revere's lawyer when he took Leonard and Kinsley to court later that same year. This research indicated that Sam Briggs sold land to the colony of Massachusetts in 1776 for the development of a gunpowder mill. The land included specific water rights to ensure that gunpowder production took place at steady levels: the gunpowder mill could order any upstream dams to open their gates in times of water scarcity. Massachusetts sold this land after the war and it passed through several other hands before Revere purchased it, with the water privilege transmitting verbatim through all the deeds. Revere's 1801 deed from Leonard and Kinsley included a similar water right, granting him the right to "commanding the water thro any gates that are, or may be constructed between the Slitting Mill aforesaid and the Forge Dam on Taunton Road so called, whenever the same may be necessary for the use of the Slitting Mill Privilege." According to these terms, Revere theoretically had the right to control the water flow through dams upstream of his own property, as far as the Forge Dam.

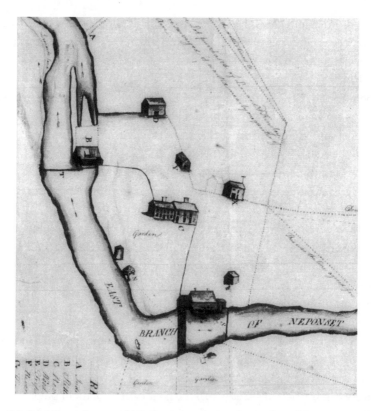

Figure 8.3. Paul Revere's map of his Canton property, from plan of Revere property on Neponset River in Canton. From Revere Family Papers, 1746–1964, microfilm edition, 15 reels (Boston: Massachusetts Historical Society, 1979), reel 15. This undated map was probably produced as one piece of evidence for Paul Revere's 1804 water rights lawsuit against Leonard and Kinsley. The map highlights the importance of the Neponset River to Revere's operations: two of his buildings abut the river and use dams to direct its flow into their waterwheels. Revere uses letter labels and dotted lines to attempt to explain boundaries of his property and the way he made use of each building. Courtesy of the Massachusetts Historical Society.

Revere recognized that this privilege was quite ambiguous. What if Leonard and Kinsley erected a dam beyond the Forge Dam? Also, was the Forge Dam in the privilege the same as the current dam on Taunton Road? He attempted to prove that the Forge Dam corresponded to the primary Leonard and Kinsley dam that caused all his problems, and he argued that he had the right to order this dam to release water to him during the day. Unfortunately he ran into treachery from Leonard and Kinsley on two separate charges. First, he discovered an "erasure" on the deed in the place where a precise description of the dam's location should have been. Revere might be able to command the "Forge Dam," but the "Forge Dam" was now unspecified, and Leonard and Kinsley could indicate a lower dam on the river. In outrage, Revere exclaimed, "Ask any indifferent man in the Town where the Dam is on Taunton Road: they point to the one above." Second, Leonard and Kinsley added language to the deed that prevented Revere from raising the river flow above a hole they drilled in a rock along the riverbed. He claimed they drilled this hole and inserted the new language "after the bargain was made & before the deed was delivered." As in many of his earlier interactions with customers or suppliers, Revere placed immense importance upon the concept of honor, which Leonard and Kinsley forfeited when they attempted to cheat him. Revere also blamed his son's absence and his own naivety for the problems, implying that his son would not have been fooled the way he was. On both of these occasions Revere signed deeds based upon verbal agreements and later learned that the specific wording deprived him of important assets.

He finally had his day in the Massachusetts Supreme Judicial Court during the 1804 October session. The court case transcript confirmed, "it became important to ascertain the place of the forge-dam mentioned in the deed—the parties not agreeing to its location." Testimony was called from Mr. Robbins, a source Revere claimed was allied to Leonard and Kinsley, and indeed, he upheld their definition of the dam. The court ruled against Revere on every count: the right to command all gates between his mill and the forge dam did not apply to the dam itself; he failed to prove definitively that his operations were harmed by the ironworks' operation of the dam; and he had "no right to command the defendants to open the gates" of any dam, but at best could open the gates himself. The judges viewed his lawsuit as an unfair attempt to injure his competitors and defended everyone's right to manipulate the river on their own property. Revere had to make two sets of payments to Leonard and Kinsley in addition to paying his own legal costs.[65]

This matter was far from over. In 1808 Leonard and Kinsley threatened to sue Revere for building his own dam and raising its height until it impeded their operations. The pair claimed the dam was "considerable higher" than his water right allowed and "does much damage by causing the back water to flow against our forge, grindstone, and gristmill wheels."[66] Revere apparently reacted to his earlier failed lawsuit by taking the water situation into his own hands, though not in the exact manner suggested. If he could not control the rate of water release from the upstream mill, he could still build a high dam to catch the water before it reached his waterwheels, and then release it as he saw fit. But he might not have stopped there.

Revere's papers include an undated deposition from Abner Crane that was later dated "1808?" and took place no earlier than 1807. Abner Crane was related to Elijah Crane, a neighbor of Revere's whose land was flooded when he raised the dam. According to Crane, Revere had the right to a certain quantity of water as measured by a hole driven into a stone along the riverbed. Crane measured the height and location of the hole in 1803 and again in 1806, and noticed that the hole had been raised. The original location showed signs of tampering and the new hole was smaller.[67] Leonard and Kinsley probably sent this deposition to Revere in preparation for their lawsuit. Did Abner have an incentive to provide false testimony, or did Revere or his son actually move the hole to increase their water right and rebel against what they saw as underhanded dealings by Leonard and Kinsley? We cannot answer these questions from the available evidence, but at any rate the two parties resolved this issue without the need for additional lawsuits. Considering the low flow of the Neponset River, the movement of new industries to the Canton area, and the increasing scale of Revere's business, we can safely assume that competition for water did not end in 1808.

Revere's riparian lawsuits spilled into other land-use disputes. On February 17, 1808, the selectmen of the town of Canton sent Revere a letter announcing that they would build a road through his property "within a few hours," leaving him no opportunity to protest. In a follow-up letter Revere hinted at a possible motivation for this sudden action: "The petitioners do it from wiked hearts, their whole view is to hurt us and not help themselves." These petitioners happened to be the Cranes, the same family whose land Revere flooded when he built his dam, and who testified that he moved the hole in the rock. Revere believed the Cranes "have aimed at our ruin since Jonathan Leonard returned from Kent to Canton. We know their views and we believe that their intention is to drive us from Canton, that they may purchase our Estate for less than nothing."[68]

In his usual thorough manner, Revere analyzed the situation. For starters, the letter was dated the 17th but was sent on the 19th by Jonathan Leonard's servant, intentionally awaiting a time when both Revere and his son were in Boston. The proposed road "would reduce the value of our Mills & Works to less than one half they cost us; for were it a Common highway it would lead every kind of Creature by our works, and our Stock of Manufactures would not be safe. Besides, the passage way between the house and river is so narrow that a Horse or Ox passing by in the night allways disturbs the family." In this letter Revere also mentioned some of his future plans for the land the road would occupy. He needed to enlarge his gardens, which were already too small for his family's needs. He planned to build a new forge building, possibly to replace the destroyed Boston foundry. And finally he expected to build "a large dwelling House" for his workmen. Revere concluded that he had to "improve every inch of ground on that side of the river," and asked if it was fair to attack two men "who spend more real money in the Town of Canton than any five men in it." This exchange revealed many elements of Revere's philosophy, such as the need for privacy and security, his frustration with "every kind of Creature" who might travel public highways, the value of his contributions to the town's economy, and the respect he deserved as a result. America might have been moving toward a classless society dominated by the "middling" sort, but Revere held old-fashioned values, and at the end of his life felt he had earned certain privileges.[69] And perhaps he had. Unlike the water regulation cases, Revere decisively won this battle. His angry protest and defense of his right to develop his own property permanently forestalled the new road construction. Although the water control lawsuits represented a version of a property dispute, a river does not look or act like property. Judges had a hard time believing Leonard's control of the dam could prove injurious since the same water would eventually arrive at Revere's stretch of the river. The road through Revere's land was entirely different. Anyone could understand the intrusion presented by a public highway, and the local government refused to impose this burden upon an unwilling landowner.

Throughout his legal battles over Revere's use of the river, the key environmental issue was not one of supply or demand, but one of control and competition. Even if the river had a larger volume, the different users would still interfere with each other's operations by regulating the river flow to suit their own schedules. Revere and his competitors used both technological and legal tools in their attempt to harness this natural power source, and the harsh language and threatening actions taken by both sides leave no doubt about

the importance of this issue. The river was a form of property, a natural resource essential for growing industries. The party that lost control would lose
its livelihood.

Revere's transition between the worlds of crafts and industry is more evident
during his Canton copper-rolling operations than ever before, as he juxta-
posed old and new methods in all aspects of his business. He adopted many
managerial advances, such as double entry accounting and billing, credit and
cash discount policies, warrantees, using a factor to solicit work and raw materials in distant markets, and quantitatively dividing company assets and ownership via a formal partnership agreement. At the same time he still allowed
his laborers many of the privileges of skilled craftsmen, such as control over
their time and free board; avoided incorporating or requesting patents; allowed fellow copper rollers access to his techniques and facilities; relied upon
verbal agreements with laborers and in important transactions; and continued
to conflate his personal sense of honor with the affairs and reputation of his
business. Liminality also applied to his technological practices: he moved toward standardization on several of his product lines and increased his use of
machinery but still fell short of interchangeable output, partly because of his
continuing reliance on skilled labor.

Revere made many choices concerning methods and technologies throughout this period, and moved forward quickly on some issues while hesitating
on others. His career illustrates his willingness to adopt new technologies if
conditions seemed right, so his decision to remain with older methods either
represented ignorance of new possibilities or approval of the status quo. In
assessing his choices we must remove value judgments from the classification
of new and old techniques. In many cases older practices made more sense
and served him better than alternatives we now consider more advanced. For
example, his old-fashioned manner of dealing with workers probably averted
the labor disputes faced by other manufacturing concerns, and he had little to
lose from treating potential competitors as colleagues. Becoming "modern" or
"industrial" meant nothing to him; having a dependable workforce, effective
technology, and large quantities of high-quality output meant the world.

As he looked back on his life's work, Revere held a new appreciation of
his position in the shop, and in the larger workings of society. Although he
continued to support the importance of the mechanical arts he saw his own
role, the manager and technical pioneer, as the most essential responsibility.

Skilled laborers knew how to master complicated processes and equipment in order to complete a job, but Revere took on those responsibilities and more. He designed the processes, he purchased the equipment, he hired and trained employees, he sought out sources of business and raw materials, and he co-ordinated all activity within his shop in order to finish each job, grow the business, and pioneer new product lines. Here lay his true reward, something he considered more valuable than a comfortable income at the end of the working day. Revere's copper mill allowed him to take on a leadership role in his manufactory and help his community by providing goods that no other American could create. He could be a Patriot once again, and this time he didn't have to ride a horse.

Conclusion

Shortly before his retirement, Paul Revere wrote "Cantondale," a poem that extolled the values that guided his life while celebrating his daily routine and rural home. The very existence of this poem tells us something about Revere's otherwise poorly documented final years. He retained at least some of his earlier views of the societal role and expected behavior of gentlemen, a title he had awarded himself many years earlier, and decided that he now had the leisure, the intellectual capability, and perhaps even the social obligation to share his views with others through verse. As a man of great wit and a lifelong subscriber to numerous periodicals, he undoubtedly respected persuasive writing and occasionally penned formal works such as his historical account of his Midnight Ride, addresses to organizations during his Masonic leadership period, and even his love poem to Miss Rachel Walker on the eve of Revolution. Now, as a soon to be retired manufacturer, he could finally share his philosophical views via some reflective prose.

Revere's poetic attempt overwhelmingly conveys his great satisfaction with this final stage of his long and eventful life. He begins by setting the stage:

Figure 9.1. Gilbert Stuart (1755–1828), oil on panel portrait *Paul Revere*, 1813. Gift of Joseph W. Revere, William B. Revere, and Edward H. R. Revere. Photograph © 2010 Museum of Fine Arts, Boston image number 30.782). This oil painting portrays a 78-year-old Paul Revere as a prosperous and alert businessman. The painting was commissioned by Joseph Warren Revere as part of a set, accompanied by a portrait of Rachel Revere, who died later that year. This formal portrait casts Revere as the patriarch of a successful business dynasty, a very different individual from the artisan who had posed for a Copley portrait forty-five years earlier.

Not distant far from Taunton road

In Canton Dale is my abode.

My Cot tho small, my mind's at ease,

My Better Half, takes pains to please,

Content sits lolling in her chair

And all my friends find welcome there
When they git home they never fail
To praise the charms of <u>Canton Dale</u>.

Several stanzas follow in a similar vein, laying out some of Revere's views. He praises the value of hard work, friends, and peaceful repose, and even seems to contradict himself by praising exercise ("I <u>exercise</u> prefer to <u>wealth</u>") before describing how he loves to recline in his cot. After attacking cheats and flattery and lauding wise and generous men, he describes at some length his daily schedule, befitting a manufacturer operating amid the symbols of rural life:

Around my <u>Cot</u>, at break of day,
The <u>robin</u> pipe's his artless lay;
The <u>yellow-Bird</u>, with pleasing note,
Sings sweet, and trills his little throat.
Near to my Couch, congenial guest,
The <u>Wren</u> has wove Her mosey Nest,
Her hopes in safe repose to dwell,
Nor ought suspects the silvian dell.
At early morn I take my round,
Invited first by <u>hammer's</u> sound;
The <u>Furnace</u> next; then <u>Roleing-Mill</u>;
'Till Breakfast's call'd, my time doth fill;
Then round my Acres (few) I trot,
To see what's done and what is not.
Give orders what ought to be done.
Then sometimes take my <u>Dog</u> and <u>Gun</u>.
Under an aged spreading <u>Oak</u>,
At noon I take a favorite <u>Book</u>.
To shun the heat and feed the Mind,
In elbow chair I sit reclined.

Whether intentionally or not, Revere's poem repeatedly juxtaposes images of nature and industry without identifying any links between them. Revere wakes each morning to birdsongs and proceeds to inspect his hammer, furnace, and rolling mill. He issues instructions to his workers, and then goes hunting with his dog and gun before reading under a tree. At the end of his poem, he states:

Or ere the Sun sinks in the West
Or tunefull birds skim to their nests
To walk thro <u>Groves</u>, and <u>grass'y Fields</u>
Contemplating what <u>Nature yealds</u>.[1]

Revere's romantic imagery, common in the late eighteenth century, labels him a supporter of the pastoral ideal. He preferred the gardenlike aspects of nature to its wilder components while keeping civilization's comforts reassuringly close at hand.[2] Unlike late nineteenth-century writers such as Mark Twain, Henry David Thoreau, and Herman Melville who viewed machinery and technology as threats to pastoral fulfillment, Revere prominently included industrial elements in his catalogue of Cantondale's delights. To Revere, his Canton property represented a "middle landscape," a harmonious compromise between wild nature and civilization. This compromise extended to his daily life, which included a blend of productive activity and contemplative leisure that combined the ideals of the hard-working self-made artisan and those of the intellectual gentleman. It is tempting to apply the themes of his Cantondale poem to his final years and conclude that he found happiness on his rural property, surrounded by a harmonious mixture of nature and technological activity that satisfied him by simultaneously being both productive and relaxing.

Perhaps Revere wrote "Cantondale" in anticipation of his life's final stage, using it to cast his journey and his destination in a balanced, pleasant, and philosophical light. On March 1, 1811, at age 76, Paul Revere officially retired, closing the book on an incredible career that began more than 60 years earlier when he first donned an apprentice's apron in his father's silver shop. For a man who defined himself by his achievements, retirement undoubtedly represented a difficult and momentous transition. But at least he knew his business would live on under the capable supervision of Joseph Warren, his trusted son and partner.

To implement an orderly succession of business ownership, Revere prepared new Articles of Agreement to formally establish a partnership between his son Joseph Warren and two of his oldest and most capable grandsons, Paul Revere III and Thomas Eayres Jr. This partnership inherited all of Revere's company's assets, totaling more than $40,000, and maintained the name Paul Revere and Son. This asset total greatly understated the firm's real worth as it failed to account for the value of property and equipment, and also excluded certain personal property that belonged to the business. Joseph Warren became the new senior partner of this family venture, responsible for four-sixths of all expenses and entitled to a similar percentage of all profits, while the other two controlled one-sixth of the business and profits each. Revere loaned Joseph Warren the use of all his property and in exchange Joseph Warren agreed to pay him $900 a year and reserved to his father the exclusive use of his house.[3] This arrangement smoothly accomplished many goals: Joseph Warren continued as director and unquestioned majority owner while the two younger men learned the trade, received leadership experience, and shared the prestige of ownership. Paul Revere's share of the property continued to see productive employment in the business, and Revere received rent from its use, helping to support his final years in an era that lacked welfare, social security, or pensions. Paul Revere and Son's new partnership united Joseph Warren with two members of the next generation, including the oldest son of Paul's oldest son, paving the way for continued family involvement as other grandchildren matured. The thought of a thriving family business must have pleased Revere, who firmly established his role as family patriarch and the founder of a successful manufacturing dynasty.

Compared to the voluminous correspondence describing his working career, there is very little information about Revere's retirement activities. The Revere Family Papers include letterbooks that extend beyond his retirement date, but they exclusively consist of draft letters written by Joseph Warren, the company's new head, to a variety of suppliers and clients. Revere's records no longer include any personal correspondence, perhaps because the family focused upon business records when preserving items for the archives. Other information sources paint a sketchy picture of Paul Revere's final years, consistent with his lifelong activities and interests. Revere's retirement years were nothing if not comfortable: he lived in a spacious three-story brick house on Charter Street in Boston's North End for some of the year, and spent his summers in the large Canton house on the company property, which included a weathervane and other items manufactured by Revere's own hand. He shared both domiciles with many members of his family: unmarried daughter Harriet always remained at his side; daughter Maria and her husband lived with him in Boston until they could afford to live on their own; Joseph Warren shared the Canton address; and clusters of grandchildren surrounded him everywhere. He remained a die-hard Federalist, keeping himself informed by subscribing to various Federalist newspapers while also paying dues to the

Boston Library until his death. Revere attended church each week and drew attention via his adherence to old-fashioned gentlemanly attire—knee breeches, long stockings, ruffled shirts, and cocked three-cornered hats—which made him look, in one boy's words, as if he "always wore small-clothes." Revere no longer held leadership positions in charitable or public organizations but certainly remained in touch with larger events. Demonstrating his public spirit in September 1814, 79-year-old Paul Revere became the first of 150 "Mechanics of the Town of Boston" to sign a petition pledging their services to help Governor Caleb Strong prepare defenses in anticipation of a British attack. The governor took them up on this offer and the mechanics, along with all of Boston's public and private students, built Fort Strong and thereby fortified Noddle's Island. A letter of thanks from Bishop Cheverus indicates that Revere paid his workers their usual salary so they could work at the fortifications.[4]

Revere suffered two profound losses when his oldest son Paul Jr. and his beloved wife Rachel both died in 1813. These deaths underlined the mortality surrounding every stage of his life: only five of his sixteen children outlived him, and he now had to carry on without the charming company of his wife, immortalized as his "Better Half" in his Cantondale poem and captured in a Gilbert Stuart painting a little more than a month before her death. Revere hopefully drew some comfort from the thriving family that would outlive him: daughter Mary from his first marriage, and Joseph Warren, Harriet, Maria, and John from his second, along with at least fifty-one grandchildren.[5]

Paul Revere died on May 10, 1818. His last will and testament reveals the great trust he placed in Joseph Warren, since he appointed him the guardian for all grandchildren under the legal age. Revere left $4,000 to each of his five surviving children and approximately $10,000 to those of his grandchildren whose parents had died. In a later codicil to the will, Joseph Warren also received the money previously given to Revere's three daughters with instructions to pay them the interest from that money in quarterly installments. Revere did not immediately grant all his business property and equipment to Joseph Warren, but rather asked him to purchase it from the estate over a period of four years, at "lawful" interest. In addition, after the estate made all Revere's bequests to his children and grandchildren, Joseph Warren received the remainder of his father's personal property and money.[6] Relishing his role as the family patriarch, Revere did his best to distribute his money to his family in a fair manner that sustained everyone while maintaining enough capital to continue profitable manufacturing operations.

The will also allowed him to demonstrate and reinforce his view of each child's position in the family. Paul Revere's three surviving daughters received a lifetime of interest payments from their inheritance, but Revere left the management and disbursement of that money to Joseph Warren, the anointed successor to the family's patriarchal role. Oldest daughter Mary had married carpenter Jedidiah Lincoln in 1797 and eventually had seven children. Youngest daughter Maria had married businessman Joseph Balestier in 1814 and traveled with him to Singapore in the 1830s while he served as an American diplomat. And middle daughter Harriet never married, but instead lived with her brothers and helped to care for different family members. Perhaps in recognition of the constant support she offered on behalf of the family, Paul Revere left her all his household furniture.

Youngest child John Revere's inheritance complemented the valuable education he had already received. Paul Revere shared his silver shop with Paul Jr. and his manufacturing activities with Joseph Warren, but spared no expense in grooming John for a higher societal role. Paul sent his youngest son to the prestigious Boston Latin School and then Harvard University and the University of Edinburgh, which awarded him a medical degree in 1811. Dr. John Revere's long and distinguished career would have swelled his father with pride, as it included research, medical practice, and professorial appointments at different elite universities that eventually earned him a reputation "as one of the best and most learned professors of medicine in the United States."[7] Revere followed a common trend among successful artisans and businessmen who frequently apprenticed their older sons into their own trade as helpers while setting up the youngest son in a more elite calling. Perhaps he lived vicariously through John, who met the obligations of gentlemanly status with far greater ease than his father ever managed.

Joseph Warren, his father's right-hand man in all business matters, received his personal share of Revere's inheritance and so much more. Revere delivered his real estate, equipment, and company capital to his son in a manner that helped him carry on the business while managing the family's overall fortunes. Exercising the option to purchase his father's share of the company, Joseph Warren oversaw the operations of Revere and Son from the implementation of the will until his own death in 1868. He could not have been better prepared to lead the manufactory. In his youth he worked under his father's supervision for many years and finally assumed overall management of the Boston foundry while his father ran the Canton operation. Immediately after becoming a full partner, he traveled to Britain and Europe in search of the latest manufacturing technologies and principles of factory management, which he implemented upon his return. Joseph Warren shared his father's ability to grasp both the technical and managerial details of his operations, and although he had big shoes to fill, the company thrived under his energetic and versatile management.

Revere and Son received an early boost during the War of 1812, when naval demands increased the company's copper sales to an unheard-of three tons of copper sheathing a week. Playing off of this wartime momentum, Joseph Warren continued expanding the company in the years after his father's death, gradually bringing it closer to the massive manufacturing operations he had witnessed during his overseas travels. He made a series of strategic decisions that kept the firm profitable in spite of America's turbulent economy, for example, de-emphasizing and eventually ceasing the production of bells in favor of larger quantities of standardized products such as sheets and bolts, more amenable to machine production methods. He continued casting ordnance thanks to large ongoing military contracts, and started moving that complex process closer to the ideal of standardization as well. In 1828 he took a major step in advancing his firm by chartering it as a corporation, the Revere Copper Company. Joseph Warren used the process of incorporation to facilitate a merger with James Davis & Son, a family-operated brass foundry located in Boston, and he remained president of the expanded joint enterprise in recognition of the Revere Company's larger assets as well as his greater experience. He also used his growing societal influence to deftly fashion a technological solution to his high shipping costs: according to the Canton Historical Society, Joseph Warren and the Revere family played a critical role in routing a line of the Boston and Providence Railroad through Canton in the 1830s, also arranging for a spur to connect their own facility to the main line. In 1844 Joseph Warren also spearheaded the establishment at Point Shirley, Massachusetts, of a large manufacturing complex that could smelt and refine copper, to maximize the supply available for manufacturing. This vast complex included eight blast furnaces and eight new reverberatory furnaces and used the "German" process to refine copper ore imported from abroad via a fuel-intensive but labor-saving system.[8] A public servant like his father, Joseph Warren served several terms in the state legislature and on Boston's board of aldermen. He died in October 1868 at age 91, in Canton. The Revere Copper Company continued operating under the supervision of other family members and underwent a series of corporate mergers and name changes that eventually joined it with the descendent of Levi Hollingsworth's company in a massive merger of six copper companies in 1928, culminating in the creation of Revere Copper & Brass in 1929. Versions of Revere's firm survive today as Revere Copper and Brass Incorporated and the Revere Ware line of cookware.[9]

Paul Revere's legacy transcends the corporate progeny of his business. His lifelong career trajectory offers many insights into early American manufacturing and business and serves as a microcosm of the great changes sweeping across the new republic. At last we can return to the questions that initiated this study. Looking at the larger American scene, what can we conclude about the birth of industrial capitalism? And returning to Revere, which factors helped him accomplish his goals, and how did these goals evolve throughout his life?

Industrial Dawn: Proto-industry Revisited

While enjoying the pleasures of his Cantondale cot in the 1810s, Paul Revere proudly watched his son run the family firm in a business climate that bore little resemblance to the colonial conditions that inaugurated his own career. America had come a long way, with new political institutions favoring the centralization of political power and widespread democracy; evolving cultural values in which economic worth and demonstrated merit displaced the value of familial status; and a thriving economy that rewarded competent, well-placed entrepreneurs while increasing the disparity between rich and poor. In spite of the attention focused on the most dramatic new industries or societal institutions, America still retained many values of the colonial society that made these changes possible. Throughout the early 1800s, older elements such as craft practices, barter exchange, and agrarian traditions coexisted with applications and ideals commonly associated with the age of industrial capitalism.

Even though profit-seeking entrepreneurship had a long history by Paul Revere's time, the progress of American industrial capitalism faced many barriers. America's shortages of investment capital, cutting-edge technology, and skilled labor slowed the rise of large-scale manufacturing and capital accumulation. The young republic's long and slow transitional period stood in stark contrast to the experience of Britain, which dwarfed all other nations' industrial progress thanks to advantages such as large reserves of skilled craftsmen, well-developed infrastructure, and naval and mercantile dominance. Until the late nineteenth century British technological advancement and manufacturing output outstripped America's, but its strong guilds and deep-seated craft traditions engendered great pride in unique quality items and craft conservatism, containing the seeds of obsolescence. Insightful British observers noticed in the 1840s that many American shops used fewer specialized skilled workers by substituting machinery for labor. While America still lagged well behind Britain in terms of capital, technology, and labor, its thriving economy, abundant natural resources, and favorable legal environment enabled investors and manufacturers to expand their operations and steadily close the gap as they shifted from craft to industrial production methods.[10]

Proto-industry emerged as a middle state between craft and industrial practices, a transitional period in which each manufacturer borrowed and combined methods for managing capital, labor, technological, and environmental resources in order to best suit the ever-changing local conditions. Appendix 2 portrays many of the connections between these four factors, illuminating the limitations and challenges facing the early manufacturers who struggled to overcome America's capital and labor shortages, technological backwardness, and limited access to trade networks and resources. Appendix 3 recaps the primary craft and industrial options available to Revere and his contemporaries.

Paul Revere's example exposes the fallacy of rigid dichotomies, because he comfortably integrated both craft and industrial practices throughout his career. He engaged in machine use, the employment of salaried laborers, subcontracting, and double entry accounting as a supposedly preindustrial artisan, and relied upon verbal contracts, room and board arrangements, information exchanges with fellow manufacturers, and family funding and labor sources when operating as a nineteenth-century manufacturer. Even taking into account the many times he became an early adopter of cutting-edge technologies and managerial strategies, Revere never abandoned all of his colonial methods and principles, nor was he alone in this regard. A broader study of Revere's fellow manufacturers turns up other individuals who changed with the time as he did, alongside large numbers of artisans whose adherence to earlier practices resulted in downsizing or unemployment, as well as firms such as blast furnaces that started their operations in a nearly industrial manner from an early point. As exceptional as Revere may have been, being both a forward-looking innovator and a prominent leader of the craft system, every American manufacturer lived in this transitional realm to some degree and we cannot fully appreciate the complexity of early American manufacturing—or of early American history—without appreciating the turbulence and dynamism of proto-industry.

Entrepreneurial manufacturers such as Revere repeatedly crossed what we consider the boundary between crafts and industry, not as an intentional business strategy, but simply because it made sense at the time. Revere's society combined old and new elements and his business followed suit. Everything in early America took place on a small scale: undersized urban centers limited the size of labor forces and markets; incomplete transportation networks raised the price of non-local raw materials and made it harder to compete in distant markets; information disseminated at a slow rate; and small, tightly knit communities knew each other intimately. Traditional craft elements had evolved in a similar European setting and fundamentally suited this small-scale environment, enabling artisans to strengthen bonds with workers and clients while ensuring that their labor and raw materials produced only the goods that customers wanted to purchase. Newer industrial practices responded to the growing population, expanded market economy, and improved networks of information and transportation by helping manufacturers expand their operations: more laborers, more and bigger machinery, more raw material inputs, stronger power sources, and greater output for more customers. In a tumultuous economic climate featuring technological advances and shifting societal conditions, success depended on one's ability to adopt or pioneer new processes while maintaining continuity with familiar methods.

The value of a proto-industrial approach remains clear today: even our globalized twenty-first-century economy, productive and specialized beyond Paul Revere's wildest dreams, offers examples of craft practices that often appear as solutions to contemporary challenges. Skilled craftspersons still ply their trade and create unique items according to their customers' specifications, often with hand tools and traditional methods that might even predate Revere's. Many trades feature versions of an apprenticeship period, allowing new members an opportunity to work alongside a mentor and forge a personal connection while mastering elusive skills. Family-run businesses thrive in many settings and maintain close reciprocal ties with their communities. "Green" firms minimize their environmental impacts by relying upon efficient and sustainable processing of local resources. And even the most revolutionary segments of high-tech industries acknowledge the merit of certain craft practices each time a startup company offers its workers control over their time, workplace perks, and the opportunity to work alongside the owner of the company with minimal managerial oversight. Industrial and craft practices coexist quite effectively, and proto-industry still holds an important position in a dynamic, heterogeneous society.

Tools of the Trade: Components of Revere's Success

Revere's career makes a little too much sense when viewed in its entirety. His different business endeavors, and specifically his many successes, seem to fit perfectly into a seamless plan. After all, he started as a well-positioned silversmith and built a solid reputation, patron network, and stockpile of capital. Ironworking gave him the chance to perfect his casting skills while earning a steady income from cast iron sales. Better yet, he constructed a versatile ironworking furnace that opened the door to the fields of bell and cannon casting, which further augmented his income while also familiarizing him with copperworking and government contracting. Malleable copper production such as bolt and spike drawing easily followed, and introduced him to new metallurgical techniques and a much larger scale of production. After mastering that line of work he had amassed all the technical and organizational qualifications required for copper rolling, his highest-profile and most advanced endeavor. All in all, a most excellent strategy.

Of course this simplification is false. Revere did not possess oracular powers and could not chart such a fortunate string of events, but he does deserve credit for fostering the components of success throughout his career. Ambition remained one of his defining characteristics and at every stage of his personal and professional life he sought ways to improve his social position, expand his business, extend his professional network, or learn something new. Ambition, while obviously important in this dynamic, opportunity-rich age, did not guarantee success in any endeavor, let alone in a series of endeavors. Revere combined ambition with competence: his powerful set of skills and attitudes allowed him to exploit any prospect that arose. His abilities had the greatest impact in three areas where his fellow Americans lacked experience: management, capital collection, and technical capability.

Revere's managerial savvy set him apart from most of his contemporaries in colonial and post-revolutionary America. More than 90 percent of the population worked on small farms and relied primarily upon barter and credit interactions within close kinship and community networks. Urban craft shops had more access to the expanding market economy but remained small and closely tied to traditional practices. Despite the short supply and high value of skilled labor many craft shops failed, often because of managerial incompetence. The apprenticeship tradition focused on technical art and mystery and not upon accounting, pricing, advertising, budgeting, inventory control, short- and long-range planning, or many other important business elements. As a result, large numbers of artisans found themselves unprepared to change with the times.

Revere displayed his sharp business skills in many ways, beginning with the careful recordkeeping practices that make him such a rewarding historical subject. His methods certainly changed over time, becoming more comprehensive, more quantitative, and better organized, but even his most primitive and incomplete early records reveal his attempt to document major transactions and periodically assess his financial status. Careful recordkeeping epitomized his desire to scrutinize and improve his operations: he never accepted the status quo, and looked for new ways of expanding his business or increasing his efficiency. Revere's analytical approach allowed him to make well-informed changes to his methods by gathering data, seeking trends, and making projections. After recording all his shop transactions, he periodically computed profits and losses; he also tracked his interactions with each client or in each branch of his business and made regular tallies of stock on hand. This approach reaped many benefits, ranging from his short-term ability to set prices or purchase raw materials to long-term strategic choices about adding new product lines or expanding his business.

Revere also developed an effective managerial leadership style, built upon many years of skilled labor and his immersion in craft traditions. Revere saw himself as a gentleman manager and tried to separate himself from the actual labor process wherever possible, but still sympathized with his workers and knew how to maintain their motivation. In spite of new machinery that changed the form and pace of the labor cycle, he continued offering his workers perks such as room and board subsidies, some control over the days and hours they worked, competitive salaries, and the ability to serve different roles in his shop and apply their own judgment. He had the natural ability to blend old and new managerial principles, offering some of the earlier benefits accorded to skilled craftsmen while maintaining control over the work process and enforcing enough discipline to maintain output quality and quantity. His background in craft production also made him an expert in the technological aspects of manufacturing, which allowed him to identify knowledgeable workers, teach them to improve, and recognize the challenges and burdens of this type of labor. Revere's managerial excellence sprung from his intimate understanding of the labor and technology that underlay all manufacturing.

Second, Revere's success hinged upon his ability to procure capital funding throughout his career. In colonial and post-Revolutionary America, debilitating capital shortages often inhibited the establishment of new operations or the expansion of existing ones. The majority of artisans began their careers as apprentices and upon the completion of their craft education possessed few resources other than their tools. After serving as journeymen for several thrifty years they might amass enough money to purchase their own shops, but even the most ambitious entrepreneurs often curtailed or canceled plans due to a lack of funding. A would-be manufacturer's ability to raise and deploy capital was second to nothing: success allowed a shop to expand and thrive while failure led, quite frequently, to bankruptcy. Revere used his networking abilities to raise capital from numerous sources at every stage of his adult life. He cultivated a wide network of family and friends and demonstrated a great facility for converting acquaintances into customers, investors, or skilled laborers. When some of his friends eventually assumed influential positions in the army or government they further helped him secure contracts and loans.

Revere's largest and most important capital funding came from government contracts and support in the final stages of his career. Most Americans had limited interactions with the federal government, perhaps through the U.S. Postal Service, army recruiters, or articles in partisan newspapers. The same Revolutionaries who resented Britain's incursion upon their freedom prevented the new American government from following a similar trajectory by restricting its size and domestic influence. At the same time, many nationalist politicians, including most members of the Federalist Party, hoped to use the government to support enterprises that augmented America's self-sufficiency and prosperity. The American government in general, and the War and Navy departments in particular, sponsored technological development since Washington's administration at the very birth of the new republic. Contractors needed a good degree of initiative to find ways to receive help in the form of loans or contracts from such a small and limited national organization, particularly in its earliest years when even government officials lacked a clear sense of the scope of their authority. Revere had a good degree of initiative, and his persistence paid off.

Even after he failed to become the director of the mint or a customs inspector, Revere still believed the government had a place for him. When he shifted his full attention to manufacturing endeavors, his role became clear. At that point, all that remained was years of frustrating correspondence and administrative delay as he patiently convinced the government to hire him for different jobs and then pay him the agreed-upon fees. No private contract gave him this much trouble, but no private contract could even remotely compete with the government's appetite and deep pockets. In particular, the navy's $10,000 loan provided him with tangible capital, the most vital requirement for entrepreneurial success in early America. Most entrepreneurs relied upon family wealth or alliances with moneyed men, and future businessmen usually formed corporations to raise their funds, but Revere and a small number of contemporaries made the government work for them.[11]

And finally, Revere also benefited from his immense technological versatility. To phrase this differently, Revere excelled at transferring technologies from other fields and locations into his own operations. Although the concept of technology transfer does not truly apply to a well-established field such as silverworking, even at that early point in his career Revere's communal practices showed the importance of cooperation within the local culture. Individual practitioners undoubtedly competed with one another for clients, but as a whole they used the division of labor principle to work together. Subcontracting allowed them to divide work and maximize their ability to deploy skilled labor to satisfy the largest number of customers.

American ironworking also involved cooperation between firms, and in this case it did involve technological sharing. The different establishments within the ironworking field developed symbiotic relationships: blast furnaces needed forges and fineries to buy their products; finishing industries depended upon furnaces for raw materials; and all firms actively shared patterns, recommended responsible workers, and answered one another's questions. In a nation with abundant natural resources and growing consumer demand, the efficiency, specialization, and support that resulted from a larger network of cooperative shops greatly outweighed any concerns about competition. In spite of the crippling scarcity of intellectual capital, technical practitioners saw more to gain from sharing their knowledge than from hoarding it. As an ironworker, bell caster, and cannon founder, Revere interacted with his fellow practitioners through letters, visits, and purchases of tools and patterns. Some would-be rivals such as James Byers became his close friends, and within a remarkably short time he managed to enter each new field, add to the community's expertise, and share his knowledge with others in return.

Revere's culminating experience with technology transfer occurred during his copperworking career. He was one of the first Americans to make malleable copper into bolts and spikes, and the very first American to roll it into sheets. He employed knowledge from all his earlier endeavors, as well as some astute reverse engineering, to help reproduce the high-quality British goods that soon became his primary competition. His activities centered on reinventing British techniques by applying lessons from casting and silverworking, and he received all the help he could through the illegal importation of British iron rollers, the primary technological element unavailable in America. The issue of cooperation arose in later years when two other American copper rollers entered the field. Once again, he seemed unconcerned with the threats posed by potential competitors, allowed others to visit his rolling mill, and offered extensive technical advice and diagrams to help Levi Hollingsworth and H. M. Salomon.

In the political sphere, American statesmen had recently created a new government that from the moment of its inception attempted to improve the British model upon which it was based. America's manufacturers lacked this luxury. Finding themselves years behind their British counterparts, cut off from British processes and equipment, and lacking Britain's surplus of cheap laborers, supply networks, and investment capital, American entrepreneurs had a lot of catching up to do. Paul Revere underwent the technical research and experimentation process not once, but many times. In mastering all his technical endeavors, what he truly excelled at was the learning process, and his example shows how he and other Americans learned new trades. In the late eighteenth century neither Revere nor any of his countrymen could compete with or surpass British technology, but by emulating British processes and combining relevant expertise from some of the fields already practiced in America, these innovators advanced America's technological infrastructure to a point where future manufacturers could break new ground. Contemporaries appreciated the efforts of these technical pioneers: Benjamin Franklin, no stranger to the concept of innovation and self-improvement, believed that men who invent "new Trades, Arts, or Manufactures, or new Improvements in Husbandry, may properly be called Fathers of their Nation."[12] And John Adams astutely defined emulation as "imitation and something more—a desire not only to equal or resemble but to excel."[13] History typically rewards inventors, but emulators also deserve their moment in the sun, particularly in the early years of nations and enterprises. Revere's career reminds us of the importance of all aspects of technology transfer: at times an individual, a technical community, or even a state has to copy or borrow, learning all the while, until the opportunity for a great leap forward presents itself. Revere never pushed his country past Britain's mastery of technology, but he closed the gap substantially during his career. Future manufacturers would continue the trajectory that Revere had started, and would see farther. After all, they stood on the shoulders of giants.

Revere's success in shifting between technical fields also drew upon his command of technology and metallurgical processes at both a practical and a theoretical level. He used methodical research to expand his repertoire and he soon mastered the scientific and technical vocabularies of the time, limited though they were. His passionate correspondence with scientists in America and England reveals the extent of his scientific interests, which certainly infringed upon his personal time. He also loved new machinery and tools, and understood them well enough to envision new uses for existing devices, identify flaws, and postulate solutions. Revere combined some of the traits of scientists and engineers, and always strove to understand why as well as how something worked.

Even though Revere appreciated the perfection of technical processes, he definitely did not want to restrict himself to a laboring role. As a master craftsman Revere wore all the hats: he was a teacher, manager, promoter, designer, producer, and researcher. As his career progressed he tried to separate these job duties. He still wanted to manage, learn new processes, and instruct his men, but for most of his career he strove to eliminate manual labor from his job description. Apparently manufacturing and financial success represented only part of his larger objective.

The Pursuit of Happiness: Revere's Goals and Identity

Summarizing a person's goals is a risky business because a lifetime of decisions never fits a single framework. Revere's goals evolved quite naturally in accordance with changing external conditions and in response to the success or failure of his different endeavors. With this in mind, several primary patterns emerge from all his career shifts and business practices, best told in the form of one final narrative, from the perspective of Paul Revere.

In the beginning, Revere's priorities, concerns, and actions followed the well-traveled trail of most young middle-class artisans. His apprenticeship inaugurated him into an ancient tradition of skilled labor and exclusive knowledge. He worked hard under his father's supervision and did his best to support his family after his father died. Silverworking verified his consummate skill as an artisan, and his pride was justified by his intelligence, perception, and love of learning. This pride gave him the confidence to undertake new endeavors in an ambitious manner. His colonial upbringing separated his personal outlook from that of the post-Revolutionary generation: as a pre-Revolutionary silversmith he interacted with the gentry on a regular basis and observed and envied their privileges. Therefore, he did not aspire to success in his trade or to the amassing of status among artisans. The gentry dominated the world of his youth, and Revere hoped to stand in their midst.

Revere's elite status in the ranks of Revolutionary artisans and his high visibility among upper-class buyers of silver items led to a critical role in the Revolutionary War. For one exciting night's ride he became the commander and instigator, taking charge in the center of events by waking the countryside, rallying support, and outmaneuvering the enemy. The Revolution encouraged him to think bigger and he tried to achieve a more prominent role in the society he served so diligently. Throughout the rest of the war he performed other important services, interacted with many of the traditional leaders of society, and hungered for more. After paying his dues in the Patriot movement he expected some form of recognition when victory was won. But his hopes and dreams shattered against an uncrossable barrier: colonial gentry privileges almost always resulted from one's birthright and he could never gain them by getting his hands dirty.

Fortunately for Revere, changing times and national events opened new possibilities. During the post-Revolutionary period, hierarchical social stratification became a hated reminder of British oppression and restriction. A growing meritocracy arose in which one's skills and ambition replaced the importance of birthright. Societal stratification now depended on economic measures that gave primary status and power to the wealthy, supported by an almost Darwinian justification of their privilege. Post-Revolutionary ideology offered Revere a chance to enter the upper crust of society on his own terms by earning his place in their ranks. Meritocracy accorded with republican principles, in which society elected its most prominent and most talented members for public service. Revere subscribed to the notion of a meritocracy but not to any version of democracy: the gifted deserved power and status, and Revere counted himself in this elite assemblage. And societal service should certainly include fitting rewards. Revere and fellow Federalists wanted to do their part to improve America's political, military, and economic position in a manner that also bettered their own circumstances. In fact, Revere honestly believed he could best help America by becoming a successful businessman and attaining positions of wealth and authority that allowed him to expand his influence and share his growing expertise with society as a whole.

At first he tried to become a merchant, funding his efforts with proceeds from his silver shop and early casting endeavors. This failed for many reasons: he did not have a good sense of purchasing habits; his timing was poor; and he lacked the necessary personal contacts, reputation, and investment capital. He tried to arrange a government appointment, but this also failed because of insufficient connections and reputation. As a card-carrying Federalist he became completely frustrated: the qualified men should rule and he knew he possessed the required qualifications, but nobody believed him.

In the process of arriving at this dead end he had an epiphany. Somehow the means to the end became the end; the manufacturing activities that funded his quest for status became his source of status. These pursuits succeeded in part from the growth of his operations and the profound changes that transformed the nature of craftsmanship. With expanding markets and increasing consumerism, small craft shops started giving way to larger and more visible manufactories managed and owned by proto-industrial entrepreneurs. As a manufacturer Revere made money and provided a valuable public service without getting his hands dirty. He was the leader, he commanded respect, and he used his mind to solve problems. It helped that he devoted most of his efforts to advanced technical issues that perfectly suited his talents.

The culmination of this epiphany occurred when he received the opportunity to roll copper. At last all the pieces fell into place. Copper rolling related to the U.S. Navy, a federal institution intimately associated with the national welfare. Copper rolling also related to America's prestige: if Revere succeeded he would strike a blow for America's technological independence and help build a fleet to assert and defend its political independence. The $10,000 loan from Stoddert became an embodiment of his important contribution. In addition to representing a vital financial benefit, the loan signified his mandate from government, tangible proof that the nation's leaders needed and sought his help. He certainly focused on the financial aspects of the loan at first, but in the years to come he repeatedly mentioned it as an example of his manufactory's importance. He was becoming a man of affairs and finally played a high-profile role in national events by providing a vital, heroic service that no other American could manage.

With his success, Revere believed the good times had only just begun. Anticipating admittance into the ranks of the Federalist elite he expected good treatment such as the continuation of major government contracts and certainly a monopoly, since he was the only American to provide copper sheeting. Unfortunately, at this moment the Federalists permanently lost their grip on power and the Jeffersonians immediately distanced themselves from manufacturing, from government spending, from a strong navy, and from Revere's business. Revere fought them at every step: he wanted his loan, he wanted his manufactory to provide copper for all the government's needs, he wanted a protective tariff that provided special protection from foreign competition, and he could not understand why he never received these well-earned rewards. He won the battle to receive the funding Stoddert had promised, but lost the war. The Jeffersonians saw the loan as a strictly monetary transaction, and Revere as just another government contractor. He would never become a high-profile public servant or government leader. The Midnight Ride lay in his past, and he would not have another opportunity to be a hero.

But he was a manufacturer, and in the end that meant something after all. For starters, labor had lost its servitude-based connotations in favor of the growing image of the "self-made man." No one, friend or foe, could mistake Revere for a "mere" laborer by this point: he owned property, managed a large workforce, contributed to his community, and produced high-tech goods that approached or equaled the quality of their English counterparts. More important than his net worth was the magnitude of his accomplishments, the technological secrets he wrested from Britain to strengthen his country, the size of his business and volume of its output. Rather than hurting his fame, the appearance of other copper rollers enhanced it: he could accurately consider himself the founder of what we now call a national industry as others followed in his footsteps, sought his advice, built upon his achievements, and continued his work. On the eve of his retirement he even accepted a new ideal, that of Cantondale, which helped him find contentment in the pastoral harmony connecting his thriving operations with their natural surroundings. Property ownership, a manufacturing legacy, societal prestige, and reclining under an oak tree represented different forms of success and happiness. At the culmination of his last and greatest ride, Paul Revere could appreciate them all.

Acknowledgments

I first started thinking about Paul Revere in the final weeks of 1996 thanks to the advice of my astute mentor. Merritt Roe Smith, my advisor in MIT's graduate program in the History and Social Study of Science and Technology, wanted to steer me to an interesting research project . . . nothing too major, just a nice scholarly paper topic that might fulfill a Ph.D. graduation requirement and get me some research and writing practice. He told me to take a look at Paul Revere "and do a little digging" because he knew there was more to this story than meets the eye. Fifteen years later, the digging has evolved into a full-scale excavation: that paper became a journal article, then a dissertation, and finally this book. What looked like an interesting new view of Paul Revere's life proved to be a lens for examining a much larger picture—one that encompassed early American artisan and social communities, entrepreneurship and enterprise formation, and proto-industrialization. As I came to understand Revere, he served as my guide through a world of early American transformations and eventually became a key part of my professional identity. At the end of this journey I find myself unexpectedly sad to leave it behind, and forever thankful to everyone who made the experience so rewarding.

To start at the beginning, I must offer my deepest thanks to MIT, and in particular to the generous support offered by the HASTS program and the Dibner Institute during my Ph.D. research. My debts to individuals at MIT are too numerous to recount here; so many people at MIT make it their job to ask penetrating questions, locate resources, and demonstrate how intense work can also be a lot of fun. I am particularly indebted to fellow students David, Rebecca, Greg, Tony, Kendall, Brendan, Eden, Ben, Rachel, Tim, and Nina, and also to TEAM-mates Pete, Kris, Carlos, Sarah, Doug, April, Eric, Cara, and Jake, for minimizing the midnight rider jokes but maximizing the midnight camaraderie. I must also thank the faculty at MIT, particularly Leo Marx, Deborah Fitzgerald, Wyn Kelley, Roz Williams, and Peter Purdue, who offered great assistance to my research while also trying their best to educate me. And most of all, I am in lifelong debt to my dissertation committee, Pauline Maier, Harriet Ritvo, and Roe Smith, whose wise advice and unfailing support continue to pay dividends in my research, my teaching, and everything else. I can truly identify with Paul Revere when I think of the amazing apprenticeship I experienced, and the craft masters who taught me the "art and mystery" of history.

Continuing in Revere's footsteps, I have been most fortunate to live and work near Boston, where I could benefit from two of our nation's finest historical organizations. From literally the first day of my research I received generous support from the Paul Revere Memorial Association. I cannot praise this organization enough, specifically their farsighted president Nina Zanneri and their insightful research director Patrick Leehey, both of whom I now proudly consider my friends. Whenever we discuss Paul Revere I half expect him to stroll into the room and pull up a chair, and anyone who visits the Paul Revere House in Boston can experience this feeling firsthand—walk through the door, talk to the staff, and enter a different world. I have also had the great pleasure and honor of receiving funding and scholarly support from the Massachusetts Historical Society (MHS), my home away from home during the main years of my research. Indeed, without the patient advice of so many wonderful MHS employees I would still be reading microfilm there today. My friends and colleagues at the MHS are far too numerous to mention but I cannot fail to specifically thank Conrad White, Len Travers, Beth Krimmel (still my fairy godmother), Bill Fowler, Brenda Lawson, and Elaine Grublin. And of course I must draw attention to the indescribable Peter Drummey, the Western Hemisphere's greatest research librarian, the man who can describe my research better than I can and cite scholarly information on any topic imaginable. If you think I exaggerate, have coffee with Peter.

Since 2001 I have had the great opportunity of working at the Franklin W. Olin College of Engineering in Needham Massachusetts, getting the "rarer than once in a lifetime" chance to help build a college and embrace its innovative spirit. I must gratefully acknowledge the generous research support provided to me by Olin's research and innovation program, support that enabled me to speed my progress and hire inspired and insightful student researchers such as Kathy King, Juliana Connelly, Lauren Cagle, and Laura Firstenberg. Heartfelt thanks must also flow to Olin's first eight cohorts of students, the classes of 2006 through 2013, for all of their questions, comments, ideas, and goodwill whenever I subjected them to draft chapters or random Revere analogies. The Olin library has enthusiastically offered ideas and support of all kinds from the very moment it came into being, and I am greatly in debt to all of the wonderful past and present librarians, particularly my great allies David Ware and Dee Magnoni. Olin's faculty and staff are more like a family than anything else, and I wish I could use the next eleven pages of this book to properly thank each of them for the many ways that they have inspired me, supported me, and made me excited about coming to work each day. I must call attention to those who directly helped me on this project, the wonderful Gillian Epstein, and Professors Lynn Andrea Stein, Debbie Chachra, and Mike Moody, who is already dearly missed. A special thank you goes to Professor Caitrin Lynch for her cheerful and penetrating insights on my draft chapters, and for putting up with bizarre email questions and jokes at all hours of the day and night. Finally, I profoundly thank Professor Jon Stolk, the co-instructor of our "Stuff of History" course, who became my partner in crime the moment we met on our shared first day of work and who continues to help me understand and appreciate the connection between materials science and the history of technology. I am indebted to Jon for so many reasons, perhaps most of all because he named Olin's rolling mill "Big Paul."

I also extend sincere thanks to everyone at the Johns Hopkins University Press for such generous aid throughout the preparation of this manuscript. The experienced editorial advice of Bob Brugger, the constant guidance and support of Josh Tong, the marketing savvy of Becky Brasington Clark, the thorough and professional copyediting of Maria denBoer, and the competent services of so many other members of this press have been invaluable. My work also depended upon the brilliant editing of my dynamic friend Professor Sara Pritchard, who maintained perfect penmanship and much-needed humor throughout all of my chapters. And I would not be where I am today without the enthusiasm and support of all of my friends and grant partners at the Savannah-Chatham school district and Georgia Historical Society. Candy Lowe and Leah Colby have been my advisors and support team for years now, and I wish I could properly thank them for believing in me, opening doors on my behalf, teaching me the power of southern charm, and for carrying Revere's message into the lucky schools throughout the Savannah-Chatham district.

And finally, most important, I thank my family, old and new, parents and siblings, in-laws and outlaws, near and far. You made this work, and make everything else that I do possible. I dedicate this book, along with my infinite and eternal love, to Tristan, Chiara, Lorelei, and Marybeth.

Appendix 1

Major Events in the Narratives of Paul Revere and America

The following captures some of the major events and milestones facing Paul Revere and America throughout his lifespan. The left column indicates events in Revere's life, while the right column indicates selected major events in early American history pertinent to his story.

Year	Event	Year
1734	Born in Boston	
1748	Begins his silversmith apprenticeship with his father	
1754	Inherits and begins operating his father's silver shop	
	French and Indian War begins	*1754*
1757	Marries Sarah Orne	
1760	Birth of Paul Revere Jr. (apprentice and silverworking partner)	
	Parliament passes Stamp Act; colonial resistance intensifies	*1765*
	Boston Massacre	*1770*
1773	Sarah Orne dies; Paul Revere marries Rachel Walker	
	Parliament passes Tea Act; Boston Tea Party takes place	*1773*
1775	Midnight Ride; other full-time Revolutionary activities	
	Battles of Lexington and Concord	*1775*
	Revolutionary War begins	
	Declaration of Independence signed	*1776*
1777	Birth of Joseph Warren (later partner and successor)	
1779	Return to silverworking after the war	
	British surrender at Yorktown	
	effectively ends the Revolutionary War	*1781*
1781	Failed attempt to become a merchant	
	Treaty of Paris formally establishes American independence	*1783*
1788	Begins iron founding	
	U.S. Constitution takes effect	*1789*
	George Washington begins his first presidential term	
1792	Begins bell casting	
1794	Begins cannon casting	
1795	Begins working with malleable copper to produce bolts and spikes ·	
	John Adams begins his presidential term	*1797*

(continued)

Year	Event	Year
	Beginning of America's "Quasi-War" with France	*1798*
	Department of the Navy formed and led by Benjamin Stoddert	
1801	Begins copper sheet rolling in Canton mill	
	Thomas Jefferson begins his presidential term	*1801*
1804	Forms partnership with Joseph Warren: Paul Revere and Son	
	Moves all operations and equipment to Canton	
1811	Retires, Joseph Warren becomes manager	
	War of 1812 begins	*1812*
1818	Dies May 10, 1818	

Appendix 2

Four Proto-industrial Production Factors and Major Linkages

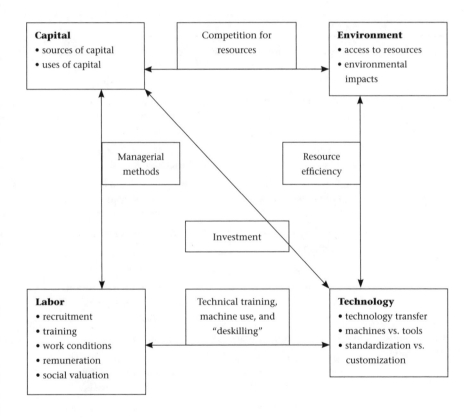

Appendix 3

Prevalent Craft and Industrial Practices in the Proto-industrial Period

	Craft (Pre-industry)	Industry
Labor	Lengthy apprenticeship training period emphasizes "art and mystery" of the trade	Brief and pragmatic on-the-job training often centers on machine operation
	Social obligations connect master and apprentices	Formal contracts emphasize monetary salary and worker responsibilities
	Skilled, versatile, autonomous laborers	Less skilled, specialized machine operators
	Skilled laborers are the primary source of knowledge and expertise	Expertise is embedded in machinery or in scientific and technical manuals
	Wages, usually supplemented with room and board	Exclusively wage labor
	Small workforce	Large workforce
	"Vertical" connections between masters and apprentices in each trade	"Horizontal" connections between journeymen or all laborers; unions
Capital and Management	Modest capital requirements	Substantial capital needs
	Rely upon local and internal funding sources such as family	Receive funding from large external sources: banks, government loans, etc.
	Owners work alongside employees	Separate managers from laborers
	Sole proprietorship or partnership	Corporation
	Minimal and primitive recordkeeping	Double entry accounting
	Word of mouth advertising, rely upon patrons and reputation	Use advertising and factors or marketing agents

	Craft (Pre-industry)	Industry
Technology	Hand tool use	Machinery makes use of external power sources
	"Bespoke" work, customized products and processes for each client	Standardized and inter-changeable output
	Rely upon craft training and employment of skilled experts for technological learning	Learn from reverse engineering, industrial espionage, or scientific published sources
	Knowledge shared among members of a trade as an "art and mystery"	Intellectual property and patent protection rewards a solitary inventor with a short-term monopoly
Environment	Manufacturers produce local, small-scale environmental impacts	Larger, widespread environmental impacts
	Manufacturers depend upon local resources, constrain and adapt their operations to work within parameters set by the market or natural rhythms	Manufacturers draw upon wider resources and implement methods such as stockpiling and long-term procurement contracts to control and guarantee resource flows
	Environmental conflicts are often local; addressed via self-correction or community mediation	Litigation more common; larger scale of impacts affects more stakeholders

Appendix 4
Selected Revere Engravings

Title	Date	Description
A View of the Year 1765	1765	An allegorical cartoon represents colonial representatives battling a dragon that symbolizes the Stamp Act.
A View of the Obelisk	1766	Representation of the large and ornate obelisk erected by the Sons of Liberty on the Boston Common to celebrate the repeal of the Stamp Act. A fire accidentally destroyed the obelisk, making this engraving extremely valuable as the surviving memory of its many patriotic messages.
A Warm Place—Hell	1768	A political caricature and companion piece to Revere's Liberty Bowl, this image portrays the punishment of the seventeen representatives who sided with the crown and voted to rescind a patriotic circular letter.
The Bloody Massacre	1770	The most famous of Revere's prints, this colorful and complex image portrays seven redcoats firing upon and killing members of an unarmed group of Bostonians during the Boston Massacre. This image controversially shows a smoking gun firing at the crowd from a window of "Butcher's Hall."
Landing of the Troops	1770	A detailed rendition of Boston's docks and buildings portrays a number of British warships surrounding the Boston harbor, conveying rows of troops marching onto Long Wharf.

Appendix 5
Furnace Startup Expenses for 1787–1788

The listing below is sorted from most to least expensive. Each entry combines all similar expenses from Revere's ledger. For example, ten different listings for nail purchases are combined into the "Nails" row. The largest expense, more than 53 pounds for iron, is not a startup cost in the strictest sense since Revere undoubtedly used this iron as casting material and not to build the furnace. However, he may have used iron for practice casts, and its inclusion in the startup costs makes more sense if we assume that he could not sell some of the output of these early attempts.

Expense	Pounds	Shillings	Pence
Iron (six entries)	53	17	1
Laborers' wages and board (twenty-seven entries)	29	8	5
Coal and carting (three entries)	14	9	4
Hay (eight entries)	12	9	7
Stones and transport (three entries)	7	19	
Nails (ten entries)	6	6	11
Sand (five entries)	4	7	
Stone lime (two entries)	4	1	
Clay	3	9	9
Trips to Halifax, Marblehead, Menotomy, and Providence (four entries)	3	4	4
Drink, liquor (five entries)	2	16	2
Merchantable boards	1	1	
Wooden bands		18	
Cattle use (two entries)		17	
Whisking stones		15	
Rings for truck wheels		12	
Stock Lock & Bolt (two entries)		8	2
Hire and transport of windlass and tub		8	
Brick molds		6	
Slates		5	4
Letters (two entries)		3	
Trucking		3	
Weigh cannon		3	
Hooks and hinges		2	6
Total	148	10	7

Appendix 6
April 1796 Payments to Faxon

The tasks listed above the first subtotal required large equipment and specialized skills. The two largest charges above the first subtotal correspond to the boring and turning work on Revere's first two contracts: ten howitzers for the federal government, and twelve smaller cannon for Massachusetts. The second subtotal includes an additional charge of more than 430 pounds for drawing copper bolts. *Note:* Some of Revere's superscript notations have been removed, and parenthetical comments were added by the author.

1795 Paul Revere to Elib Faxon	Debit (pounds–shillings–pence)
To boring, turning, & filing 10 howitzers @ £7 10s	75–0–0
10 Howitzers	
To cutting of 3 Sprues	1–16–0
To cleaning 5 Howitz's @ 12	3–0–0
To turning & boring a Cillinder	4–10–0
Aug. To Turning & boring boring & filing 12 pieces	
of cannon @ £3 each	36–0–0
To finishing a Mortar	0–18–0
To Carting a Cillinder	0–7–6
(subtotal)	£ 121–11–6
To drawing copper bolts (detailed numbers hard	
to read)	430–2–9
(total)	£ 551–14–3
Credit By his account of this date	281–16–4
Ballance	
Boston April 2, 1796 Rec'd ballance in full	£ 269–17–11

Appendix 7
Revere's Second Letter to Benjamin Stoddert, February 26, 1800

Dear Sir,

In Decem 1798 I had the honor of addressing a letter to you respecting Malleable Copper, wherein I mentioned that I manufactored either old or new copper into Bolts, Spikes, or any Matereals that was wanted in Ship Building.

In consequence of some conversations with Col. Humphries, since his arrival in Boston, I have been induced to trouble you again on that subject. I learn by that Gentleman, that there are no persons in either Philadelphia or New York that can make copper so Malable that it can be drawn in Bolts, Spikes, +c under the Hammer.

Within the last 16 months I have Manufactored into Bolts, Spikes, +c, upward of 10000 thousand weight for the Ships Boston & Essex. And upward of 13000 lb for the Merchants of Boston & Salem. I melt it into Pigs of 250 lbs each & draw it hot down to the size I want, some of it I have done into 10 penny nails.

Col Humphries tells me you have it much at heart to finish all the Ships built for our government with copper from mines in the United States and that there is a mine in Maryland & another in the Jerseys which produce a large quantity of Copper Ore but they cannot find any person that can smelt it so as to make it Malleable. I have never tryed, but from the experiments I have made I have no doubt I can do it & if Government will send me ten or fifteen hundred weight of Oar & will pay my expenses I will Build a Furnace on purpose this furnace is different from a Common Air Furnace & will endeavor to perfect myself in it. I have two sons who are concerned with me, if I make my self Master of the Business I will teach them. If these matters are worthy of your Notice, you know my character from his Excellency, the President of the United States, & from the Hon. Harrison G. Otis, member of Congress, to whom I am personally known.

If Government are not provided with bolts & spikes for the Ship which is to be built here, I shall be glad to supply one, or two, tons; which I suppose will be the most I shall be able to git old copper for. But if Government can purchase old copper sufficient for the whole, I will undertake to make all the Bolts & Spikes & cast work for her. The wrought work such as Bolts, Spikes, Dove tails, Brace Nails, +c for which I find the copper at 50 cents pr pound the cast work at 41 cents. I will allow 26 cents pr pound for all copper turned in by Government.

Col. Humphries has been at my works & after examining the Bolts & Spikes +c. He had the frankness & goodness to say that He saw my letter to you & that he really believed He was the cause why there were no attention paid to it, for he then thought that no person in America could do it & but few in Europe.

Appendix 8
Employee Salaries, 1802–1806

Name	Years of Employment	Salary
Willaby Dexter (foreman)	1802–1803	$45/month raised to $49/month
Jeremiah Dexter (foreman)	1802–1806	$1/day raised to $2/day
Amos Fales	1802–1804	$14/month
Jeremiah Vase	1802–1803	$11/month
Joseph Pettier	1802–1803	$1/day
John Story	1802–1805	$10.50/month raised to $16/month
Asa Smith	1803–1807	$7/month raised to $12/month
Otis Withington	1804	Approx. $14/month
Jason Clap	1804–1806	$150/year raised to $16/month
Epparim Leonard	1804	$17/month
Enos Withington	1804–1806	$17/month raised to $18/month
Benaiah Wilder	1805	$16/month raised to $18/month
Thomas Pattersole	1805–1806	$15/month
Isaac Bosworth	1806	$1/day
Timothy Allen	1806	$1/day
Josiah Johnson	1806	Approx. $.66/day

Appendix 9

Typical Stages in the Growth of a Large Technological System

Stage	Description
Invention	An inventor-entrepreneur develops a "radical" invention suitable for development into a larger system. The invention might only exist as an idea at this early stage.
Development	The inventor-entrepreneur expands and modifies the original invention, adding "conservative" inventions that render it more suitable to real-world conditions. The invention becomes socially constructed at this point, because developers give it economic, political, and social characteristics.
Innovation	Managers become more prominent, and help the inventor-entrepreneur combine all system components into a coherent whole that starts functioning. Innovation might include the need to hire workers, advertise, produce goods, offer services, or perform any other vital tasks.
Transfer	System developers now expand the system by opening up new facilities or branching into new markets. Transfer entails a reevaluation of the system because it was initially developed with its original environment in mind.
Growth, Competition, and Consolidation	Growing systems come into contact with other technological systems, and also need to optimize interactions with external constituencies such as regulatory or policy-making bodies, natural resource suppliers, and sources of funding. Growing systems attempt to maximize their efficiency and reduce uncertainty by integrating vital external components into the system.

Notes

Introduction

1. Carl Bridenbaugh, *The Colonial Craftsman* (New York: New York University Press, 1950), p. 87.

2. Appendix 1 recaps some of the highlights from Paul Revere's life, intertwined with the larger national context.

3. This freedom and the rapidity of change must not be overstated. America remained completely within the British mercantile sphere well into the 1800s, and American cycles of prosperity and depression were directly related to European wars and Britain's willingness to retract credit or flood the American market with inexpensive goods. In 1775 British America still operated in a preindustrial manner that combined almost medieval technologies with a colonial economy. See John F. Kasson, *Civilizing the Machine* (New York: Hill and Wang, 1979); Carrol W. Pursell Jr., *Technology in America* (Cambridge, Mass.: MIT Press, 1981); Eric Hobsbawm, *The Age of Revolution* (New York: Vintage Books, 1962); and John McCusker and Russell Menard, *The Economy of British America* (Chapel Hill: University of North Carolina Press, 1985), esp. pp. 70, 327.

4. The nationwide transition between craft and industrial labor is discussed in many excellent works, including (but by no means limited to) Thomas C. Cochran, *Frontiers of Change: Early Industrialism in America* (New York: Oxford University Press, 1981); Kasson, *Civilizing the Machine*; Walter Licht, *Industrializing America* (Baltimore: Johns Hopkins University Press, 1995); Jonathan Prude, *The Coming of Industrial Order* (Cambridge: Cambridge University Press, 1983); and Merritt Roe Smith, *Harpers Ferry Armory and the New Technology* (Ithaca: Cornell University Press, 1977).

5. For example, mill villages, initially organized around textile production in New England, usually offered a mix of paternalism and economic opportunity to the many families involved in the putting-out process, a system allowing merchants to coordinate the division of labor. Industrial single-product cities such as Lowell (textiles) or Lynn (shoes) featured large outlays of capital, extensive mechanization, collection of labor into factories or huge workshops, and modern business management practices. And diversified manufacturing centers also appeared in urban areas not dominated by a single production line, featuring a mix of smaller factories, home manufactures, and artisan shops that produced great product variety in numerous work settings. Licht, *Industrializing America*, pp. 22–34.

6. According to proto-industrial theory, agricultural productivity also increased as a result of this shift to manufacturing because more laborers left their marginal farms to engage in manufacturing, and population grew thanks to lowered ages of marriage among laborers with higher incomes. Sheilagh Ogilvie and Markus Cerman, "The Theo-

ries of Proto-industrialization," in *European Proto-industrialization*, ed. Sheilagh Ogilvie and Markus Cerman (Cambridge: Cambridge University Press, 1996), pp. 1–5. The theory's initial formulation can be found in Franklin F. Mendels, "Proto-Industrialization: The First Phase of the Industrialization Process," *Journal of Economic History* 32, no. 1 (March 1972): 241–261, and additional modifications are presented in Gay L. Gullickson, "Agriculture and Cottage Industry: Redefining the Causes of Proto-Industrialization," *Journal of Economic History* 43, no. 4 (December 1983): 831–850.

7. As a predictive theory proto-industry has many failings: for example, its projected impacts such as factory expansion and population growth do not always occur. Regardless of the theory's dubious ability to explain European industrialization, its applicability to America is far weaker. American farmers had access to large quantities of land, especially in the early years, and these farmers engaged in land clearing, animal raising, or other high-paying jobs in addition to household manufacturing. Therefore, the impact of household manufacturing on their transition toward factory labor is small. These farmers had little trouble producing enough food for the growing factory population, and also had the income to buy manufactured goods from the start. Kenneth Pomeranz, *The Great Divergence: China, Europe, and the Making of the Modern World Economy* (Princeton: Princeton University Press, 2001), pp. 288–289. Proto-industrial weaknesses are described in D. C. Coleman, "Proto-Industrialization: A Concept Too Many," *Economic History Review* 36, 2nd series (August 1983): 443–447; S. R. Epstein, "Craft Guilds, Apprenticeship, and Technological Change in Preindustrial Europe," *Journal of Economic History* 58, no. 3 (September 1998): 706; Kenneth Morgan, *The Birth of Industrial Britain* (New York: Longman, 1999), pp. 33–38; Sheilagh Ogilvie and Markus Cerman, "Proto-industrialization, Economic Development and Social Change in Early Modern Europe," in *European Proto-industrialization*, ed. Sheilagh Ogilvie and Markus Cerman (Cambridge: Cambridge University Press, 1996), pp. 228–229.

8. Appendix 2 illustrates some of the connections that unify the capital, labor, technology, and environmental factors.

CHAPTER ONE: Artisan, Silversmith, and Businessman (1754–1775)

1. The phrase "and the world he lived in" is taken, with great respect, from the title of Esther Forbes's seminal work, *Paul Revere and the World He Lived In* (Boston: Houghton Mifflin, 1969). Some information about Revere's arrangement with Copley was taken from Deborah Anne Federhen, "Paul Revere, Silversmith: A Study of His Shop Operation and His Objects" (master's thesis, University of Delaware, 1988), p. 34, and also from Revere's daybooks in "Waste Book and Memoranda (1761–1783)," Revere Family Papers (hereafter RFP), microfilm edition, 15 reels (Boston: Massachusetts Historical Society, 1979), reel 5, vol. 1.

2. The terms *silversmith* and *goldsmith* were completely interchangeable at this time. Because all American artisans using this title worked far more with silver than gold, I use the term *silversmith* throughout this book.

3. Patrick M. Leehey, "Reconstructing Paul Revere; an Overview of his Ancestry, Life, and Work," in *Paul Revere—Artisan, Businessman, and Patriot* (Boston: Paul Revere Memorial Association, 1988), pp. 21–23.

4. The "Bumpkin" quote is taken from a January 12, 1775 letter from Paul Revere's cousin John Rivoire to Revere, in "Loose Manuscripts 1746–1801," reel 1, RFP. John

Rivoire is quoting an earlier letter from Paul in this letter, so apparently Paul offered this reason for his father's name change. Also see Leehey, "Reconstructing Paul Revere," pp. 19–21. To distinguish him from his son Paul, the "midnight rider" and subject of this book, the name Apollos Rivoire will be used throughout this book even though he was referred to as Paul Revere after a certain point.

5. The North End was the poorest section of town in 1771 and declined further by 1790. Allan Kulikoff, "The Progress of Inequality in Revolutionary Boston," *William and Mary Quarterly* 28, no. 3 (July 1971): 397–398; Russell Blaine Nye, *The Cultural Life of the New Nation, 1770–1830* (New York: Harper Torchbooks, 1960), p. 126.

6. These agricultural improvements included policies, practices, and technologies: the adoption of the enclosures policy shifted land ownership from commonly to privately held fields; new crop rotation practices allowed farmers to replenish their soils without losing productivity; the development of the seed drill gave farmers the ability to seed fields more efficiently; and improvements to plowing technology, including the use of a lighter and stronger iron plow to better lift and break up soil.

7. The preceding analysis represents a distillation of the first chapter of Lawrence A. Peskin's *Manufacturing Revolution: The Intellectual Origins of Early American Industry* (Baltimore: Johns Hopkins University Press, 2003). Note that the term *mercantilism* was not used until the late 1700s, but British economists advocated some aspects of mercantilist economic theory centuries earlier.

8. Margaret Ellen Newell, "The Birth of New England in the Atlantic Economy," in *Engines of Enterprise: An Economic History of New England*, ed. Peter Temin (Cambridge, Mass.: Harvard University Press, 2000), pp. 11, 59, 61; and Kulikoff, "The Progress of Inequality," p. 376.

9. These categories contain a surprising number of gradations and overlaps: for example, most farmers also produced goods such as textiles or even practiced an artisan trade, and many hired laborers did so on a temporary basis while saving their earnings to buy land or start a trade of their own.

10. Marketplaces, or the locations for the buying and selling of goods and services, existed in some of the earliest human societies, and a market economy can be said to appear when a number of marketplaces fold into one another, creating price convergence, division of labor, and increasing returns. Capitalist attitudes and conditions existed throughout colonial America, but remnants of pre-capitalist practices still exerted an influence. For example, some colonists still used the concept of a "just price," or a price for goods and services determined by notions of fairness rather than profit maximization. In addition, many economic relations depended on personal relationships, and unpaid labor (indentured servants, apprentices, or slaves) affected labor rates as well. The analysis of early market growth and the corresponding spread of capitalism is explored in many sources, including Winifred Barr Rothenberg, *From Market-Places to a Market Economy* (Chicago: University of Chicago Press, 1992), p. 243; Winifred Barr Rothenberg, "The Invention of American Capitalism: The Economy of New England in the Federal Period," in *Engines of Enterprise: An Economic History of New England*, ed. Peter Temin (Cambridge, Mass.: Harvard University Press, 2000), pp. 78–79; Eric Foner, *Tom Paine and Revolutionary America* (New York: Oxford University Press, 1976), p. 26; Winifred Barr Rothenberg, quoted in Edwin J. Perkins, "The Entrepreneurial Spirit in Colonial America: The Foundations of Modern Business History," *Business History Review* 63, no. 1 (Spring 1989): 160; Richard Lyman Bushman, "Markets and Composite

Farms in Early America," *William and Mary Quarterly* 55, no. 3 (July 1998): 353, 354, 356, 363–365; Howard S. Russell, *A Long, Deep Furrow: Three Centuries of Farming in New England* (Hanover: University Press of New England, 1976), pp. 33, 57; Stuart Bruchey, *The Roots of American Economic Growth, 1607–1861* (New York: Harper, 1965), p. 44; Paul B. Hensley, "Time, Work, and Social Context in New England," *New England Quarterly* 65, no. 4 (December 1992): 546–551; Jan de Vries, "The Industrial Revolution and the Industrious Revolution," *Journal of Economic History* 54, no. 2 (June 1994): 254–260; and Gordon S. Wood, *The Radicalism of the American Revolution* (New York: Vintage Books, 1991), p. 135.

11. *Artisan* was the most popular term in the seventeenth and early eighteenth centuries, *craftsman* was most common in the late eighteenth century, and *mechanic* gained ascendancy after that. This study uses the term *artisan*, although the sources often use other terms. Alison Burford, *Craftsmen in Greek and Roman Society* (London: Thames and Hudson Ltd., 1972), p. 13; Thomas J. Schlereth, "Artisans and Craftsmen: A Historical Perspective," in *The Craftsman in Early America*, ed. Ian Quimby (New York: W. W. Norton, 1984), p. 37; James R. Farr, *Artisans in Europe: 1300–1914* (Cambridge: Cambridge University Press, 2000), pp. 3–6.

12. Paul A. Gilje, "Identity and Independence: The American Artisan," in *American Artisans: Crafting Social Identity*, ed. Howard B. Rock, Paul A. Gilje, and Robert Asher (Baltimore: Johns Hopkins University Press, 1995), p. xii; Schlereth, "Artisans and Craftsmen," p. 39; S. R. Epstein, "Craft Guilds, Apprenticeship, and Technological Change in Preindustrial Europe," *Journal of Economic History* 58, no. 3 (September 1998): 686, 701; Farr, *Artisans in Europe*, pp. 240–241.

13. *Iliad*, book 12, lines 342–345. Quote taken from Robert Fagles translation (New York: Penguin Classics Deluxe Edition, 1998), p. 335.

14. Burford, *Craftsmen in Greek and Roman Society*, pp. 67–89. The development of science in classical Greece contributed to the gendering of craftwork. Even though women performed highly skilled tasks such as weaving, this work was not held in the same regard as men's work. Many of the vital services provided by women—child raising, education, weaving, cooking—exited the domain of crafts. Richard Sennett, *The Craftsman* (New Haven: Yale University Press, 2008), p. 23.

15. Farr, *Artisans in Europe*, pp. 33–34, 37–40, 260–263, 264–268.

16. Guilds formed in the thirteenth century as part of an arrangement with local governments: in exchange for receiving quasi-legal authority to control local entrance to trades (among other benefits), guilds assumed responsibility for maintaining quality standards. Pressures from widening commercial activity weakened guilds in the sixteenth century, which caused England to react with the Statute of Artificers (1562), which backed guild practices with the authority of national law. Walter Licht, *Industrializing America* (Baltimore: Johns Hopkins University Press, 1995), pp. 48–49.

17. This emphasis upon honor applied to guilds as well, which often attempted to expand their social role by serving political, social, and administrative functions. Some guilds received authority from local government to collect taxes or rents or to help members with legal issues. Responsibility accompanied this authority: guilds enforced societal organization by publicly advocating the values of paternalism, hierarchy, and discipline. Guilds often took on the role of charitable societies or fraternities comprised of like-minded individuals seeking to perform good works. W. J. Rorabaugh, *The Craft Apprentice* (New York: Oxford University Press, 1986), pp. 4–5; Epstein, "Craft Guilds,"

pp. 684–687, 691–692; and Farr, *Artisans in Europe*, pp. 5–6, 20–21, 33–35, 228–230, 261–263.

18. The silversmiths' guild, known as the Worshipful Company of Goldsmiths of the City of London, was exceptionally powerful. More than five hundred years old by the time of the first American colonies, it maintained a complete monopoly in Britain, and oversaw all aspects of the trade. It had no authority in the colonies. Graham Hood, *American Silver, A History of Style, 1650–1900* (New York: Praeger Publishers, 1971), p. 15; Farr, *Artisans in Europe*, pp. 89–90; Epstein, "Craft Guilds," pp. 696, 701; Thomas Max Safley and Leonard M. Rosenband, *The Workplace before the Factory*, ed. Thomas Max Safley and Leonard M. Rosenband (Ithaca: Cornell University Press, 1993), pp. 4–5.

19. According to the 1790 census (America's first official population tally), only 5.1 percent of America's population, or a total of 201,655 individuals, lived in urban settings, compared to 3,727,559 rural residents. The urban population would remain under 10 percent until the 1840 census. The census defined an urban center as one with 2,500 or more inhabitants; 24 such centers existed in 1790. Most Americans did not emphasize book-learning, scientific knowledge, or artistic finesse, but did gain familiarity with construction and repair skills as well as general mechanical knowledge. American craftsmen had to learn to do whatever needed doing with any tools at hand. Christine Daniels, "'WANTED: A Blacksmith who understands Plantation Work': Artisans in Maryland, 1700–1810," *William and Mary Quarterly* 50, no. 4 (October 1993): 747; Gilje, "Identity and Independence," p. xii; Thomas C. Cochran, *Frontiers of Change: Early Industrialism in America* (New York: Oxford University Press, 1981), pp. 10, 14, 54.

20. Artisans thrived in the middle and northern colonies, and fared worse in the rural South, where foreign imports and the lure of agriculture discouraged both masters and apprentices. One author defines artisans as those who derived half or more of their income from nonfarm pursuits, and believes they constituted 10 to 15 percent of all colonial-era households, with a higher percentage in cities: one-third of all New York City households were headed by artisans according to this definition. Edwin J. Perkins, *The Economy of Colonial America*, 2nd ed. (New York: Columbia University Press, 1988), p. 116. Bridenbaugh estimates that artisans constituted 18 percent of the total colonial-era laboring population, representing the second largest occupational class next to farmers, and he claims that artisans may have comprised between one-third and one-half of employed urban dwellers. Carl Bridenbaugh, *The Colonial Craftsman* (New York: New York University Press, 1950), pp. 1, 30–32, 96, 127. Looking at smaller sample populations, one regional study contends that by the time of the Revolution, one-fourth of Connecticut men practiced some form of artisanal activity. Newell, "The Birth of New England in the Atlantic Economy," p. 62. Philadelphia had the largest number of crafts and craftsmen: 934 of 3,432 property owners were artisans in 1774. This number does not include all artisans, merely those heads of families who possessed real estate. Carl Bridenbaugh, *Cities in Revolt: Urban Life in America, 1743–1776* (London: Oxford University Press, 1955), p. 272.

21. Kulikoff, "The Progress of Inequality in Revolutionary Boston," p. 378; Bridenbaugh, *The Colonial Craftsman*, p. 127; Daniels, "Artisans in Maryland," pp. 753, 760–765.

22. The government eventually attempted to take on some of the duties of guilds such as quality control and prosecution of frauds. Bridenbaugh, *The Colonial Craftsman*, pp. 1, 126–127, 146; Rorabaugh, *Craft Apprentice*, pp. 8–9.

23. Foner, *Tom Paine and Revolutionary America*, pp. 28–29.

24. Richard Hofstadter, *America at 1750* (New York: Vintage Books, 1973), pp. 132–134; Bridenbaugh, *The Colonial Craftsman*, 155–156; Bridenbaugh, *Cities in Revolt*, p. 137; Nye, *Cultural Life of the New Nation*, pp. 132–133.

25. Wood, *Radicalism*, pp. 21–26; Leehey, "Reconstructing Paul Revere," p. 15; Rorabaugh, *Craft Apprentice*, p. 5; Gary B. Nash, *The Urban Crucible* (Cambridge, Mass.: Harvard University Press, 1986), pp. 163–166; Nye, *Cultural Life of the New Nation*, pp. 105–106, 109, 111, 115.

26. Gordon S. Wood, "The Enemy is Us: Democratic Capitalism in the Early Republic," *Journal of the Early Republic* 16, no. 2 (Summer 1996): 301; Wood, *Radicalism*, pp. 21–26, 195; Gordon S. Wood, *Revolutionary Characters: What Made the Founders Different* (New York: Penguin, 2006), pp. 14–15; Nash, *The Urban Crucible*, pp. 163–166. The colonies lacked a true gentry or aristocracy in the English sense. American gentlemen would hardly draw attention in comparison to the much greater authority and wealth commended by the four hundred or so noble families comprising England's hereditary and titled aristocracy. The wealthiest of Americans would, at best, be considered minor gentry in England, and most would simply fall under the category of gentlemen. Had they possessed such an incredibly wealthy and enduring hereditary class, colonial politics might have been far more stable: for example, Virginia had the closest approximation of a gentry class and had the most stable government. Wood, *Revolutionary Characters*, p. 12; Wood, *Radicalism of the American Revolution*, pp. 119–121.

27. Separation between colonial social classes was far from absolute. In Philadelphia, for example, artisans and merchants lived near each other and although most individuals carried out their transactions with their equals, a fairly frequent number of social and business interactions took place between these groups, including occasional intermarriages. Even though Americans had a large gap between rich and poor it was smaller than Europe's gap, and the chance to rise in society was larger in the United States. Wood, *Radicalism of the American Revolution*, p. 122; Charles S. Olton, "Philadelphia's Mechanics in the First Decade of Revolution, 1765–1775," *Journal of American History* 59, no. 2. (September 1972): 313; Kulikoff, "Progress of Inequality in Revolutionary Boston," pp. 376, 388, 390–391; Wood, *Revolutionary Characters*, p. 12; Nye, *Cultural Life of the New Nation*, pp. 105–107.

28. This negative view of manual labor extends to the Roman Empire, if not earlier. While Saint Augustine did claim that labor helped mankind fulfill God's plan for them, he also associated work with hierarchical servitude, and later European philosophers throughout the middle and early modern ages continued to cast physical labor as social control that helped the lower classes overcome idleness and vice. In colonial America, even a famous and respected painter such as John Singleton Copley was regarded by many of his wealthy clients merely as a skilled laborer, which infuriated him. These attitudes dramatically changed after the Revolution when the "useful arts" were viewed as a vital productive element in a successful enterprising society. Quote from Wood, *Radicalism*, p. 37. See also Bridenbaugh, *The Colonial Craftsman*, p. 96; Wood, *Radicalism*, p. 23; Wood, *Revolutionary Characters*, p. 15; Gilje, "Identity and Independence," p. xiii; Farr, *Artisans in Europe*, pp. 11–13.

29. Office holding was often cited as a societal obligation or burden; for example, Benjamin Franklin and others had to give up their lucrative business interests in order to be taken seriously as candidates. Adam Smith cited the British landed gentry as ideal

rulers because their assured income put them above the demands and influences of the marketplace. Alexander Hamilton concurred in *Federalist No. 35*, contending that lawyers and other scholarly professionals rose above the selfish interests of the marketplace and therefore deserved positions of leadership. Wood, *Revolutionary Characters*, pp. 16–17, 20, 237–238.

30. Historian Gordon Wood contends that Revolutionary-era artisans had difficulty "establishing their self-esteem and worth in the face of the age old scorn in which their mean occupations were held." Wood, *Radicalism*, p. 278; Billy G. Smith, *The "Lower Sort": Philadelphia's Laboring People, 1750–1800* (Ithaca: Cornell University Press, 1990), pp. 138–142.

31. In its strongest incarnation, the artisan perspective took an almost Marxist stance, equating an artisan's skill and service with the property that most non-artisans accepted as the determinant of wealth and status. Ronald Schultz, *The Republic of Labor: Philadelphia Artisans and the Politics of Class, 1720–1830* (New York: Oxford University Press, 1993), p. 5; Ronald Schultz, "The Small Producer Tradition and the Moral Origins of Artisan Radicalism in Philadelphia, 1720–1810," *Past and Present*, no. 127 (May 1990): 87–89, 99; Foner, *Tom Paine and Revolutionary America*, p. 40; Bridenbaugh, *The Colonial Craftsman*, pp. 165–173; Gilje, "Identity and Independence," p. xii; Howard B. Rock, *Artisans of the New Republic* (New York: New York University Press, 1984), p. 8.

32. As cities achieved greater populations and concentrations of wealth, craft trades split into subcategories with increasing specialization. For example, blacksmiths split into at least a dozen different sub-trades, including ornamental ironworkers, coppersmiths, whitesmiths, or anchor forgers. Bridenbaugh, *Cities in Revolt*, pp. 272–273; Wood, "The Enemy is Us," p. 300; Rorabaugh, *Craft Apprentice*, p. 6; Barbara McLean Ward, "Boston Goldsmiths, 1690–1730," in *The Craftsman in Early America*, ed. Ian Quimby (New York: W. W. Norton and Co., 1984), p. 129; Nash, *The Urban Crucible*, pp. 235–239.

33. Nash, *The Urban Crucible*, pp. 235–239; Kulikoff, "Inequality in Revolutionary Boston," p. 387; Bridenbaugh, *The Colonial Craftsman*, p. 162; Farr, *Artisans in Europe*, pp. 113–114; Sharon V. Salinger, "Artisans, Journeymen, and the Transformation of Labor in Late Eighteenth-Century Philadelphia," *William and Mary Quarterly* 40, no. 1 (January 1983): 63; Foner, *Tom Paine and Revolutionary America*, pp. 28–29, 32.

34. In general, these "wealthiest practitioners" had net worths valued in the tax records 225 percent higher than the average wealth in each group. Kulikoff, "Progress of Inequality in Revolutionary Boston," pp. 385–388; Jeannine Falino, "'The Pride Which Pervades thro every Class': The Customers of Paul Revere," in *New England Silver & Silversmithing*, ed. Jeannine Falino and Gerald W. R. Ward (Boston: University Press of Virginia, 2001), p. 159.

35. McLean Ward, "Boston Goldsmiths," pp. 131–138; Kathryn C. Buhler, *American Silver from the Colonial Period through the Early Republic in the Worster Art Museum* (Worster: Worster Art Museum, 1979), pp. 10–11; Kathryn C. Buhler, *American Silver in the Museum of Fine Arts* (Meriden: Museum of Fine Arts, 1972), p. 39; Hood, *History of Style*, p. 74.

36. Leehey, "Reconstructing Paul Revere," pp. 19–20. No records reveal whether he borrowed money to pay this indenture fee and buy the initial equipment and raw materials needed to start a silversmith shop. Perhaps he paid it from his journeyman income.

37. Janine E. Skerry, "The Revolutionary Revere: A Critical Assessment of the Silver

of Paul Revere," in *Paul Revere—Artisan, Businessman, and Patriot* (Boston: Paul Revere Memorial Association, 1988), pp. 43–46.

38. Skerry, "The Revolutionary Revere," p. 45; Stephanie Grauman Wolf, *As Various as their Land* (New York: HarperCollins, 1993), p. 184; McLean Ward, "Boston Goldsmiths," p. 147; Farr, *Artisans in Europe*, pp. 52–53.

39. Leehey, "Reconstructing Paul Revere," p. 25; Wolf, *As Various as their Land*, pp. 124–128; Bridenbaugh, *The Colonial Craftsman*, p. 130; Jayne E. Triber, *A True Republican* (Amherst: University of Massachusetts Press, 1998), p. 13. Many schools required students to fill out preprinted pages of mathematical exercises in "copybooks" that would later be retained as a reference guide. Although Revere's surviving records do not include a copybook, he might have used one to learn to add, subtract, multiply, and divide in a manner consistent with the complex and often inconsistent currency and measurement units of his day. Patricia Cline Cohen, *A Calculating People: The Spread of Numeracy in Early America* (Chicago: University of Chicago Press, 1982), pp. 120–123.

40. In America, merchants, lawyers, doctors, and clergymen also acquired their skills through versions of apprenticeships (although they were not formally called apprenticeships), studying and practicing under an established mentor for a fixed amount of time. Apprenticeship periods varied from 2 to 12 years in different countries, and apprentices began their training between ages 10 and 20. France and Germany typically adopted shorter terms than England, and certain trades required less training than others. After 1563, England's Statute of Artificers required at least seven years of apprenticeship, in part to correct the chronic unemployment problem. Exceptions in apprentice policies often took place even within the same trade: some apprentices never became masters while others did so without finishing their terms. Farr, *Artisans in Europe*, p. 35; Epstein, "Craft Guilds, Apprenticeship, and Technological Change," p. 689; Ian Quimby, "Some Observations on the Craftsman in Early America," in *The Craftsman in Early America*, ed. Ian Quimby (New York: W. W. Norton, 1984), p. 5; Edmund S. Morgan, *American Slavery, American Freedom; the Ordeal of Colonial Virginia* (New York: W. W. Norton & Company, 1975), p. 66; Ruth Schwartz Cowan, *A Social History of American Technology* (New York: Oxford University Press, 1997), p. 47; Bernard Elbaum, "Why Apprenticeship Persisted in Britain but not in the United States," *Journal of Economic History* 59, no. 2 (June 1989): 339.

41. Henry J. Kauffman, *The Colonial Silversmith: His Techniques & His Products* (New York: Galahad Books, 1969), p. 43.

42. Bridenbaugh, *The Colonial Craftsman*, pp. 130–133; Jack Larkin, *The Reshaping of Everyday Life, 1790–1840* (New York: HarperPerennial, 1989), p. 282; John Marshall Phillips, *American Silver* (New York: Dover, 2001), p. 14; Rorabaugh, *Craft Apprentice*, p. 11; Deborah A. Federhen, "From Artisan to Entrepreneur: Paul Revere's Silver Shop Operations," in *Paul Revere—Artisan, Businessman, and Patriot* (Boston: Paul Revere Memorial Association, 1988), p. 72; Bruce Laurie, *Artisans into Workers* (New York: Hill and Wang, 1989), p. 36; Howard B. Rock, *The New York City Artisan, 1789–1825* (Albany: State University of New York Press, 1989), pp. 193–194.

43. Increasing numbers of Revolutionary-era Americans opposed blatant signs of class distinctions. For example, servants did not like to wear livery, and people of all classes expected the right to eat or wear what they pleased if they could afford it. All men expected to be equal before the law. Nye, *Cultural Life of the New Nation*, p. 108.

44. Bridenbaugh, *The Colonial Craftsman*, pp. 130–133; Rorabaugh, *The Craft Appren-*

tice, p. 4; Wood, *Radicalism of the American Revolution,* pp. 53, 162; Salinger, "Artisans, Journeymen, and the Transformation of Labor," p. 70; Schultz, *The Republic of Labor,* p. 40. Many historians have explored and debated the relationship between early American labor shortages and other factors, such as wages, apprenticeships, and societal views toward manual labor. This connection, known as the scarce labor or labor scarcity hypothesis, played some role in the development of early manufacturing but was far from the only one. The labor scarcity hypothesis is discussed in H. J. Habakkuk, *American and British Technology in the Nineteenth Century* (Cambridge: Cambridge University Press, 1962); Peter Temin, "Labor Scarcity and the Problem of American Industrial Efficiency in the 1850s," *Journal of Economic History* 26 (September 1966); Peter Temin, "Labor Scarcity in America," *Journal of Interdisciplinary History* 1 (Winter 1971).

45. Silversmiths used a hammer and punch to stamp their maker's mark (usually their initials or last name) on the bottom of all large finished pieces. This practice is discussed in more detail below. Rorabaugh, *Craft Apprentice,* p. 8; Federhen, "Artisan to Entrepreneur," p. 73; Laurie, *Artisans into Workers,* pp. 35–36; Schultz, *The Republic of Labor,* p. 6.

46. For example, while 40 to 50 percent of families in mid-eighteenth-century England depended on wage labor for support, only one in three American workers earned wages, and even fewer in the skilled trades. Most American apprentices in the colonial era soon graduated to master status, so the number of journeymen stayed relatively small. The number of journeymen gradually increased in Philadelphia and New York City, but remained less numerous than masters until wage labor gained prevalence between 1800 and 1830. These laborers never constituted a dispossessed or permanent "proletariat" in the modern sense because they often possessed skills and hoped that they might improve their position by entering into business for themselves someday. Gary J. Kornblith, "The Artisanal Response to Capitalist Transformation," *Journal of the Early Republic* 10 (Fall 1990): 316–318; Olton, *Philadelphia's Mechanics,* p. 315; Smith, *Lower Sort,* pp. 197–200; Salinger, "Artisans, Journeymen, and the Transformation of Labor," p. 75; Perkins, *Economy of Colonial America,* p. 116; John J. McCusker and Russell B. Menard, *The Economy of British America 1607–1789* (Chapel Hill: University of North Carolina Press, 1991), p. 246; Bridenbaugh, *The Colonial Craftsman,* p. 146; Rorabaugh, *Craft Apprentice,* pp. 8–9.

47. A surprising number of the daily rituals and breaks centered around alcohol, including alehouse socializing at lunch or after work and drinks consumed in the place of work. One common tradition was "Saint Mondays," in which many workers treated each Monday as a holiday that helped them recover from the weekend's excesses. Newly hired journeymen were expected to treat the entire shop to drinks on their first day, and apprentices often had the task of making runs to the tavern throughout the day to keep the shop jug full, facing occasional complaints that they "robbed the mail" on the way back to the shop. Foner, *Tom Paine and Revolutionary America,* pp. 36–37; Rock, *Artisans of the New Republic,* pp. 296, 299–301; Laurie, *Artisans into Workers,* p. 37; Laura Rigal, *The American Manufactory: Art, Labor, and the World of Things in the Early Republic* (Princeton, N.J.: Princeton University Press, 2001), p. 33; Robert B. Gordon, *American Iron, 1607–1900* (Baltimore: Johns Hopkins University Press, 1996), p. 2.

48. Sennett, *The Craftsman,* p. 38.

49. Paul Revere to John Rivoire (his cousin), October 6, 1781, in "Loose Manuscripts 1746–1801," reel 1, RFP.

50. Schultz, *Moral Origins of Artisan Radicalism*, p. 87; Falino, "The Customers of Paul Revere," p. 153; Triber, *A True Republican*, p. 21; Federhen, *Paul Revere, Silversmith*, p. 2.

51. Federhen, "Artisan to Entrepreneur," p. 65; Falino, "The Customers of Paul Revere," p. 164.

52. *Iliad*, book 23, lines 823–829, Robert Fagles translation, p. 582.

53. According to one rough estimate, approximately three hundred silversmiths practiced their craft in Boston, New York, and Philadelphia throughout the entire colonial period, and another two hundred practiced their craft in all other areas. Colonial Williamsburg Foundation Staff, *The Silversmith in Eighteenth Century Williamsburg* (Williamsburg, Va.: Colonial Williamsburg Foundation, 1956), p. 21. A more specific estimate states that Connecticut alone featured 13 practicing silversmiths prior to 1750, and 125 new ones from 1750 to 1790. Newell, "Birth of New England in the Atlantic Economy," p. 62; Phillips, *American Silver*, pp. 24–25.

54. Hood, *History of Style*, pp. 11, 58, 92, 127, 163; Hermann Frederick Clarke, "The Craft of Silversmith in Early New England," *New England Quarterly* 12, no. 1 (March 1939): 68–69.

55. Phillips, *American Silver*, pp. 11–13; *The Silversmith in Eighteenth Century Williamsburg*, pp. 14–15, 22; James A. Mulholland, *History of Metals in Colonial America* (Tuscaloosa: University of Alabama Press, 1981), p. 86; Bridenbaugh, *The Colonial Craftsman*, p. 86; Federhen, *Paul Revere, Silversmith*, p. 4. Silversmiths served a vital role in medieval Europe, as society's assayers and wealth estimators. In the absence of a single authoritative test of metal purity the silversmith's experience, judgment, and reputation earned society's trust and became a measuring stick and weapon against counterfeiters. Sennett, *The Craftsman*, p. 61.

56. McLean Ward, "Boston Goldsmiths," p. 126; Bridenbaugh, *The Colonial Craftsman*, p. 143; Hood, *History of Style*, p. 11; Clarke, "The Craft of Silversmith in Early New England," pp. 68–9; Skerry, "Revolutionary Revere," p. 43; Mulholland, *History of Metals in Colonial America*, pp. 87–88.

57. Many colonists, ranging from enlisted soldiers to political or economic leaders, voiced their annoyance with British attitudes. These complaints emphasized how the British commanders expected deference from their soldiers, as well as nearly all colonists, and treated them as inferiors. Revere would have bristled at this, in light of his later writings on the subject of meritocracy and deference. James A. Henretta and Gregory H. Nobles, *Evolution and Revolution: American Society 1600–1820* (Lexington, Mass.: D. C. Heath & Co., 1987), p. 115; Triber, *A True Republican*, pp. 23, 204. Revere's silver shop records for this period are conveyed in many entries in "Waste Book and Memoranda (1761–1783)," reel 5, vol. 1, RFP. See also Federhen, *Paul Revere, Silversmith*, p. 14.

58. Phillips, *American Silver*, pp. 14–16; Mulholland, *History of Metals in Colonial America*, pp. 88–89; *The Silversmith in Eighteenth Century Williamsburg*, pp. 21–22; Kauffman, *The Colonial Silversmith*, p. 42.

59. *Sterling silver* is an English term used to describe an alloy of silver and copper. The addition of copper to sterling silver adds a richer color to the silver while increasing its durability and workability; however, too much copper lowers the perceived value of the metal. English regulatory agencies (such as guilds and assaying offices) rigidly enforced the sterling standard of 92.5 percent silver and 7.5 percent copper. The quantity of silver and copper in American products was more variable than their English counterparts

because of the lack of regulatory mechanisms. One set of tests performed on a variety of early American spoons reveals copper contents between 10.5 and 14.5 percent. Phillips, *American Silver*, p. 20; Kauffman, *The Colonial Silversmith*, pp. 27–30.

60. Mulholland, *History of Metals in Colonial America*, pp. 88–89; Kauffman, *The Colonial Silversmith*, p. 31; *The Silversmith in Eighteenth Century Williamsburg*, pp. 24–25; Edwin Tunis, *Colonial Craftsmen* (Baltimore: Johns Hopkins University Press, 1965), pp. 82–83.

61. Kauffman, *The Colonial Silversmith*, p. 30; Mulholland, *History of Metals in Colonial America*, pp. 83, 88–89; Hood, *History of Style*, pp. 18–19; Tunis, *Colonial Craftsmen*, p. 84; *The Silversmith in Eighteenth Century Williamsburg*, pp. 24–26.

62. Tunis, *Colonial Craftsmen*, p. 82; *The Silversmith in Eighteenth Century Williamsburg*, pp. 24–26; Kauffman, *The Colonial Silversmith*, p. 15.

63. Mulholland, *History of Metals in Colonial America*, pp. 83, 88–89; Hood, *History of Style*, pp. 18–19; Tunis, *Colonial Craftsmen*, p. 84; *The Silversmith in Eighteenth Century Williamsburg*, pp. 24–26; Kauffman, *The Colonial Silversmith*, p. 62.

64. Phillips, *American Silver*, p. 19.

65. Gordon, *American Iron*, p. 2.

66. Skerry, "Revolutionary Revere," p. 47; Federhen, "Artisan to Entrepreneur," p. 66; Phillips, *American Silver*, p. 14. Benjamin Burt (1729–1805) was also the son of a silversmith and was older than Revere. Burt and Revere greatly outstripped the production recorded by four other contemporaries in the same period, that is, 47, 56, 90, and 105 silver objects. Some of these other silversmiths died early or lost business after siding with the British in the Revolution, and Burt failed to change with the times, sticking with traditional items. After the war he failed to expand and adapt the way Revere did: Revere made many new teapots, sugar bowls, and creamers, and Burt made few. Falino, "The Customers of Paul Revere," pp. 154–155.

67. The letter states, "I Cannot pay you for the Board of my dr Child till my Returne or the Returne of the Vessel which will be about 3 Months . . . This, I hope, will not be the means of my poor Childs Suffering." Josiah Collins to Paul Revere, November 22, 1774, "Loose Manuscripts 1746–1801," reel 1, RFP. This is a poignant example of the risk involved in trade endeavors and the interconnectedness of personal credit relationships—one late payment or bad debt could easily ripple throughout society. Federhen, *Paul Revere, Silversmith*, pp. 27–29.

68. Federhen, "Artisan to Entrepreneur," p. 69; Federhen, *Paul Revere, Silversmith*, p. 15.

69. Although (contrary to modern rumors) Revere never worked on George Washington's dentures, Patriot statesman John Jay was one of his early customers and the two corresponded in the following years. *The Silversmith in Eighteenth Century Williamsburg*, p. 5; Skerry, "Revolutionary Revere," p. 47; Kauffman, *The Colonial Silversmith*, p. 36; Federhen, *Paul Revere, Silversmith*, pp. 5, 26–27.

70. The most prevalent form of advertising in colonial America was the trade card, a small engraved paper that listed a craftsman's name, address, product line, and any distinguishing symbols that might appear on that shop's signboard. Craftsmen often pasted these cards on cases used to hold the objects they created. Silvio A. Bedini, *Thinkers and Tinkers: Early American Men of Science* (New York: Scribner, 1975), pp. 224–225.

71. Bridenbaugh, *The Colonial Craftsman*, p. 99; Clarence S. Brigham, *Paul Revere's*

Engravings (Worster: American Antiquarian Society, 1954), pp. 4, 133; many entries in "Waste Book and Memoranda (1761–1783)," reel 5, vol. 1, RFP (see in particular May 3, 1774; July 9, 1774; May 2, 1781).

72. The quote describing Revere's teapot is taken from Phillips, *American Silver*, p. 101. Also see Daniels, "Artisans in Maryland," p. 748; Gloria L. Main, "The Standard of Living in Southern New England, 1640–1773," *William and Mary Quarterly* 45, no. 1 (January 1988): 125.

73. The exception to this generalization is the lowest "unpropertied" class, which actually grew in numbers and faced increasingly desperate conditions as society became more stratified. Kulikoff, *Progress of Inequality in Revolutionary Boston*, pp. 380–381; Schultz, *The Republic of Labor*, pp. 38–39; Smith, *Lower Sort*, pp. 197–200.

74. Wolf, *As Various as their Land*, p. 65; Falino, "The Customers of Paul Revere," pp. 152–153; Main, "The Standard of Living in Southern New England," pp. 127–129; Kulikoff, *Progress of Inequality in Revolutionary Boston*, p. 375.

75. Falino, "The Customers of Paul Revere," pp. 153–154, 156–157.

76. Phillips, *American Silver*, pp. 101, 108; Skerry, "Revolutionary Revere," pp. 48–49.

77. After Parliament repealed most of the Townshend duties in 1770, the tax on tea was not removed, and the Tea Act of 1773 made matters even worse by granting a monopoly on colonial tea sales to the East India Company. Although upper-class women were fairly active in the pre-Revolutionary tea boycott, several letters of complaint written to colonial newspapers illustrate that this was no small sacrifice. If men were allowed to continue drinking imported rum, argued some women, why must they lose their own social ritual? Loyalist Peter Olivier joked that, although most ladies agreed to only drink tea they already possessed when they needed it to forestall sickness, "they were cautious enough to lay in large Stocks before they promised; & they could be sick just as suited their Convenience or Inclination." Wolf, *As Various as their Land*, pp. 80–81, 85; Federhen, *Paul Revere, Silversmith*, pp. 9–11.

78. Federhen, "Artisan to Entrepreneur," pp. 67–68; Skerry, "Revolutionary Revere," p. 51; Falino, "The Customers of Paul Revere," p. 174. Dr. Paine's tea service was a gift to his new wife, Lois Orne, a distant relative of Revere's wife Sarah Orne.

79. Federhen, "Artisan to Entrepreneur," p. 70; Federhen, *Paul Revere, Silversmith*, pp. 19–22, 25, 30–31.

80. Revere made his most famous creation, one of the most famous pieces of silver ever designed, to commemorate an important Patriot event. In June 1768, the Massachusetts General Court disobeyed a direct order from Secretary of State Hillsborough by voting not to rescind a circular letter protesting the Townshend Act. Fifteen members of the Sons of Liberty who took part in this vote commissioned Revere to make a large engraved bowl celebrating the event. The result, called the Sons of Liberty Bowl (or merely Liberty Bowl), is a punch bowl bursting with engraved patriotic slogans and images such as a liberty cap, the Magna Carta, and three images relating to British reformer John Wilkes. Revere took the bold step of placing his maker's mark squarely at the center of the piece, exposing himself to charges of sedition. As a complementary finishing touch to this political statement Revere designed and engraved *A Warm Place—Hell*, a copperplate print depicting the seventeen house members who voted to rescind as they sadly marched toward the gaping jaws of hell, funded by fifteen patrons whose names feature prominently around the bowl. Edwin J. Hipkiss, "A Note on the Origin of the Paul Re-

vere Liberty Bowl," Museum of Fine Arts pamphlet, February 16, 1949; Brigham, *Paul Revere's Engravings*, p. 136; Triber, *A True Republican*, pp. 64–65; McLean Ward, "Boston Goldsmiths," p. 130; Falino, "The Customers of Paul Revere," p. 172.

81. He earned a high of more than 294 pounds in 1762, and recorded an income of only 11 pounds in 1770. However, this low figure is extremely dubious. Revere did not record income from dentistry and other sources, and in general his records are incomplete, particularly when he was occupied with Revolutionary activities such as in 1770. "Waste Book and Memoranda (1761–1783)," reel 5, vol. 1, RFP; described and analyzed further in Triber, *A True Republican*, p. 209.

82. Perkins, *The Economy of Colonial America*, p. x ; Triber, *A True Republican*, pp. 35, 71; Federhen, "Artisan to Entrepreneur," p. 69; Federhen, *Paul Revere, Silversmith*, p. 15; Nash, *The Urban Crucible*, pp. 207–208.

83. Quote taken from Elbridge Henry Goss, *The Life of Colonel Paul Revere* (Boston: Plimpton Press, 1902), p. 110. See also Triber, *A True Republican*, pp. 90–92; Forbes, *Paul Revere*, pp. 183–186.

84. Revere's uncles Thomas and Nathaniel Hitchborn were boat builders as were his cousins Robert and Thomas Jr. Cousin Benjamin was a lawyer and cousin William was a hatter. In-laws, such as cousin-in-law Nathaniel Fosdick, also offered him their business. Federhen, *Paul Revere, Silversmith*, p. 32.

85. Falino, "The Customers of Paul Revere," pp. 169–173; Federhen, "Artisan to Entrepreneur," pp. 74–75.

86. Federhen, *Paul Revere, Silversmith*, p. 33; Steven C. Bullock, *Revolutionary Brotherhood: Freemasonry and the Transformation of the American Social Order, 1730–1840* (Chapel Hill: University of North Carolina Press, 1996), pp. 95–97; Triber, *A True Republican*, p. 83.

87. Triber, *A True Republican*, p. 85; Larkin, *Reshaping of Everyday Life*, p. 44; Falino, "The Customers of Paul Revere," pp. 153, 160; Skerry, "Revolutionary Revere," p. 47; Federhen, "Artisan to Entrepreneur," p. 74; Federhen, *Paul Revere, Silversmith*, pp. 29–31.

88. Silver cups were indispensable status symbols in upper-class households, often prominently displayed in sideboards, which were eventually known as cupboards. Clarke, "The Craft of Silversmith," p. 74.

89. Federhen, "Artisan to Entrepreneur," pp. 69, 75; Federhen, *Paul Revere, Silversmith*, pp. 16–17. Revere probably chose to import forks and knives because they required more labor than spoons.

90. Bridenbaugh, *The Colonial Craftsman*, pp. 1, 126–127.

91. Farr, *Artisans in Europe*, pp. 58–59; Daniels, "Artisans in Maryland," pp. 754–755; Wood, *Radicalism*, pp. 64, 66–67; Bridenbaugh, *The Colonial Craftsman*, p. 154; Larkin, *Reshaping of Everyday Life*, p. 37; Jonathan Prude, *The Coming of Industrial Order* (Cambridge: Cambridge University Press, 1983), p. 11; Federhen, *Paul Revere, Silversmith*, p. 7.

92. Naomi R. Lamoreaux, "Rethinking the Transition to Capitalism in the Early American Northeast," *Journal of American History* 90, no. 2 (September 2003): 3–4.

93. Nathan Rosenberg and E. Birdzell Jr., *How the West Grew Rich* (New York: Basic Books, 1987), pp. 126–127.

94. Bridenbaugh, *The Colonial Craftsman*, p. 129; Larkin, *Reshaping of Everyday Life*, pp. 43–44; Lamoreaux, "Rethinking the Transition to Capitalism," p. 6.

CHAPTER TWO: Patriot, Soldier, and Handyman of the Revolution (1775–1783)

1. Robert Morris and John Dickinson letter to Oswell Eve, November 21, 1775, in *Letters of Delegates to Congress, 1774–1789*, vol. 2 (September 1775–December 1775), ed. Paul H. Smith et al. (Washington, D.C.: Library of Congress, 1976–2000). This resource is also available online at the Library of Congress website at http://memory.loc.gov/ammem/amlaw/lwdg.html.

2. Samuel Adams letter to Elbridge Gerry, January 2, 1776, in *Letters of Delegates to Congress*, vol. 3 (January 1776–May 1776), ed. Smith et al.

3. Carl Bridenbaugh, *Cities in Revolt: Urban Life in America, 1743–1776* (London: Oxford University Press, 1955), pp. 251–252. Britain's debt in 1763 surpassed 122 million pounds, with yearly interest payments on this debt exceeding 4 million pounds. In the first few years after the war this debt increased by 7 million pounds a year, and the people of England were already crushed by a massive tax burden Parliament imposed in earlier years to fund the war. Robert Middlekauff, *The Glorious Cause: The American Revolution, 1763–1789* (New York: Oxford University Press, 1985), p. 57.

4. One Boston merchant's club in the 1760s organized the Society for Encouraging Trade and Commerce within the Province of Massachusetts Bay. Similar groups soon formed in the other large colonial towns. Over the years these groups stayed in touch, and they eventually coordinated common actions such as non-importation movements. Bridenbaugh, *Cities in Revolt*, p. 282.

5. Many artisans became economic nationalists, deeply resenting merchants who brought cheap British manufactured goods to local markets. Eric Foner, *Tom Paine and Revolutionary America* (New York: Oxford University Press, 1976), pp. 33–34; Lawrence A. Peskin, *Manufacturing Revolution: The Intellectual Origins of Early American Industry* (Baltimore: Johns Hopkins University Press, 2003), pp. 28–29; Middlekauff, *Glorious Cause*, pp. 27–31; T. H. Breen, *The Marketplace of Revolution: How Consumer Politics Shaped American Independence* (New York: Oxford University Press, 2004), p. xv.

6. Middlekauff, *Glorious Cause*, pp. 77–83; John F. Kasson, *Civilizing the Machine* (New York: Hill and Wang, 1979), p. 9; Peskin, *Manufacturing Revolution*, pp. 28–29.

7. Quote taken from Pauline Maier, Merritt Roe Smith, Alexander Keyssar, and Daniel Kevles, *Inventing America* (New York: W. W. Norton, 2003), p. 171; Maier, *Resistance to Revolution*, p. 297. During this time period, resistance groups realized that they needed a broader base of popular support to properly justify their ability to act on behalf of "the people." This led to membership changes, modified ideologies, and new types of outreach activities. Pauline Maier, *From Resistance to Revolution* (New York: W. W. Norton, 1991), pp. 58–59, 85–89; Jayne E. Triber, *A True Republican* (Amherst: University of Massachusetts Press, 1998), pp. 43–44; Middlekauff, *Glorious Cause*, pp. 91–94.

8. Artisans played an essential role throughout the colonial resistance movement, and commanded more economic, political, and social status than in any other nation at this time. Artisan dominance of Patriot organizations reflected late eighteenth-century demographics: artisans, shopkeepers, and tradesmen comprised about two-thirds of each urban setting. Although many artisans advocated for American nationalization and independence, this was surely not unanimous. Howard B. Rock, *Artisans of the New Republic* (New York: New York University Press, 1984), p. 20; Edwin J. Perkins, *The Economy of Colonial America*, 2nd ed. (New York: Columbia University Press, 1988), p. 122.

9. Charles S. Olton, "Philadelphia's Mechanics in the First Decade of Revolution, 1765–1775," *Journal of American History* 59, no. 2 (September 1972): 312, 320–321; Gordon S. Wood, *The Radicalism of the American Revolution* (New York: Vintage Books, 1991), pp. 244–247; Bridenbaugh, *Cities in Revolt*, pp. 282–283; Foner, *Tom Paine and Revolutionary America*, pp. 33–34, 41, 61–62; Maier, *Resistance to Revolution*, pp. 97–100.

10. Parliament repealed the Stamp Act in 1766 for various reasons, including the dismissal of the unpopular Prime Minister George Grenville and lobbying efforts by British merchants affected by colonial non-importation initiatives, although it saved face by passing a Declaratory Act confirming its right to implement laws that held authority over its colonies "in all cases whatsoever."

11. Breen, *Marketplace of Revolution*, pp. xvi–xviii; Peskin, *Manufacturing Revolution*, p. 42. Americans formed a number of communal associations starting in the early days of colonial resistance, in hopes of increasing the quantity of native manufactures. For example, in Boston the old Manufactory House used premiums to lure key students to a revitalized spinning house; New Yorkers initiated the Society for the Promotion of Arts, Agriculture, and Economy; and Philadelphia started a linen manufactory. Bridenbaugh, *Cities in Revolt*, p. 268; Kasson, *Civilizing the Machine*, p. 9.

12. Alfred Young, *The Shoemaker and the Tea Party: Memory and the American Revolution* (Boston: Beacon Press, 2000), pp. 54–55; Olton, "Philadelphia's Mechanics," pp. 317–319. Many artisans continued to advocate aggressively for what they perceived as class interests during and after the war. For example, artisans won four of the ten elected city offices in Philadelphia in 1770, and soon afterward the Philadelphia Committee of Nineteen (the organization taking the lead in the resistance movement) invited members from religious associations, mechanics, and ethnic representatives to its membership. Pennsylvania's rewritten constitution was influenced by artisans and included statements of universal male suffrage, short terms of office, and other features compatible with artisan ideology. Wood, *Radicalism of the American Revolution*, p. 245; Ronald Schultz, "The Small Producer Tradition and the Moral Origins of Artisan Radicalism in Philadelphia 1720–1810," *Past and Present*, no. 127 (May 1990): 99–100; Peskin, *Manufacturing Revolution*, pp. 31–33, 40.

13. Samuel Adams's efforts played a large part in keeping Boston government centered on the Town Meeting model, which remained until 1822, well after he died. Boston's artisan organizations continued to take a respectful and constructive tone in their approaches to merchants after the war. New York City artisans generally had more political influence than those in Boston or Philadelphia, as merchants and other gentry courted them. *Cities in Revolt*, pp. 6–7; Russell Blaine Nye, *The Cultural Life of the New Nation, 1770–1830* (New York: Harper Torchbooks, 1960), p. 125; Gary Kornblith, "Artisan Federalism: New England Mechanics and the Political Economy of the 1790s," in *Launching the "Extended Republic": The Federalist Era*, ed. Ronald Hoffman and Peter J. Albert (Charlottesville: University Press of Virginia, 1996), pp. 255–258; Sean Wilentz, *Chants Democratic* (New York: Oxford University Press, 1984), pp. 63–64.

14. Organization was facilitated by the actions of printers as well as mariners who spread word to different locations. Communication between the cities and regions improved steadily up through the Revolution, epitomized by the impressive U.S. Postal Service that included packet boats and non-crown options for patrons concerned with privacy. Most cities handled thousands of pieces of mail a year. Bridenbaugh, *Cities in Revolt*, pp. 282–283, 289–290.

15. Revere to John Rivoire, July 1, 1782, in "Loose Manuscripts 1746–1801," Revere Family Papers (hereafter RFP), microfilm edition, 15 reels (Boston: Massachusetts Historical Society, 1979), reel 1.

16. Triber, *A True Republican*, pp. 81, 85; David Hackett Fischer, *Paul Revere's Ride* (New York: Oxford University Press, 1994), pp. 22, 25.

17. Rallying song lyrics taken from Elbridge Henry Goss, *The Life of Colonel Paul Revere* (Boston: Plimpton Press, 1902), p. 128. Also see Triber, *A True Republican*, pp. 93–95; *Annals of the Massachusetts Charitable Mechanics Association* (Boston: Press of Rockwell and Churchill, 1892), p. 16; Edwin Griffin Porter, *Rambles in Old Boston* (Boston: Cupples, Upham, and Company, 1887), p. 98.

18. Goss, *The Life of Colonel Paul Revere*, p. 641. Revere might have joined the caucus earlier; their records begin in 1772.

19. Triber, *A True Republican*, pp. 62, 88; Goss, *The Life of Colonel Paul Revere*, pp. 635–642; Fischer, *Paul Revere's Ride*, p. 20.

20. Quotation taken from Revere's transcript of his Midnight Ride, submitted to Jeremy Belknap in 1798. This deposition is available in "Loose Manuscripts 1746–1801," reel 1, RFP, and online at the Massachusetts Historical Society website: http://www.masshist.org/cabinet/april2002/reveretranscription.htm. Revere's leadership role in this organization is also mentioned in Middlekauff, *Glorious Cause*, p. 265.

21. The Central Intelligence Agency's Center for the Study of Intelligence (CSI) describes this organization as "the first Patriot intelligence network on record." The CSI rightly describes this operation as "amateurish," illustrated by the fact that they often met in the same location, The Green Dragon Tavern, and allowed their group to be infiltrated by a British spy, Dr. Benjamin Church. Refer to the CSI website at https://www.cia.gov/library/ for more information.

22. Quotation found in Triber, *A True Republican*, p. 101. Triber, *A True Republican*, pp. 95–98; Fischer, *Paul Revere's Ride*, pp. 26–27.

23. Clarence S. Brigham, *Paul Revere's Engravings* (Worster: American Antiquarian Society, 1954), pp. 22–29, 43–47, 52–57, 79–81; Triber, *A True Republican*, pp. 79–80.

24. After the Revolution he was chosen as a health officer of Boston and coroner of Suffolk County, and he also helped found the Massachusetts Mutual Fire Insurance Company, among other minor service roles. Triber, *A True Republican*, pp. 99–101, 124, 126; Fischer, *Paul Revere's Ride*, pp. 16, 20; Bridenbaugh, *Cities in Revolt*, pp. 6–7; Kornblith, "Artisan Federalism," pp. 257–258.

25. David Hackett Fischer makes this point in an even more forceful manner in *Paul Revere's Ride*, pp. 301–302. Artisan organizations reappeared in the 1790s in the form of Democratic-Republican societies. Wood, *Radicalism*, pp. 244, 276.

26. The most authentic retelling of Paul Revere's ride takes place in his own words. Revere wrote a deposition for the Massachusetts Provisional Congress in 1775 and corrected it that same year. He went into even more detail in 1798 in a long letter to Jeremy Belknap, corresponding secretary of the Massachusetts Historical Society. All quotes in the following sections, unless otherwise noted, are taken from Revere's letter to Jeremy Belknap, "Loose Manuscripts 1746–1801," reel 1, RFP. The most careful reconstruction and interpretation of the Midnight Ride takes place in Fischer, *Paul Revere's Ride*, particularly chapters titled "The Mission," "The Warning," and "The Capture." Jayne Triber offers a concise and shrewd recap and analysis of the Midnight Ride in *A True Republican*, pp. 102–105.

27. Quotation found in Arthur Tourtellot, *Lexington and Concord: The Beginning of the War of the American Revolution* (New York: W. W. Norton, 2000), p. 114.

28. This information concerning the early interpretations of Revere's Midnight Ride is taken from the beginning of David Hackett Fischer's comprehensive historiographical essay included in pp. 327–341 of *Paul Revere's Ride*.

29. Tourtellot describes the battle on Lexington Green in *Lexington and Concord*, pp. 131–138, and the book deals with other aspects of the events of April 18–20 as well.

30. Letter quoted in Goss, *Life of Colonel Paul Revere*, p. 263; located in "Revere Family Papers II," reel 4, RFP. Also see Triber, *A True Republican*, pp. 110, 115.

31. Triber, *A True Republican*, pp. 110–111. Revere was not the only person forced into a different line of work. During the Revolution, many skilled artisans suffered major business disruptions due to the presence of armies in their towns or the decreased demand for manufactured items resulting from the distressed economy. Large numbers of these artisans entered the military or served related roles, such as privateering, working in armories, or surveying. Silvio A. Bedini, *Thinkers and Tinkers: Early American Men of Science* (New York: Scribner, 1975), pp. 238–243.

32. Artillery expertise was in such short supply in the Continental army that many gunners depended upon British treatises for basic instructions regarding the use and maintenance of heavy ordnance. Bedini, *Thinkers and Tinkers*, pp. 259–261. Washington's quote taken from Nye, *Cultural Life of the New Nation*, p. 106. George Washington came to realize that artisans, other members of the middle classes, and even foreigners attempted to pass as gentlemen, hoping to use the military as a vehicle for social advancement. James Martin and Mark Lender, *A Respectable Army: The Military Origins of the Republic, 1763–1789* (Wheeling, Ill.: Harlan Davidson, 2006), pp. 106–107; Gordon S. Wood, *Revolutionary Characters: What Made the Founders Different* (New York: Penguin, 2006), p. 19.

33. Revere's commissions to the Massachusetts militia and then to the artillery regiment and his promotion to the rank of lieutenant colonel are all located in "Loose Manuscripts 1746–1801," reel 1, RFP.

34. Many of Revere's military exploits and reproductions of some of his correspondence are reproduced in Goss, *Life of Colonel Paul Revere*, pp. 277–289.

35. Bedini, *Thinkers and Tinkers*, pp. 232, 298. Undated entries in "*Journal and Commonplace Book, 1777–1801,*" reel 14, vol. 51.1, RFP.

36. Revere's letter to Lamb quoted in Goss, *Life of Colonel Paul Revere*, p. 280. Samuel Adams's response to Revere is located in Samuel Adams to Paul Revere, in *Letters of Delegates to Congress*, vol. 7 (May 1, 1777–September 18, 1777), ed. Paul H. Smith et al., online at http://memory.loc.gov/ammem/amlaw/lwdg.html. See also Triber, *A True Republican*, pp. 124–128.

37. August 1778 letter from Paul Revere to Rachel Revere, "Revere Family Papers II," reel 4, RFP; also reproduced in Goss, *Life of Colonel Paul Revere*, pp. 305–307. Expedition details are also covered in Triber, *A True Republican*, pp. 130–133.

38. An overview of the Penobscot expedition is discussed in Triber, *A True Republican*, pp. 134–135, but the most authoritative and detailed source by far is George E. Buker, *The Penobscot Expedition* (Annapolis: Naval Institute Press, 2002). Chapters 3 and 4 cover the American assault, and chapter 5 describes the retreat.

39. Quote taken from Buker, *The Penobscot Expedition*, p. 148. The full treatment of Revere's trials and vindication is contained in great detail in chapter 8; see also Triber, *A*

True Republican, pp. 136–139. The investigating committee initially found Revere guilty of refusing to allow General Wadsworth the use of a boat under Revere's command, and dereliction in duty during the retreat when Revere marched his men from the Penobscot River back to Boston without specific orders. When Revere disputed these charges again in 1782, the court-martial found that although he did initially refuse the use of his boat, he almost immediately reversed his decision and gave it to the general. And the court threw out the second charge on the grounds that the entire expedition was in disarray during the retreat, and no orders could be effectively given or received.

40. Patriots became particularly dependent upon gunpowder imports, which accounted for more than 90 percent of all powder used during the war, as described in Orlando W. Stephenson, "The Supply of Gunpowder in 1776," *American Historical Review* 30, no. 2 (January 1925): 277. The American army's lack of scientific and technical expertise, particularly as applied to surveying, weapons manufacture, and the planning of fortifications, became so prominent that General Washington himself attempted to correct the deficit. He first petitioned John Hancock for relief, saying, "In a former part of my Letter I mentioned the want of Engineers. I can hardly express the Disappointment I have experienced on this Subject, the Skill of those we have being very imperfect." Benjamin Franklin later requested help from French engineers, some of whom joined the Continental army to help design defensive fortifications, construct maps, and generally supervise engineering activities. At the start of the war, the Continental army often had to do without these services. Bedini, *Thinkers and Tinkers,* pp. 245–246. See also Gary M. Walton and James F. Shepherd, *The Economic Rise of Early America* (New York: Cambridge University Press, 1980), pp. 179–180; Peskin, *Manufacturing Revolution,* pp. 57–59; Kasson, *Civilizing the Machine,* pp. 11–12; York, *Mechanical Metamorphosis,* pp. 64–65.

41. Goss, *Life of Colonel Paul Revere,* pp. 411–415; Brigham, *Paul Revere's Engravings,* pp. 213–225. During this period Revere and his family shared a crowded Watertown house with Henry Knox and his wife, which might have helped Revere receive contracts from Knox in the future.

42. Brigham, *Paul Revere's Engravings,* pp. 225–234, 237; Triber, *A True Republican,* p. 115; Goss, *Life of Colonel Paul Revere,* pp. 415–417. Some of Revere's copper plates are still preserved in the Massachusetts State House.

43. Quoted in Goss, *Life of Colonel Paul Revere,* pp. 397–398.

44. Stephenson, "The Supply of Gunpowder in 1776," pp. 271–274. Quote taken from p. 274. Peskin, *Manufacturing Revolution,* pp. 52–53.

45. Stephenson, "The Supply of Gunpowder in 1776," p. 276.

46. Quoted in Goss, *Life of Colonel Paul Revere,* pp. 400–401.

47. Brigham, *Paul Revere's Engravings,* pp. 19–21, 122–123; Goss, *The Life of Colonel Paul Revere,* p. 404; Triber, *A True Republican,* p. 119. Goss and other authors imply that Revere helped to set up and possibly operate the mill, but Triber correctly points out that the source materials do not credit Revere for any service other than designing it.

48. Quote taken from Goss, *Life of Colonel Paul Revere,* pp. 405–406; general information in Triber, *A True Republican,* p. 127; Forbes, *Paul Revere and the World He Lived In,* pp. 321–322; and Robert B. Gordon, *American Iron, 1607–1900* (Baltimore: Johns Hopkins University Press, 1996), p. 202.

49. Wood, *Radicalism of the American Revolution,* pp. 247–248, 176–177; Peskin, *Manufacturing Revolution,* pp. 57–59; Foner, *Tom Paine and Revolutionary America,* p. 27; Merrill Jensen, *The New Nation* (New York: Vintage Books, 1950), pp. 184, 219–220; Walton

and Shepherd, *Economic Rise of Early America*, p. 181; Nettels, *Emergence of a National Economy*, p. 25.

50. Jensen, *The New Nation*, pp. 85–89, 129–130; Middlekauff, *Glorious Cause*, p. 548.

CHAPTER THREE: Mercantile Ambitions and a New Look at Silver
(1783–1789)

1. Paul Revere to John Rivoire (his cousin), October 6, 1781, in "Loose Manuscripts 1746–1801," Revere Family Papers (hereafter RFP), microfilm edition, 15 reels (Boston: Massachusetts Historical Society, 1979), reel 1.

2. Revere to John Rivoire, October 6, 1781, "Loose Manuscripts 1746–1801," reel 1, RFP.

3. Adams quote taken from Samuel Adams letter to John Adams, July 2, 1785, cited in Allan Kulikoff, "The Progress of Inequality in Revolutionary Boston," *William and Mary Quarterly* 28, vol. 3 (July 1971): 375. Also see Jeannine Falino, "'The Pride Which Pervades thro every Class': The Customers of Paul Revere," in *New England Silver & Silversmithing*, ed. Jeannine Falino and Gerald W. R. Ward (Boston: University Press of Virginia, 2001), p. 152; Stuart Bruchey, *The Roots of American Economic Growth, 1607–1861* (New York: Harper, 1965), pp. 200–201; Kulikoff, *Progress of Inequality*, pp. 399, 406, 408, 409; Gordon S. Wood, *The Radicalism of the American Revolution* (New York: Vintage Books, 1991), pp. 227, 341; Drew R. McCoy, *The Elusive Republic* (New York: W. W. Norton, 1980), p. 107; Gordon S. Wood, "The Enemy is Us: Democratic Capitalism in the Early Republic," *Journal of the Early Republic* 16, no. 2 (Summer 1996): 301.

4. Sons of artisans enjoyed more social mobility than many other groups, with some becoming merchants. Kulikoff, "The Progress of Inequality," p. 406.

5. Edwin J. Perkins, *The Economy of Colonial America*, 2nd ed. (New York: Columbia University Press, 1988), p. 123; Thomas Cochran and William Miller, *The Age of Enterprise: A Social History of Industrial America* (New York: Harper Torchbooks, 1961), p. 3; Carl Bridenbaugh, *Cities in Revolt: Urban Life in America, 1743–1776* (London: Oxford University Press, 1955), pp. 274–275; Kulikoff, "The Progress of Inequality," p. 406; Barbara McLean Ward, "Boston Artisan Entrepreneurs of the Goldsmithing Trade in the Decades before the Revolution," in *Entrepreneurs: The Boston Business Community, 1700–1850*, ed. Conrad E. Wright and Katheryn P. Viens (Boston: Massachusetts Historical Society, 1997), p. 29.

6. Deborah Anne Federhen, "Paul Revere, Silversmith: A Study of His Shop Operation and His Objects" (master's thesis, University of Delaware, 1988), pp. 55–57.

7. Ward, "Boston Artisan Entrepreneurs," pp. 25–27, 37; Patrick M. Leehey, "Reconstructing Paul Revere; an Overview of his Ancestry, Life, and Work," in *Paul Revere—Artisan, Businessman, and Patriot* (Boston: Paul Revere Memorial Association, 1988), p. 30; Deborah A. Federhen, "From Artisan to Entrepreneur: Paul Revere's Silver Shop Operations," in *Paul Revere—Artisan, Businessman, and Patriot* (Boston: Paul Revere Memorial Association, 1988), p. 85; Federhen, *Paul Revere, Silversmith*, p. 35.

8. Federhen, *Paul Revere, Silversmith*, p. 58.

9. November 13, 1783 letter to Frederick Geyer, "Letterbook 1783–1800," reel 14, vol. 53.1, RFP.

10. June 30, 1783 letter to Frederick Geyer, "Letterbook 1783–1800," reel 14, vol. 53.1, RFP.

11. Leehey, "Reconstructing Paul Revere," p. 30; June 30, 1783 letter to Frederick Geyer, "Letterbook 1783–1800," reel 14, vol. 53.1, RFP.

12. Revere's quote taken from November 11, 1785 letter to Frederick Geyer, "Letterbook 1783–1800," reel 14, vol. 53.1, RFP. See also letters written to Frederick Geyer on December 25, 1783, April 15, 1784, and August 5, 1784, and Perkins, *The Economy of Colonial America*, p. 126.

13. The tremendous quantity of circulating currency did help America to shift from the barter to a market economy amid widespread economic turbulence. Paul A. Gilje, "The Rise of Capitalism in the Early Republic," *Journal of the Early Republic* 16, no. 2 (Summer 1996): 174; Curtis P. Nettels, *The Emergence of a National Economy, 1775–1815* (New York: Holt, Rinehart, and Winston, 1962), p. 25; John McCusker and Russell Menard, *The Economy of British America* (Chapel Hill: University of North Carolina Press, 1985), pp. 366–370; Gary M. Walton and James F. Shepherd, *The Economic Rise of Early America* (New York: Cambridge University Press, 1980), pp. 181–185.

14. Most craftsmen and shopkeepers considered the extension of credit to their customers a risky but essential component of their business. Two New York bakers estimated that they lost between 2.5 and 5 percent of their income to uncollected debts, which proved very costly over time. Similarly, payments in bank notes could prove worthless in the years of bank uncertainty. Howard B. Rock, *Artisans of the New Republic* (New York: New York University Press, 1984), pp. 163–165; Bridenbaugh, *Cities in Revolt*, pp. 275–276.

15. Quoted in Merrill Jensen, *The New Nation* (New York: Vintage Books, 1950), pp. 190–191.

16. Massachusetts tried to rapidly repay its debt with high taxes, and the state's tax burden increased fivefold or more. Taxpayers had to sell property and equipment to pay taxes and auctions frequently took place, leading to the Shays' Rebellion protest in 1786. Even though this rebellion was suppressed, it achieved its aims by causing lawmakers to reduce taxes. Winifred Barr Rothenberg, "The Invention of American Capitalism: The Economy of New England in the Federal Period," in *Engines of Enterprise: An Economic History of New England*, ed. Peter Temin (Cambridge, Mass.: Harvard University Press, 2000), p. 73. Also see Walton and Shepherd, *Economic Rise of Early America*, pp. 199–200; Jensen, *The New Nation*, pp. 185–187; Nettels, *Emergence of a National Economy*, pp. 47–48, 60–61; Wood, *Radicalism of the American Revolution*, p. 249; Ronald Schultz, "The Small Producer Tradition and the Moral Origins of Artisan Radicalism in Philadelphia 1720–1810," *Past and Present*, no. 127 (May 1990): 105; Ronald Schultz, *The Republic of Labor: Philadelphia Artisans and the Politics of Class 1720–1830* (New York: Oxford University Press, 1993), pp. 116–117.

17. Federhen, "Artisan to Entrepreneur," p. 85.

18. A Massachusetts protective tariff passed in July 1785, prohibiting the import of 58 types of goods entirely and taxing others at 5 to 15 percent. Following this victory the Association of Tradesmen and Manufacturers took their cause to the rest of the nation, drafting a circular letter that they first sent to colleagues in all seaport cities, who passed it on to other areas. The circular letter helped to draw America's mechanics closer together in a common cause. Lawrence A. Peskin, *Manufacturing Revolution: The*

Intellectual Origins of Early American Industry (Baltimore: Johns Hopkins University Press, 2003), pp. 71–73, 82.

19. Mechanic ideology can be best described as neo-mercantilism, advocating four main elements: a positive balance of trade, an active state to direct the economy, a balanced and self-sufficient American economy, and a concern for the public welfare over individual gain. Peskin, *Manufacturing Revolution*, p. 75. Also see Gary Kornblith, "Artisan Federalism: New England Mechanics and the Political Economy of the 1790s," in *Launching the "Extended Republic": The Federalist Era*, ed. Ronald Hoffman and Peter J. Albert (Charlottesville: University Press of Virginia, 1996), pp. 258–259, 263–267; Schultz, *Moral Origins of Artisan Radicalism*, p. 85.

20. *Annals of the Massachusetts Charitable Mechanics Association* (Boston: Press of Rockwell and Churchill, 1892), p. 10. This anecdote is also mentioned in Edwin Griffin Porter, *Rambles in Old Boston* (Boston: Cupples, Upham, and Company, 1887), pp. 98–99. These oft-quoted words are probably the creation of Daniel Webster, who related this story in an 1843 speech in Andover.

21. Jayne E. Triber, *A True Republican* (Amherst: University of Massachusetts Press, 1998), pp. 156–158; Peskin, *Manufacturing Revolution*, pp. 87–88; Silvio A. Bedini, *Thinkers and Tinkers: Early American Men of Science* (New York: Scribner, 1975), pp. 261–262.

22. April 26, 1789 and January 24, 1791 letters from Fischer Ames to Paul Revere, "Loose Manuscripts 1746–1801," reel 1, RFP. Also see Triber, *A True Republican*, pp. 160–162. This appointment eventually went to scientist and statesman David Rittenhouse, a friend of Secretary of State Thomas Jefferson.

23. Leehey, "Reconstructing Paul Revere," p. 31; Triber, *A True Republican*, pp. 160–161.

24. The specific time and mechanism of the origin and spread of a market economy in America is a complex topic that is still a matter of some scholarly debate. Many of the capitalist conditions listed above appeared haltingly or in unusual ways. For example, paper money first appeared during the war, but vanished after the passage of the new Constitution. By this point, bank notes (greatly in excess of the assets of the issuing banks) had become quite prominent. This subject is covered in great detail throughout the Summer 1996 issue of the *Journal of the Early Republic* (vol. 16, no. 2). Also see Bruce Laurie, *Artisans into Workers* (New York: Hill and Wang, 1989), pp. 15–19; Gilje, "Rise of Capitalism," pp. 176–177; Wood, *Radicalism of the American Revolution*, p. 135.

25. The high-volume production of standardized forms of goods started taking place in other trades, such as hat, pin, and pottery making in Europe by the 1700s, prior to the use of machinery. Self-reinforcing cycles of production and demand led to the expansion of industrialization at greater levels after 1789, as illustrated in the following chapters. Paul A. Gilje, "Identity and Independence: The American Artisan," in *American Artisans: Crafting Social Identity*, ed. Howard B. Rock, Paul A. Gilje, and Robert Asher (Baltimore: Johns Hopkins University Press, 1995), p. xiv; Gilje, "Rise of Capitalism in the Early Republic," p. 162; Jonathan Prude, "Capitalism, Industrialization, and the Factory in Post-revolutionary America," *Journal of the Early Republic* 16, no. 2 (Summer 1996): 242–245; Peskin, *Manufacturing Revolution*, pp. 61- 63; James R. Farr, *Artisans in Europe, 1300–1914* (Cambridge: Cambridge University Press, 2000), pp. 55–56; Robert B. Gordon and Patrick M. Malone, *The Texture of Industry* (New York: Oxford University Press, 1994), pp. 87, 227, 237; Otto Mayr and Robert C. Post, eds., *Yankee Enterprise: The*

Rise of the American System of Manufactures (Washington: Smithsonian Institution Press, 1981), p. xii; Brooke Hindle and Steven Lubar, *Engines of Change: The American Industrial Revolution, 1790–1860* (Washington: Smithsonian Institution Press, 1986), p. 153; David S. Landes, *The Unbound Prometheus: Technological Change and Industrial Development in Western Europe from 1750 to the Present* (New York: Cambridge University Press, 2003), pp. 1–2.

26. Taken from Table E in Federhen, *Paul Revere, Silversmith*, p. 92.

27. Federhen, *Paul Revere, Silversmith*, pp. 36–37.

28. Federhen, *Paul Revere, Silversmith*, p. 6.

29. Federhen, *Paul Revere, Silversmith*, pp. 48–54, 77–79.

30. Silver merchandise recorded in an undated entry (between August 30 and September 17, 1783), "Boston Wastebook 1783–1797," reel 5, vol. 2, RFP. Also see Federhen, *Paul Revere, Silversmith*, pp. 58, 81; Walter Licht, *Industrializing America* (Baltimore: Johns Hopkins University Press, 1995), pp. 16–17; Gilje, "Rise of Capitalism in the Early Republic," pp. 162–165; Winifred Barr Rothenberg, *From Market-Places to a Market Economy* (Chicago: University of Chicago Press, 1992), pp. 243–244; Rothenberg, "Invention of American Capitalism," pp. 107–108; Rock, *Artisans of the New Republic*, p. 242; T. H. Breen, *The Marketplace of Revolution: How Consumer Politics Shaped American Independence* (New York: Oxford University Press, 2004), pp. 120–121; W. J. Rorabaugh, *The Craft Apprentice* (New York: Oxford University Press, 1986), p. 68; Billy G. Smith, "The Vicissitudes of Fortune" in *Work and Labor in Early America*, ed. Stephen Innes (Chapel Hill: University of North Carolina Press, 1988), pp. 246–247; Jonathan Prude, *The Coming of Industrial Order* (Cambridge: Cambridge University Press, 1983), pp. 90, 109–110, 201.

31. John Marshall Phillips, *American Silver* (New York: Dover, 2001), p. 109; Graham Hood, *American Silver, A History of Style, 1650–1900* (New York: Praeger Publishers, 1971), pp. 162, 166; Janine E. Skerry, "The Revolutionary Revere: A Critical Assessment of the Silver of Paul Revere," in *Paul Revere—Artisan, Businessman, and Patriot* (Boston: Paul Revere Memorial Association, 1988), p. 53.

32. The colonial consumption of tea increased to incredible levels after the war, aided by the opening of Chinese trade to American ships in 1784, and Revere's business records illustrate his ability to tap into this trend. In addition to the 54 teapots he produced after the war, Revere made 50 pairs of sugar tongs, 58 creampots, 18 sugar bowls, and 21 teapot stands, among other. Falino, "Customers of Paul Revere," pp. 157–158; Federhen, *Paul Revere, Silversmith*, pp. 44–45.

33. Federhen, *Paul Revere, Silversmith*, pp. 63–65.

34. Skerry, "Revolutionary Revere," p. 53; Federhen, "Artisan to Entrepreneur," p. 77.

35. Artisans using mechanized equipment often found ways to maintain their older methods of skilled work while exploiting power machinery's assets and thereby avoiding much of the physical toil. Gordon and Malone, *Texture of Industry*, p. 354.

36. Karl Marx famously dissected the differences between tools and machines in his masterwork, *Capital*. Marx contended that machines include one or more tools among their components, and guide those tools with power sources and transmission devices. Marx analyzes how machines have the ability to displace and supersede human workers, and usher in the industrial revolution. This argument is advanced in Karl Marx, *Capital*, vol. 1, chapter 15, particularly in section one. Dirk J. Struik, *Yankee Science in the Making* (New York: Dover Publications, 1991), pp. 175–176; Lindy Biggs, *The Rational Factory*

(Baltimore: Johns Hopkins University Press, 1996), pp. 4–5; Landes, *The Unbound Prometheus*, pp. 1–2; Gordon and Malone, *Texture of Industry*, pp. 236, 354.

37. James A. Mulholland, *History of Metals in Colonial America* (Tuscaloosa: University of Alabama Press, 1981), pp. 83, 88–89; Hood, *History of Style*, pp. 18–19; Federhen, "Artisan to Entrepreneur," p. 77; Federhen, *Paul Revere, Silversmith*, pp. 41–42.

38. Federhen, *Paul Revere, Silversmith*, pp. 43–44.

39. Federhen, "Artisan to Entrepreneur," p. 80; Federhen, *Paul Revere, Silversmith*, pp. 49–50.

40. December 5, 1791 letter to Mess. George & William Burchell, "Letterbook 1783–1800," reel 14, vol. 53.1, RFP.

41. Federhen, *Paul Revere, Silversmith*, pp. 70–71, 73–74; Federhen, "Artisan to Entrepreneur," pp. 87–88. Information pertaining to David Moseley can be found in reel 6, vol. 8, RFP, particularly in the "Moseley Estate" records and the "Cash and Memoranda Book, 1791–1801."

42. Janson quote from Rorabaugh, *Craft Apprentice*, p. 36. Also see Rorabaugh, *Craft Apprentice*, pp. 16, 20, 55; McCoy, *The Elusive Republic*, p. 92; Bernard Elbaum, "Why Apprenticeship Persisted in Britain but not in the United States," *Journal of Economic History* 49, no. 2 (June 1989): 346; Sharon V. Salinger, "Artisans, Journeymen, and the Transformation of Labor in Late Eighteenth-Century Philadelphia," *William and Mary Quarterly* 40, no. 1 (January 1983): 70.

43. Wages were not standard. Some workers received piece rates, or day rates, or by the month. Carpenters received a day rate in the summer and a piece rate in winter when short days equated to less productivity. Wood, *Radicalism*, p. 185; Laura Rigal, *The American Manufactory: Art, Labor, and the World of Things in the Early Republic* (Princeton, N.J.: Princeton University Press, 2001), p. 14; Nettels, *Emergence of a National Economy*, pp. 265–266; Salinger, "Transformation of Labor," pp. 62, 64, 69, 72–73, 75, 83; Smith, "The Vicissitudes of Fortune," pp. 239–242, 246–247.

44. Leehey, "Reconstructing Paul Revere," p. 26; Federhen, "Artisan to Entrepreneur," p. 86.

45. Federhen, "Artisan to Entrepreneur," pp. 75, 86–88; Federhen, *Paul Revere, Silversmith*, p. 71.

46. Other craftsmen joined him in this shift: in colonial woodworking, for example, many cabinetmakers assembled products from essentially mass-produced components. Stephanie Grauman Wolf, *As Various as their Land* (New York: HarperCollins, 1993), p. 179.

47. Wood, "The Enemy is Us," p. 304; Conrad E. Wright and Katheryn P. Viens, eds., *Entrepreneurs: The Boston Business Community, 1700–1850* (Boston: Massachusetts Historical Society, 1997), pp. x–xi; Gilje, "The Rise of Capitalism," pp. 172–175.

48. Hamilton's "Federalist program" consisted of three revenue laws derived from tariffs on foreign imports and a liquor tax; a Funding Act that required the federal government to pay all federal and state debts via tax revenues in a manner that restored faith in federally issued securities and America's credit; a Bank Act that established the Bank of the United States, privately run but supported in part with federal funds, to expand the currency supply by issuing more bank notes than the coin in its possession; a Coinage Act that defined the sizes and values of new gold and silver coins; and a patent law that gave inventors fourteen years of monopolistic control over their products. Walton and Shepherd, *Economic Rise of Early America*, pp. 187–188; Nettels, *Emergence of*

a National Economy, pp. 105, 110–123; Gilje, "Rise of Capitalism in the Early Republic," pp. 176–177.

49. Rorabaugh, *Craft Apprentice*, pp. 24, 59, 63–65; Jack Larkin, *The Reshaping of Everyday Life, 1790–1840* (New York: HarperPerennial, 1989), pp. 53, 59; Gilje, "The Rise of Capitalism," p. 170; Wood, *Radicalism*, p. 185; Wood, "The Enemy is Us," p. 301; Nettels, *Emergence of a National Economy*, pp. 265–266; Bruce Laurie, "'Spavined Ministers, Lying Toothpullers, and Buggering Priests': Third-Partyism and the Search for Security in the Antebellum North," in *American Artisans: Crafting Social Identity*, ed. Howard B. Rock, Paul A. Gilje, and Robert Asher (Baltimore: Johns Hopkins University Press, 1995), pp. 99, 105–106.

50. The Massachusetts Charitable Mechanic Association deviated from prior artisan organizations by ignoring the hierarchy and divisions between separate trades, instead treating its members as individual and independent businessmen. This was a gradual process, however: even as individual artisans moved away from craft identification in their day-to-day lives they still fell into traditional craft groups for public events such as parades. Peskin, *Manufacturing Revolution*, pp. 140–144; *Annals of the Massachusetts Charitable Mechanics Association* (Boston: Press of Rockwell and Churchill, 1892), pp. 1–2; Quimby, "Some Observations on the Craftsman in Early America," pp. 7–9; Oscar Handlin and Mary Flug Handlin, *Commonwealth* (Cambridge, Mass.: Harvard University Press, 1947), p. 132.

51. *Annals of the Massachusetts Charitable Mechanics Association*, p. 291.

52. The Massachusetts Charitable Mechanics Association enrolled approximately 146 of Boston's 1,259 master craftsmen (11 percent), and its members were relatively well off for the group—80 percent of the assembly had more than $500 in taxable property, which is double the rate of nonmembers. In keeping with his love of profitable networking, Revere recorded large sales of silver to the officers and membership roster of this group. Ian Quimby, "Some Observations on the Craftsman in Early America," in *The Craftsman in Early America*, ed. Ian Quimby (New York: W. W. Norton, 1984), pp. 9, 13; Oscar and Mary Handlin, *Commonwealth* (Cambridge, Mass.: Belknap Press, 1969), p. 132; Falino, "The Customers of Paul Revere," pp. 162–163.

CHAPTER FOUR: To Run a "Furnass": The Iron Years (1788–1792)

1. Paul Revere to Messrs. Brown and Benson, November 3, 1788, "Letterbook 1783–1800," Revere Family Papers (hereafter RFP), microfilm edition, 15 reels (Boston: Massachusetts Historical Society, 1979), reel 14, vol. 53.1.

2. Robert Raymond, *Out of the Fiery Furnace: The Impact of Metals on the History of Mankind* (University Park: Pennsylvania State University Press, 1986), pp. 9–10; Stephen Sass, *The Substance of Civilization: Materials and Human History from the Stone Age to the Age of Silicon* (New York: Arcade Publishing, 1999), p. 83.

3. Iron ores might have entered copper-smelting furnaces in different ways. Some iron-rich ores could have been used as pigments. In addition, coppersmiths added "flux" materials to aid the production of copper: a flux combines with impurities in the copper ore to make them easier to remove. Hematite, one of the most frequently used fluxes, is rich in iron.

4. Raymond, *Out of the Fiery Furnace*, pp. 55–63; Sass, *Substance of Civilization*, pp. 90–93. The term *iron age* generally applies to the earliest widespread use of iron in a

specific region, and therefore takes place at different times depending on the technological proficiency of its inhabitants. Many historians approximate the starting date of the Near East's iron age in 1200 BC, with the caveat that this is a major generalization. The iron age started several centuries later in Europe and India. The iron age does not have a definitive ending point tied to any subsequent technological developments: as the final stage of prehistory it ends in each region whenever historical recordkeeping begins.

5. This analysis regrettably focuses upon the transfer of iron technology in the direction of Paul Revere, which excludes the fascinating story of Eastern ironworking, such as Chinese iron casting. Although China initially borrowed iron technology from Europe in the seventh century BC, it quickly adapted its own ironworking traditions directed toward advanced blast furnace technology. Chinese furnaces were far superior to European ones, making use of horizontal and box bellows capable of delivering a rapid and continuous stream of air to the fire, and could directly convert iron ore into cast iron by reaching temperatures above 1530 degrees C. In contrast, Europeans did not manage to cast iron until the fifteenth century. Chinese technicians learned to stir molten iron with steel rods, thereby removing carbon from the entire mass of iron and forming solid, pure steel. Steelmaking advances also took place in India around 600 BC, with the discovery of crucible steel. Workers filled small crucibles with wrought iron, wood, and plant leaves, and fired them in a charcoal pit for hours. Carbon from the plants and wood infused the iron, forming steel. This process was later borrowed by the Arabs to make the legendary Damascus steel. Raymond, *Out of the Fiery Furnace*, pp. 73–80.

6. Raymond, *Out of the Fiery Furnace*, pp. 63–67; James M. Swank, *History of the Manufacture of Iron in all Ages and Particularly in the United States From Colonial Times to 1891* (Philadelphia: Allen, Lane, & Scott, 1892), pp. 11–18, 33–44, 113.

7. David Lewis, *Iron and Steel in America* (Wilmington, Del.: Hagley Museum, 1976), p. 10; Robert B. Gordon, *American Iron, 1607–1900* (Baltimore: Johns Hopkins University Press, 1996), pp. 7, 12–14.

8. Gordon, *American Iron*, pp. 4–5, 10–11, 14; Peter Temin, *Iron and Steel in Nineteenth-Century America, an Economic Inquiry* (Cambridge, Mass.: MIT Press, 1964), p. 10.

9. Most fineries contained between two and four hearths, each requiring the full attention of a skilled laborer and a waterpower-driven hammer. In some cases a blast furnace would operate an on-site finery to produce bar iron from freshly made pig iron. Gordon, *American Iron*, pp. 14, 125; Robert B. Gordon and Patrick M. Malone, *The Texture of Industry* (New York: Oxford University Press, 1994), pp. 79–80.

10. Gordon and Malone, *Texture of Industry*, pp. 227, 242; Gordon, *American Iron, 1607–1900*, p. 1; James B. Hedges, *The Browns of Providence Plantation: The Colonial Years* (Lebanon, N.H.: University Press of New England, 1952), p. 129.

11. Eminent historian of technology Thomas Hughes describes his theory of technological systems in the essay "The Evolution of Large Technological Systems," in *The Social Construction of Technological Systems*, ed. Wiebe Bijker, Thomas Hughes, and Trevor Pinch (Cambridge, Mass.: MIT Press, 1987).

12. Lewis, *Iron and Steel in America*, pp. 17–18; Gordon, *American Iron, 1607–1900*, pp. 55–57, 62.

13. James A. Mulholland, *History of Metals in Colonial America* (Tuscaloosa: University of Alabama Press, 1981), pp. 59–61; Brooke Hindle, *Emulation and Invention* (New York: W. W. Norton and Company, 1981), pp. 6–8.

14. Gordon and Malone, *The Texture of Industry*, p. 77; Mulholland, *History of Met-*

als in Colonial America, pp. 104–109; Lewis, *Iron and Steel in America*, p. 22; Margaret H. Hazen and Robert M. Hazen, *Wealth Inexhaustible* (New York: Van Nostrand Reinhold, 1985), pp. 196–197; Gordon, *American Iron, 1607–1900*, pp. 1, 57; Gary M. Walton and James F. Shepherd, *The Economic Rise of Early America* (New York: Cambridge University Press, 1980), p. 160; Joseph E. Walker, *Hopewell Village; a Social and Economic History of an Iron-Making Community* (Philadelphia: University of Pennsylvania Press, 1966), p. 414.

15. Hindle, *Emulation and Invention*, p. 11.

16. Temin, *Iron and Steel in Nineteenth-Century America*, p. 22; Mulholland, *A History of Metals in Colonial America*, p. 146; Gordon, *American Iron*, p. 3.

17. Mulholland, *A History of Metals in Colonial America*, pp. 64–66, 70–72; Charles B. Dew, *Bond of Iron: Master and Slave at Buffalo Forge* (New York: W. W. Norton & Company, 1994); Gordon, *American Iron*, p. 1.

18. By relying upon charcoal for so long, American ironworkers avoided the difficult transition to coal fuel and missed several developments of England's industrial revolution. England's mechanization greatly accelerated after the reverberatory furnace and other innovations enabled the use of coal as a power source in lime burning, textiles, gunpowder, soap, sugar, and iron production. By the early nineteenth century, American ironworkers experimented with new technologies at both small and large works, everywhere except the South. By the mid-nineteenth century, America's ironworks had arrived at a point of technological sophistication, incorporating more efficient furnace and forge heating methods as well as the use of anthracite and bituminous coal. Hindle, *Emulation and Invention*, pp. 9–10; Gordon, *American Iron*, pp. 2–3.

19. Mulholland, *A History of Metals in Colonial America*, pp. 145–148.

20. Samuel Slater personifies the traditional, immigration-centered technology transfer process. Slater began an apprenticeship in 1782 in a British textile factory, where he received training in both the technological and managerial aspects of cloth production. In 1789 he tricked customs officials into allowing him to emigrate, a direct violation of British law. Within a few years he found a firm of merchants willing to fund his spinning mill, America's first successful textile factory. For an introduction to the subject of technology transfer, see Nathan Rosenberg, *Perspectives on Technology* (New York: Cambridge University Press, 1976); Svante Lindqvist, *Technology on Trial* (Stockholm: Almqvist & Wiksell International, 1984); David Jeremy, *Transatlantic Industrial Revolution* (Cambridge, Mass.: MIT Press, 1981); David Jeremy, *Artists, Entrepreneurs, and Machines* (Brookfield: Ashgate Publishing, 1991); J. R. Harris, *Industrial Espionage and Technology Transfer* (Brookfield: Ashgate Publishing, 1998); Neil York, *Mechanical Metamorphosis: Technological Change in Revolutionary America* (Westport: Greenwood Press, 1985), pp. 39, 158, 172; Hindle, *Emulation and Invention*, p. 6; Doron S. Ben-Atar, *Trade Secrets: Intellectual Piracy and the Origins of American Industrial Power* (New Haven: Yale University Press, 2004), pp. 8–10, 29–31. America followed a long European tradition of encouraging the immigration of skilled foreigners in possession of technological secrets. Britain owed some of its eighteenth-century technological dominance to its ability to lure foreign artisans, and often used royal cash rewards or promises of patent monopolies to do so.

21. George Washington quote from Hindle, *Emulation and Invention*, pp. 19–20. Historian of technology David Jeremy proposes this four-step model for technology diffusion in *Artists, Entrepreneurs, and Machines*, p. 5. Early Americans interested in carrying out this technology transfer process benefited from some of America's social and

economic assets, including an economic mindset geared at improvement, relatively high rates of education, a perception of social and economic mobility that inspired entrepreneurship, and the absence of institutions that inhibited technical change in other nations. These national assets aided both the invention process and the practical application of invention. Nathan Rosenberg, *Technology and American Economic Growth* (New York: M. E. Sharpe Inc., 1972), pp. 32–35.

22. Paul Revere to Messrs. Brown and Benson, November 3, 1788, "Letterbook 1783–1800," reel 14, vol. 53.1, RFP; Renee Lynn Ernay, "The Revere Furnace" (master's thesis, University of Delaware, 1989), p. 12; Gordon, *American Iron, 1607–1900*, pp. 2, 19–21.

23. Revere started his own correspondence with one of these British experts in 1791, to prepare for his bell-making career. Esther Forbes, *Paul Revere and the World He Lived In* (Boston: Houghton Mifflin, 1969), p. 392.

24. Mulholland, *A History of Metals in Colonial America*, p. 148; York, *Mechanical Metamorphosis*, p. 172.

25. Gordon, *American Iron*, p. 2. America's comparatively open technical exchange policy is also discussed in Merritt Roe Smith, "Eli Whitney and the American System of Manufacturing," in *Technology in America*, ed. Carroll W. Pursell Jr. (Cambridge, Mass.: MIT Press, 1981), pp. 57–59.

26. June 22, 1795 letter from Stephen Rochefontaine, "Loose Manuscripts 1746–1801," reel 1, RFP.

27. Hedges, *The Browns of Providence Plantation*, pp. 122–136, 140–145; Edwin J. Perkins, *The Economy of Colonial America* (New York: Columbia University Press, 1980), p. 128. Revere made four furnace-related business trips between April 11, 1787, and March 20, 1788. His first trip to Menotomy (now Arlington, Massachusetts) almost certainly concerned the search for reliable raw material supplies, as he specified in several later records that he purchased sand from there. He traveled to Providence on July 28 to visit the Hope furnace and to speak to Brown in person. The purpose of his third trip to Halifax on August 9 is unknown, and his fourth trip to Marblehead in March 1788 involved the purchase of cannon since he included a bill for weighing cannon on the same day. Marblehead was also the port used by Brown and Benson for their iron shipments. Ledger titled "1787 The Iron Furnace D. to Stock," reel 6, vol. 9, part 2, RFP.

28. Deed, June 28, 1792, "Loose Manuscripts 1746–1801," reel 1, RFP. Please note that Revere's cousin refers to him as "Esquire."

29. Revere's terminology offers several clues about the nature of these monetary relationships. His notations of Samuel's payments always contain numerous details, such as "By Cash from Sam'l Hichborn to pay Wm Little for Iron." These comments could imply that Samuel's payments were loans to help defray expenses and that Samuel might have cared about the purpose of each loan. In contrast, whenever Revere recorded a transaction with Benjamin, it simply read "By Cash from Benj. Hichborn Esq." Because the use of Benjamin's money is not recorded, this could be a straightforward payment of money, perhaps to repay an old debt. But on one November 28, 1787 receipt, Revere instructed Benjamin to pay a carting expense and charge it to the furnace, implying a working partnership between the two. Ledger titled "1787 The Iron Furnace D. to Stock," reel 6, vol. 9, part 2, RFP; "Boston Wastebook 1783–1797," reel 5, vol. 2, RFP; November 28, 1787 note/receipt from Revere to Benjamin Hitchborn, "Loose Manuscripts 1746–1801," reel 1, RFP.

30. Ledger titled "1787 The Iron Furnace D. to Stock," reel 6, vol. 9, part 2, RFP. This

ledger is summarized in Appendix 6. This ledger is a four-page subsection of Revere's "Account Book, Boston, 1783–1804," and contains many unrelated entries. This section refers to the list of expenses charged to his furnace account. Revere probably incurred additional expenses not recorded on this list, particularly if he purchased items for his silver shop or hardware store and then used them for the furnace. But in general he thought of his different business endeavors as independent entities, and recorded many transfers of cash and goods between his shops in his account books. He was also very meticulous about the smallest expenses, often recording charges of 1 shilling or less. The only items from Revere's ledger not included in the above table are cash receipts from Benjamin Hitchborn, since these may represent payments to Revere, or payments from Revere to others that happened to be reimbursed by Hitchborn.

31. The following analysis is limited by gaps in his records: Revere made one list of expenses from 1787 to 1788, and the next ledger details his sales and expenses from 1792 to 1794.

32. Technically, what Revere called a furnass should be called a foundry, although many contemporaries used *furnace* and *forge* interchangeably to describe any large heating oven. More common usage throughout the eighteenth century referred to blast furnaces as "furnaces" and foundries as "air furnaces."

33. Many bellows were water powered, but this would not be an option in downtown Boston. Steam-powered bellows were not used in America until almost the mid-nineteenth century. Some of the information used to describe the foundry is taken from the 1751 edition of the London-published *Universal Dictionary of Arts and Sciences*, cited in Henry Kauffman, *American Copper and Brass* (Camden, N.J.: Thomas Nelson & Sons, 1968), p. 57.

34. Gordon, *American Iron*, p. 195; Gordon and Malone, *Texture of Industry*, p. 38.

35. Gordon, *American Iron*, p. 195.

36. Gordon, *American Iron*, pp. 2, 195.

37. Ledger titled "1787 The Iron Furnace D. to Stock," 1792–1794 furnace records, reel 6, vol. 9, part 2, RFP.

38. Gordon, *American Iron*, pp. 4–5, 10, 195.

39. This entire section is derived from an analysis of Revere's untitled 1792–1794 wastebook, contained in reel 6, vol. 8, part 2, RFP.

40. Ledger titled "1787 The Iron Furnace D. to Stock," 1792–1794 furnace records, reel 6, vol. 9, part 2, RFP. Also see November 14, 1793 receipt from Zebulon White, "Loose Manuscripts 1746–1801," reel 1, RFP.

41. Revere to Thomas Wadsworth, July 28, 1785, in "Loose Manuscripts 1746–1801," reel 1, RFP.

42. Walter Licht, *Industrializing America* (Baltimore: Johns Hopkins University Press, 1995), p. 4.

43. September 13, 1792 and December 2, 1793 receipts from Nelson Miller, September 13, 1792 receipt to William Story, and September 13, 1792 receipt to Stanley Caster, "Loose Manuscripts 1746–1801," reel 1, RFP.

44. May 23 entry in "1793 Memoranda Book," reel 14, vol. 15.11, RFP.

45. According to Revere scholar Deborah Federhen, Mathew Metcalf might have been one of Revere's prewar silvershop apprentices or journeymen. Deborah A. Federhen, "From Artisan to Entrepreneur: Paul Revere's Silver Shop Operations," in *Paul*

Revere—Artisan, Businessman, and Patriot (Boston: Paul Revere Memorial Association, 1988), p. 73.

46. The 1792–1794 furnace records are in a ledger titled "1787 The Iron Furnace D. to Stock," reel 6, vol. 9, part 2, RFP. Drinking study taken from Howard B. Rock, *Artisans of the New Republic* (New York: New York University Press, 1984), p. 298; Howard B. Rock, *The New York City Artisan 1789–1825* (Albany: State University of New York Press, 1989), p. 47.

47. May 23 entry in "1793 Memoranda Book," reel 14, vol. 15.11, RFP.

48. 1792–1794 furnace records in ledger titled "1787 The Iron Furnace D. to Stock," reel 6, vol. 9, part 2, RFP, and March 8, 1788 receipt from Mr. Little, "Loose Manuscripts 1746–1801," reel 1, RFP. The iron ambiguity is caused by vague charges to merchants that fail to explain the type of product bought or sold.

49. November 1, 1787 letter from J. Blagge, April 11, 1788 letter from Brown and Benson, "Loose Manuscripts 1746–1801," reel 1, RFP.

50. November 3, 1788 letter to Messrs. Brown and Benson, "Letterbook 1783–1800," reel 14, vol. 53.1, RFP.

51. September 3, 1789 letter to Messrs. Brown and Benson, "Letterbook 1783–1800," reel 14, vol. 53.1, RFP.

52. E. H. Goss, *The Life of Colonel Paul Revere* (Boston: Joseph G. Cupples, 1891), p. 534.

53. Puddling eventually replaced fineries by using the reverberatory principle to reflect and concentrate heat onto the pig iron to remove impurities. Rosenberg, *Technology and American Economic Growth*, p. 77; Gordon, *American Iron*, pp. 15–16; Charles F. Carroll, *The Timber Economy of Puritan New England* (Providence: Brown University Press, 1973), p. 10; Theodore Wertime, *The Coming of the Age of Steel* (Leiden: E. J. Brill, 1961), p. 110; Paul F. Paskoff, *Industrial Evolution: Organization, Structure, and Growth of the Pennsylvania Iron Industry* (Baltimore: Johns Hopkins University Press, 1983), p. 21.

54. John R. Stilgoe, *Common Landscape of America* (New Haven: Yale University Press, 1982), p. 293; Walker, *Hopewell Village*, pp. 121, 136. When coal became the fuel of choice later in the nineteenth century, environmental impacts grew: coal and ore excavations left cavities in the ground, sulfur in coal entered wastewater and acidified streams and lakes, coal mines were messes of refuse and toxicity, air pollution intensified and was noticeable. Steam power also accelerated environmental degradation—ironworks could be located in cities and affected residents. Gordon, *American Iron*, pp. 4–5.

55. This comparison of costs and expenses represents a worst-case scenario for Revere. His recorded cash receipts end prematurely on November 12, 1793, and he might have received more income from his 1793 sales after that date. Similarly, his post-November 25 expenses include close to 100 pounds of copper, tin, iron, coal, and one unlisted purchase, clearly not used in the 1793 production. The above analysis therefore might underestimate Revere's income and overstate his expenses.

56. Revere's wastebooks were lists of all transactions that took place, arranged chronologically. His ledgers were account books sorted by client. Periodically, he transcribed the entries from wastebooks into the proper section of the ledger, allowing him to determine how much each client owed.

57. 1792–1794 furnace records, in ledger titled "1787 The Iron Furnace D. to Stock," reel 6, vol. 9, part 2, RFP.

58. Gordon, *American Iron*, p. 1.

59. Perkins, *Economy of Colonial America*, pp. 127–130.

CHAPTER FIVE: Bells, Cannon, and Malleable Copper (1792–1801)

1. Jayne E. Triber, *A True Republican* (Amherst: University of Massachusetts Press, 1998), pp. 8–9, 20, 158.

2. Edward Stickney and Evelyn Stickney, *The Bells of Paul Revere* (Bedford: n.p., 1976), pp. 4–5, 14; Triber, *A True Republican*, p. 167; Elbridge Henry Goss, *The Life of Colonel Paul Revere* (Boston: Plimpton Press, 1902), pp. 538–541; Esther Forbes, *Paul Revere and the World He Lived In* (Boston: Houghton Mifflin, 1969), pp. 385–387.

3. Edwin Griffin Porter, *Rambles in Old Boston* (Boston: Cupples, Upham, and Company, 1887), p. 330.

4. Douglass C. North, *The Economic Growth of the United States, 1790–1860* (New York: W. W. Norton, 1966), pp. 46–54; Paul A. Gilje, "The Rise of Capitalism in the Early Republic," *Journal of the Early Republic* 16, no. 2 (Summer 1996): 176–177.

5. Percival Price, *Bells and Man* (New York: Oxford University Press, 1983), pp. 1, 73, 91; John Camp, *Bell Ringing* (New York: A. S. Barnes & Company, 1974), pp. 11–13, 77.

6. Figure 5.1 taken from Forbes, *Paul Revere's Ride*, p. 391. Also see Stickney and Stickney, *Bells of Paul Revere*, p. 4.

7. Camp, *Bell Ringing*, pp. 85–86; Henry J. Kauffman, *American Copper and Brass* (Camden, N.J.: T. Nelson Publishers, 1968), p. 170.

8. Stickney and Stickney, *Bells of Paul Revere*, p. 5.

9. Malachite ore could plausibly have entered pottery kilns as a green pigment or as an impurity within the pottery clay. Early pottery workers might have noticed a metallic residue at the bottom of their kilns, and further experimentation could have showed them how to intentionally convert malachite into usable metal.

10. Robert Raymond, *Out of the Fiery Furnace: The Impact of Metals on the History of Mankind* (University Park: Pennsylvania State University Press, 1986), pp. 10–13, 16, 21; Stephen Sass, *The Substance of Civilization: Materials and Human History from the Stone Age to the Age of Silicon* (New York: Arcade Publishing, 1999), pp. 49–53. Copper smelting may have been independently discovered in China and Africa.

11. The source of tin used by these early bronze workers in the Ancient Near East is a matter of some speculation and mystery. Tin oxides occur in hard rocks such as granite, which lay well beyond the extraction capabilities of this time period. Therefore, any usable tin would have to come from "native" sources, such as a vein of fairly pure tin that might be washed out of a rock by the action of a stream. The world's major tin deposits exist far from the early centers of bronze production, necessitating the development of trade networks. Raymond, *Out of the Fiery Furnace*, pp. 31–34.

12. Raymond, *Out of the Fiery Furnace*, pp. 25–27; Sass, *The Substance of Civilization*, pp. 59–62.

13. Margaret H. Hazen and Robert M. Hazen, *Wealth Inexhaustible* (New York: Van Nostrand Reinhold, 1985), p. 92.

14. Bell metal quote taken from *A Supplement to Chambers' Cyclopedia or Universal Dictionary of Arts and Sciences* (London, 1753), quoted in Kauffman, *American Copper and Brass*, p. 169. Also see Kauffman, *American Copper and Brass*, p. 169; Stickney and Stickney, *The Bells of Paul Revere*, p. 7.

15. Arthur H. Nichols, *The Bells of Paul and Joseph W. Revere* (Boston: Newcomb & Gauss Press, 1911), p. 1; Kauffman, *American Copper and Brass*, p. 163.

16. December 7, 1789 letter from Nicholas Brown, "Loose Manuscripts 1746–1801," Revere Family Papers (hereafter RFP), microfilm edition, 15 reels (Boston: Massachusetts Historical Society, 1979), reel 1.

17. December 3, 1791 letter to Dr. Lestrom (sic), August 20, 1792 letter from Dr. Lettsom, and March 29, 1793 letter from Dr. Lettsom, all in "Letterbook 1783–1800," reel 14, vol. 53.1, RFP.

18. An additional problem with casting copper is the fact that copper takes on oxygen when it is melted, making it viscous and hard to mold. Founders often overcame this difficulty by adding lead to the melt. James A. Mulholland, *History of Metals in Colonial America* (Tuscaloosa: University of Alabama Press, 1981), p. 93.

19. Ledger titled "1787 The Iron Furnace D. to Stock," 1792–1794 furnace records, reel 6, vol. 9, part 2, RFP.

20. Kauffman, *American Copper and Brass*, pp. 171–172; Camp, *Bell Ringing*, pp. 84–86.

21. For example, in one August 30, 1802 letter to Messrs. Heywood, Flagg, and Stowell, Revere states, "Your Bell is an exact coppy from the English Bell in Christ Church in Boston" (Letterbook 1801–1806, reel 14, vol. 53, RFP). Revere's list of bells and their weights also indicates groupings of bells with similar weights: for example, in the years 1804 through 1807 he cast bells with weights of 1,015, 1,017, 1,017, 1,018, and 1,020 pounds. See "Account Book Boston 1793–1810," reel 6, vol. 9, RFP, and "Stock Book 1793–1828," reel 11, vol. 40, RFP.

22. Because the molten metal flowed downward into the mold it had to lie at a lower elevation than the furnace, probably in a large pit that could be expanded or partially filled in to bring each bell to the proper height.

23. Porter, *Rambles in Old Boston*, p. 241.

24. Paul Jr. no longer played a central role in the family businesses by the 1790s, as his primary function seems to have been the management of the silver shop, whose importance faded in comparison to the more lucrative ironwork, copperwork, and bronzework in the foundry. Paul Revere paid many of Paul. Jr.'s expenses, including his taxes, rent, and wood, for the remainder of Paul Jr.'s life. Evidence of Revere's support of his oldest son is located in several receipts in the receipt book, Boston 1780–1805 (reel 12, vol. 41, RFP) as well as notations in the wastebook and memoranda (reel 5, vol. 2, RFP).

25. Undated entry, "Stock Book 1793–1828," reel 11, vol. 40, p. 1, RFP.

26. Quoted in Forbes, *Paul Revere and the World He Lived In*, p. 390.

27. Undated entry, "Stock Book 1793–1828." reel 11, vol. 40, p. 1, RFP. Although this is undated, the reference to dollars instead of pounds and shillings implies that it took place after 1795.

28. Renee Lynn Ernay, "The Revere Furnace" (master's thesis, University of Delaware, 1989), p. 13.

29. "Memorandum of Bells cast by me Paul Revere," reel 6, vol. 9, RFP, and entry in Stock Book 1793–1828, reel 11, vol. 40, RFP.

30. Revere has many notations of bell prices throughout his records, including those in his Boston Wastebook, 1799–1803, reel 5, vol. 4, RFP; Boston Wastebook, 1804–1811, reel 5, vol. 5, RFP, and many others.

31. William H. McNeill, *The Pursuit of Power: Technology, Armed Force, and Society since A.D. 1000* (Chicago: University of Chicago Press, 1984), pp. 81–83, 86–88; Geoffrey Parker, *The Military Revolution: Military Innovation and the Rise of the West, 1500–1800* (New York: Cambridge University Press, 1996), pp. 7–9, 83–84; Albert C. Manucy, *Artillery Through the Ages* (Washington: U.S. Printing Office, 1962 reprint), pp. 3–4.

32. Venetian Senate quote taken from Parker, *The Military Revolution,* p. 10. Also see McNeill, *The Pursuit of Power,* pp. 87–89, 100; Parker, *The Military Revolution,* pp. 89–90, 128–129.

33. Knox quote taken from Mulholland, *A History of Metals in Colonial America,* p. 118. Also see Mulholland, *A History of Metals in Colonial America,* pp. 122–123; Manucy, *Artillery Through the Ages,* p. 10.

34. Quote taken from Mulholland, *A History of Metals in Colonial America,* p. 129; also see pp. 124–127; Hazen and Hazen, *Wealth Inexhaustible,* p. 204.

35. Mulholland, *A History of Metals in Colonial America,* p. 7.

36. McNeill, *The Pursuit of Power,* pp. 166–170; Kauffman, *American Copper and Brass,* p. 196; T. K. Derry and Trevor Williams, *A Short History of Technology* (New York: Oxford University Press, 1960), pp. 150, 350. Revere paid Elib Faxon more than 121 pounds to bore and finish 10 howitzers and 12 cannon in 1795. The receipt from Faxon, displayed in the following section, includes charges such as "To turning & boring & filing." Receipt dated "1795 Paul Revere to Elib Faxon," in "Loose Manuscripts 1746–1801," reel 1, RFP.

37. Revere to James Byers, May 24, 1795, "Letterbook 1783–1800," reel 14, vol. 53.1, RFP.

38. Robert B. Gordon, *American Iron, 1607–1900* (Baltimore: Johns Hopkins University Press, 1996), p. 202; Clyde Sanders and Dudley Gould, *History Cast in Metal* (Naperville, Ill.: Cast Metals Institute, 1976), p. 152.

39. Henry Knox letter to Paul Revere, March 11, 1794, "Loose Manuscripts 1746–1801," reel 1, RFP.

40. Tench Coxe letter to Paul Revere, June 16, 1794, Tench Coxe contract with Paul Revere, July 23, 1794, Henry Knox letter to Paul Revere, June 17, 1794, all in "Loose Manuscripts 1746–1801," reel 1, RFP.

41. Revere's quotes to Byers taken from letters to James Byers on October 29, 1795, and May 24, 1795, from "Letterbook 1783–1800," reel 14, vol. 53.1, RFP. Also see Tench Coxe contract with Paul Revere, July 23, 1794, "Loose Manuscripts 1746–1801," reel 1, RFP.

42. Ledger entry beginning "Memorandum of Stock in Furnace May 1799 as taken by Joshua & Jos. W. Revere," in "Account Book Boston, 1783–1804," reel 6, vol. 9b, RFP.

43. These figures are taken from Revere's tally entitled "Boston Octo 22 1795 To casting & finishing 10 brass howitzers weighing . . . ," in "Account Book Boston, 1783–1804," reel 6, vol. 9, RFP. Revere did not know the amount of leftover copper that he sent to Byers, so he wasted approximately 8,200 pounds of metal minus whatever he sent.

44. Revere letter to Tench Coxe, July 31, 1794, "Loose Manuscripts 1746–1801," reel 1, RFP.

45. Revere letter to Tench Coxe, July 31, 1794, "Loose Manuscripts 1746–1801," reel 1, RFP. Revere's many letters to different merchants and sea captains on the subject

of copper and tin purchases are collected in the "Letterbook 1783–1800," reel 14, vol. 53.1, RFP.

46. Quote about the degree of waste taken from Revere letter to Nathaniel Gorham, January 27, 1796, "Letterbook 1783–1800," reel 14, vol. 53.1, RFP. General contract information appears in the ledger beginning with "Boston New England 1794, Aug 16, Furnace for Brass Howitzers," and ledger six pages later, titled "1794 The United States," both in "Account Book Boston, 1783–1804," reel 6, vol. 9b, RFP. Revere's contract and some of his defenses of his operations are included in Tench Coxe to Paul Revere, July 21, 1794, "Loose Manuscripts 1746–1801," reel 1, RFP; in letters to Tench Coxe on November 3, 1794, to Henry Knox on November 3, 1794, and to Nathaniel Gorham, January 27, 1796, in "Letterbook 1783–1800," reel 14, vol. 53.1, RFP; and also in an undated 1796 entry in "Account Book Boston, 1783–1804," reel 6, vol. 9b, RFP.

47. Revere letter to General Knox, November 3, 1794, "Letterbook 1783–1800," reel 14, vol. 53.1, RFP.

48. Quote taken from Gordon, *American Iron*, p. 3. Revere's contract information located in April 1, 1798 ledger titled "Henry Jackson Esqu," "Account Book Boston, 1783–1804," reel 6, vol. 9b, RFP.

49. Revere to James Lawrason, June 14, 1795, "Letterbook 1783–1800," reel 14, vol. 53.1, RFP; ledgers beginning with "Boston April 16 1795," "Account Book Boston, 1783–1804," reel 6, vol. 9b, RFP.

50. James Lawrason to Paul Revere, undated letter (date given as July 9, 1794, in following letter cited here), Lawrason to Paul Revere, May 28, 1795, Revere to Lawrason, June 14, 1795, and Lawrason to Paul Revere July 17, 1795, in "Loose Manuscripts 1746–1801," reel 1, RFP, and "Letterbook 1783–1800," reel 14, vol. 53.1, RFP.

51. William Rhodes and Nathaniel Fisher to Revere, May 10, 1798, "Loose Manuscripts 1746–1801," reel 1, RFP.

52. Revere to Rhodes and Fisher, May 16, 1798, "Letterbook 1783–1800," reel 14, vol. 53.1, RFP.

53. Ledgers beginning with "Boston April 16 1795," "Account Book Boston, 1783–1804," reel 6, vol. 9b, RFP.

54. James Lawrason to Paul Revere, undated letter (date given as July 9, 1794, in following letter cited here), Lawrason to Paul Revere, May 28, 1795, Revere to Lawrason, June 14, 1795, and Lawrason to Paul Revere July 17, 1795, "Loose Manuscripts 1746–1801," reel 1, RFP, and "Letterbook 1783–1800," reel 14, vol. 53.1, RFP.

55. Revere to Rhodes and Fisher, May 16, 1798, "Letterbook 1783–1800," reel 14, vol. 53.1, RFP.

56. Revere to Nathan Fisher, August 23, 1798, and Revere to Nathan Fisher February 7, 1799, in "Letterbook 1783–1800," reel 14, vol. 53.1, RFP; ledger titled "William Rhodes & Nathan Fischer," August 30, 1798 in "Account Book Boston, 1783–1804," reel 6, vol. 9b, RFP.

57. Revere's pricing information appears in an undated 1796 entry in "Account Book Boston, 1783–1804," reel 6, vol. 9b, RFP. Other data taken from ledgers beginning with "Boston April 16 1795," "Account Book Boston, 1783–1804," reel 6, vol. 9b, RFP. Ledger dates usually correspond to the first entry in the ledger, as Revere often does not date the final transaction of each contract.

58. Elib Faxon receipt dated April 2, 1796, "Loose Manuscripts," reel 1, RFP. This receipt is reproduced in Appendix 6.

59. Otis E. Young Jr., "Origins of the American Copper Industry," *Journal of the Early Republic* 3 (Summer 1983): 118–123; Charles K. Hyde, *Copper for America* (Tucson: University of Arizona Press, 1998), p. 9; Mulholland, *A History of Metals in Colonial America*, pp. 43–47, 53.

60. The transatlantic shipment of copper ore could be very profitable thanks to the high metal content of copper ore, which often exceeded 50 percent usable metal. Iron ore, in contrast, typically contains less than 10 percent usable metal, partially explaining why blast furnaces usually smelted iron close to its source. Demand for iron also dwarfed copper demand. Mulholland, *A History of Metals in Colonial America*, pp. 49, 53, 91, 166; Kauffman, *American Copper and Brass*, p. 17; Edwin Tunis, *Colonial Craftsmen* (Baltimore: Johns Hopkins University Press, 1965), pp. 76–77.

61. Actual copper mine output is difficult to establish due to unreliable recordkeeping. The Simsbury and Shuyler mines produced most American copper, and American output plunged when they slowed or shut down. American copper output exploded in the 1840s with the discovery of vast quantities of pure copper in Michigan, which by 1850 accounted for 88 percent of America's copper production. Mulholland, *A History of Metals in Colonial America*, pp. 21, 43–47, 53; Hyde, *Copper for America*, pp. 4–6; Young, "Origins of the American Copper Industry," pp. 118–123, 135–137.

62. Thomas Cooper, in *The Emporium of Arts and Sciences*, vol. 3 (Philadelphia: Kimber & Richardson, 1814), p. 50. Also partially quoted in Hyde, *Copper for America*, p. 3.

63. Hyde, *Copper for America*, pp. 7, 10; Mulholland, *A History of Metals in Colonial America*, pp. 91–92; Kauffman, *American Copper and Brass*, p. 37; Tunis, *Colonial Craftsmen*, pp. 76–77.

64. The demand for bolts and spikes increased dramatically when shipbuilders started fastening copper sheets to the outside of ship hulls at this time to prolong hull life, a development covered at length in the following chapter. Rusting was exacerbated by a chemical reaction known as galvanic action between the iron fasteners and copper sheets. J. R. Harris, *Industrial Espionage and Technology Transfer* (Brookfield: Ashgate Publishing, 1998), p. 263.

65. Revere to unknown recipient, quoted in Goss, *The Life of Colonel Paul Revere*, p. 544, taken from a private autograph collection.

66. April 24, 1797 letter to John Brown Esq., "Letterbook 1783–1800," reel 14, vol. 53.1, RFP.

67. Revere to Harrison G. Otis, March 11, 1800, H. G. Otis Papers, Massachusetts Historical Society.

68. The concept of "strength" can be further divided according to the type of stress being resisted: tensile strength is resistance to pulling apart, shear strength is resistance to crosswise stresses, and bending strength, unsurprisingly, is resistance to bending motion.

69. Gordon, *American Iron*, p. 13. Even with respect to iron, a metal used and understood far more than copper in Revere's day, skilled workers did not fully comprehend the ramifications of their procedures or the different characteristics of metals.

70. Plastic deformation differs from elastic deformation, which, as the name implies, involves a temporary change in a substance's shape for as long as an external force is applied to it. An elastically deformed substance is able to revert back to its original shape when this force is removed. A plastically deformed substance is permanently altered and

does not return to its original shape. Elastic deformation can be envisioned as a rubber band stretching and returning to its exact starting state.

71. Harris, *Industrial Espionage and Technology Transfer*, p. 264.

72. Revere to Edward Edwards, January 20, 1800, "Letterbook 1783–1800," reel 14, vol. 53.1, RFP.

73. Revere to Jacob Sheaf, October 28, 1795, "Letterbook 1783–1800," reel 14, vol. 53.1, RFP.

74. February 7, 1796 letter to unknown recipient, "Letterbook 1783–1800," reel 14, vol. 53.1, RFP.

75. February 7, 1796 letter to unknown recipient, "Letterbook 1783–1800," reel 14, vol. 53.1, RFP.

76. Revere to Jacob Sheafe, January 7, 1799, "Letterbook 1783–1800," reel 14, vol. 53.1, RFP.

77. Paul Revere to Jacob Sheafe Esq., January 7, 1799, "Letterbook 1783–1800," reel 14, vol. 53.1, RFP.

78. Revere to Jacob Sheafe, February 13, 1799, "Letterbook 1783–1800," reel 14, vol. 53.1, RFP.

79. Revere to Harrison G. Otis, March 11, 1800, H. G. Otis Papers, Massachusetts Historical Society.

80. U.S. Office of Naval Records and Library, *Naval Documents Related to the Quasi-war Between the United States and France* (Washington: U.S. Government Printing Office, 1935–1938), vol. 3, p. 223.

81. Ledger beginning with "1795 Octo 14 United States," in "Account Book Boston, 1783–1804," reel 6, vol. 9b, RFP.

82. Thomas Cochran and William Miller, *The Age of Enterprise: A Social History of Industrial America* (New York: Harper Torchbooks, 1961), p. 8.

83. Revere to Naval Committee building the ship at Mr. Hartt's yard, August 31, 1798, and Revere to Jacob Sheafe, December 7, 1798, both in "Letterbook 1783–1800," reel 14, vol. 53.1, RFP.

84. Ledger entry beginning "The Committee for building the 32 Gun Frigate at Harts Yard," in "Account Book Boston, 1783–1804," reel 6, vol. 9b, RFP.

85. Miscellaneous ledger entries in "Account Book Boston, 1783–1804," reel 6, vol. 9b, RFP. This book is a collection of various records, not strictly arranged in chronological order. The beginning entry used in this analysis is a ledger page labeled "Furnace 1799," and the transactions list extends for many pages, although it is interspersed with wastebook entries. The following several pages of discussion all draw upon this source material.

CHAPTER SIX: Paul Revere's Last Ride: The Road to Rolling Copper (1798–1801)

1. Paul Revere to Benjamin Stoddard (sic) Esq., December 31, 1798, "Letterbook 1783–1800," Revere Family Papers (hereafter RFP), microfilm edition, 15 reels (Boston: Massachusetts Historical Society, 1979), reel 14, vol. 53.1.

2. Leonard D. White, *The Federalists: A Study in Administrative History* (New York: Macmillan, 1948), p. 159; Ian W. Toll, *Six Frigates: The Epic History of the Founding of the U.S. Navy* (New York: W. W. Norton, 2006), pp. 47–48. Humphreys is also remembered

for designing the *USS Constitution* and five other frigates. Revere and Humphreys eventually became close friends.

3. White, *The Federalists*, pp. 470–474.

4. Some government departments had started establishing their own administrative systems under the Articles of Confederation but few of these officials continued in office under the new Constitution, with the key exception of War Secretary Henry Knox. Merrill Jensen, *The New Nation* (Boston: Northeastern University Press, 1981), pp. 360–364, 373.

5. White, *The Federalists*, p. 363; Russell F. Weigley, *History of the United States Army* (New York: Macmillan Company, 1967), pp. 82–84; Richard H. Kohn, *Eagle and Sword* (New York: Macmillan Publishing, 1975), pp. 114–115, 123, 127, 187–188.

6. Massive errors in the War Department's procurement process in 1790 gave Alexander Hamilton an excuse to implement a more centralized purchasing system under the Treasury Department's control. In 1798 a congressional committee returned purchasing authority to the War and Navy departments. White, *The Federalists*, pp. 33, 148, 343–344, 360–362; Kohn, *Eagle and Sword*, pp. 111, 123–125.

7. Donald Hickey, *The War of 1812: A Forgotten Conflict* (Urbana: University of Illinois Press, 1989), p. 6. America's prosperity also resulted from the war between France and Britain, which created huge markets for goods and reduced competition among merchants.

8. Stanley Elkins and Eric McKitrick, *The Age of Federalism* (New York: Oxford University Press, 1993), p. 647; Michael A. Palmer, *Stoddert's War* (Annapolis: Naval Institute Press, 1987), pp. 4–6; Toll, *Six Frigates*, pp. 78–81. French hostility against American shipping originated in the West Indies as much as the French Directory: the Directory instructed French vessels and courts to treat neutral shipping the same way Britain treated it, but since local courts had no way of knowing what this meant, they did as they pleased. In the West Indies, this led to numerous captures and mock trials that always resulted in condemnation of American ships.

9. Elkins and McKitrick, *The Age of Federalism*, pp. 629–630; Palmer, *Stoddert's War*, pp. 7–8; White, *The Federalists*, p. 156.

10. Quoted in Palmer, *Stoddert's War*, pp. 125 and 10. Also see Elkins and McKitrick, *The Age of Federalism*, pp. 634–635; Palmer, *Stoddert's War*, pp. 12–13; Toll, *Six Frigates*, pp. 105–107.

11. The North African Barbary pirates attacked merchant ships sailing in Mediterranean waters, enslaving their crews and selling their cargo. Even the powerful European nations agreed to pay tribute to these attackers, in addition to ransoms for well-to-do captives.

12. Humphreys achieved these incredible properties through a combination of design choices: extra sails and a longer hull led to greater speed; numerous heavy cannons (forty-four guns total) produced greater firepower; and an extra thick hull made of live oak (only available in North America and Cuba) gave his ships additional durability. These modifications would normally lead to "hogging," or a sagging in the center of the wooden hull resulting from the additional stresses from the ship's increased weight and length. Hogging would greatly slow and eventually sink a ship. Humphreys mitigated this hogging via huge internal braces that used the weight of the cannons to push down against the center of the hull and prevent it from buckling inward. White, *The Federal-*

ists, p. 157; Palmer, *Stoddert's War*, pp. 125–126; Elkins and McKitrick, *The Age of Federalism*, p. 644; Hickey, *The War of 1812*, p. 91; Toll, *Six Frigates*, pp. 48–50.

13. Palmer, *Stoddert's War*, pp. 125–126.

14. U.S. Office of Naval Records and Library, *Naval Documents related to the Quasi-war between the United States and France*, vol. 1 (Washington, D.C.: U.S. Government Printing Office, 1935), pp. v–vi, 185; Leonard D. White, *The Jeffersonians: A Study in Administrative History* (New York: Macmillan, 1951), p. 291.

15. Hickey, *The War of 1812*, p. 7; Palmer, *Stoddert's War*, pp. 98–103, 130–131, 235; Elkins and McKitrick, *The Age of Federalism*, pp. 653–654, 890–891; Toll, *Six Frigates*, pp. 117–120, 128.

16. Palmer, *Stoddert's War*, pp. 125–127, 141.

17. According to Tench Coxe's *Statement of the Arts and Manufactures of the United States of America for the year 1810* (Philadelphia, 1814), hemp was produced in every state in 1810, with Kentucky providing the largest quantity. Also see U.S. Office of Naval Records and Library, *Naval Documents related to the Quasi-war between the United States and France*, vol. 2 (Washington, D.C.: U.S. Government Printing Office, 1935), pp. 129–134.

18. Palmer, *Stoddert's War*, pp. 33–35.

19. Nathan Rosenberg, "Why in America?" in *Yankee Enterprise: The Rise of the American System of Manufactures*, ed. Otto Mayr and Robert C. Post (Washington: Smithsonian Institution Press, 1981), pp. 53–54; John F. Kasson, *Civilizing the Machine* (New York: Hill and Wang, 1979), p. 22.

20. Cicero quoted in Kasson, *Civilizing the Machine*, p. 7. Also see Walter Licht, *Industrializing America* (Baltimore: Johns Hopkins University Press, 1995), pp. 2, 16; Kasson, *Civilizing the Machine*, pp. 4–8; John R. Nelson, *Liberty and Property* (Baltimore: Johns Hopkins University Press, 1987), pp. 41–45; Schultz, *The Republic of Labor*, p. 114; Drew R. McCoy, *The Elusive Republic* (New York: Norton, 1982), pp. 107–111.

21. Licht, *Industrializing America*, pp. 15–16; Kasson, *Civilizing the Machine*, pp. 16–17. American government officials, including Jefferson, encouraged the emigration of British artisans. Artisan knowledge was seen as a method of securing economic independence from England, while large-scale manufacturing inspired concern. Doron S. Ben-Atar, *Trade Secrets: Intellectual Piracy and the Origins of American Industrial Power* (New Haven: Yale University Press, 2004), pp. 118–133.

22. Quoted from Ralph Ketcham, ed., *The Political Thought of Benjamin Franklin* (New York: Hackett Publishing Company, 2003), p. 364.

23. Kasson, *Civilizing the Machine*, pp. 16–17, 20, 38, 40; Licht, *Industrializing America*, pp. 13–14.

24. Rush became the first president of the United Company of Philadelphia for Promoting American Manufactures in 1775. This company employed almost five hundred people to produce cotton, linen, and woolen products. Rush warned that dependence upon British goods would soon become subjection to British domination and even slavery, while domestic manufactures would increase independent self-reliance and efficiently employ the nonagricultural labor force. Kasson, *Civilizing the Machine*, pp. 9–10.

25. Coxe, *Statement of the Arts and Manufactures of the United States of America for the year 1810*, p. xxv.

26. Licht, *Industrializing America*, p. 2; Jacob E. Cooke, *Tench Coxe and the Early Republic* (Chapel Hill: University of North Carolina Press, 1978), pp. 156–157; Kasson, *Civilizing the Machine*, pp. 29–30; McCoy, *The Elusive Republic*, pp. 113–117; Lawrence A. Peskin, *Manufacturing Revolution: The Intellectual Origins of Early American Industry* (Baltimore: Johns Hopkins University Press, 2003), pp. 94–97.

27. Michael B. Folsom and Steven D. Lubar, *The Philosophy of Manufactures* (Chapel Hill: University of North Carolina Press, 1985), pp. xxiii, xxvi; Nelson, *Liberty and Property*, pp. 43–45; Merritt Roe Smith, "Technological Determinism in American Culture" in *Does Technology Drive History*, ed. Merritt Roe Smith and Leo Marx (Cambridge, Mass.: MIT Press, 1994), pp. 2–5; Coxe, *Statement of the Arts and Manufactures of the United States of America for the year 1810*, p. xx.

28. Licht, *Industrializing America*, pp. 13–14; Brooke Hindle, *Emulation and Invention* (New York: W. W. Norton and Company, 1981), p. 21; John R. Nelson, "Manufactures Reconsidered," in *Major Problems in the History of American Technology*, ed. Merritt Roe Smith and Gregory Clancey (Boston: Houghton Mifflin Company, 1998), pp. 132–133.

29. The most spectacular failure of the early manufacturing societies was the Society for Establishing Useful Manufactures, formed by some of Alexander Hamilton's colleagues with his support in 1791. Aided by political support and tax exemption from different state legislatures, a highly influential group of investors, a favorable charter from New Jersey, and more than $625,000 in capital from sales of its shares, it intended to create a cotton factory of unprecedented size that would eventually produce iron, glass, paper, and other products. However, it lacked technical and managerial manufacturing expertise and failed after some of its directors became bankrupt in a financial crisis of 1792. Worse yet, the Society for Establishing Useful Manufactures' high-profile connection between a for-profit institution and public officials who stood to benefit from it raised opposition to government sponsorship of industry, which suppressed the pro-manufacturing supporters for many years after its public failure. Curtis P. Nettels, *The Emergence of a National Economy, 1775–1815* (New York: Holt, Rinehart, and Winston, 1962), pp. 124–125; Brooke Hindle and Steven Lubar, *Engines of Change: The American Industrial Revolution, 1790–1860* (Washington: Smithsonian Institution Press, 1986), p. 89; Hindle, *Emulation and Invention*, pp. 16–17; Jensen, *The New Nation*, pp. 224–226, 286–287, 293, 297; Neil York, *Mechanical Metamorphosis: Technological Change in Revolutionary America* (Westport: Greenwood Press, 1985), pp. 41–42, 164–167; Licht, *Industrializing America*, pp. 17–18.

30. Maurer Maurer, "Coppered Bottoms for the Royal Navy: A Factor in the Maritime War of 1778–1783," *Military Affairs* 14, no. 2 (Summer 1950): 57, 60.

31. Quote from R. J. B. Knight, "The Introduction of Copper Sheathing into the Royal Navy 1779–1786," *The Mariner's Mirror* 59 (August 1973): 299. Also see J. R. Harris, *Industrial Espionage and Technology Transfer* (Brookfield: Ashgate Publishing, 1998), pp. 262–264; Edgard Moreno, "Patriotism and Profit: The Copper Mills at Canton" in *Paul Revere—Artisan, Businessman, and Patriot* (Boston: Paul Revere Memorial Association, 1988), p. 98.

32. Quotes and other information taken from Maurer, "Coppered Bottoms for the Royal Navy," pp. 58–60.

33. Knight, "Introduction of Copper Sheathing," pp. 299–302, 306–307; Charles K. Hyde, *Copper for America* (Tucson: University of Arizona Press, 1998), p. 10; Harris, *Industrial Espionage and Technology Transfer*, pp. 262–264.

34. The French navy went to great lengths to understand and emulate the developments taking place in the Royal Navy. Faced with their own massive and debilitating costs for replacing wooden hulls, French manufacturers attempted to duplicate the copper sheet rolling process but had great difficulty mastering it. Prior to 1783, France relied upon hammered copper plates, which were rough, often fell off because the fasteners rusted, and were not fully effective at deterring barnacles. But successful industrial espionage and the hiring of English workers allowed France to catch up. The French also depended on English-made rollers, which were increasingly hard to come by. Harris, *Industrial Espionage and Technology Transfer*, pp. 267–277.

35. Maurer, "Coppered Bottoms for the United States Navy," p. 694.

36. Knight, "Introduction of Copper Sheathing," pp. 299–300; Margaret H. Hazen and Robert M. Hazen, *Wealth Inexhaustible* (New York: Van Nostrand Reinhold, 1985), p. 92; Maxwell Whiteman, *Copper for America: The Hendricks Family and a National Industry, 1755–1939* (New Brunswick: Rutgers University Press, 1971), p. 47; Moreno, "Patriotism and Profit: The Copper Mills at Canton," p. 101.

37. Whiteman, *Copper for America*, p. 47; Moreno, "Patriotism and Profit: The Copper Mills at Canton," p. 99; James A. Mulholland, *History of Metals in Colonial America* (Tuscaloosa: University of Alabama Press, 1981), p. 161.

38. Maurer, "Coppered Bottoms for the United States Navy," p. 696.

39. The dates of early iron-rolling mills in America are unknown, but they certainly started to appear around 1810 or so. Thomas C. Cochran, *Frontiers of Change: Early Industrialism in America* (New York: Oxford University Press, 1981), p. 72; Mulholland, *A History of Metals in Colonial America*, pp. 92–93; Henry Kauffman, *American Copper and Brass* (Camden, N.J.: T. Nelson, 1968), pp. 20–23.

40. U.S. Office of Naval Records and Library, *Naval Documents related to the Quasi-war between the United States and France*, vol. 3 (Washington, D.C.: U.S. Government Printing Office, 1936), p. 194.

41. "The 74's" refer to the new 74-gun ships under construction. U.S. Office of Naval Records and Library, *Naval Documents related to the Quasi-war between the United States and France*, vol. 4 (Washington, D.C.: U.S. Government Printing Office, 1937), p. 112.

42. Nicholas Roosevelt was a member of the famous Roosevelt family and great-granduncle to President Theodore Roosevelt. Following these early manufacturing attempts, he worked on steamboat technology with Robert Fulton and eventually built and sailed the *New Orleans*, the first steamboat to operate on western rivers.

43. U.S. Office of Naval Records and Library, *Quasi-war between the United States and France*, vol. 4, pp. 118–119.

44. U.S. Office of Naval Records and Library, *Naval Documents related to the Quasi-war between the United States and France*, vol. 5 (Washington, D.C.: U.S. Government Printing Office, 1937), pp. 58–62.

45. U.S. Office of Naval Records and Library, *Quasi-war between the United States and France*, vol. 5, pp. 58–62.

46. Paul Revere to Jacob Sheafe, Esq., December 7, 1798, "Letterbook 1783–1800," reel 14, vol. 53.1, RFP; U.S. Office of Naval Records and Library, *Quasi-war Between the United States and France*, vol. 5, p. 139; Moreno, "Patriotism and Profit: The Copper Mills at Canton," p. 101.

47. Paul Revere to Benjamin Stoddard (sic) Esq., December 31, 1798, "Letterbook 1783–1800," reel 14, vol. 53.1, RFP.

48. Paul Revere to Benjamin Stoddard (sic) Esq., February 26, 1800, "Letterbook 1783–1800," reel 14, vol. 53.1, RFP.

49. Paul Revere to Harrison G. Otis, March 11, 1800, H. G. Otis Papers, Massachusetts Historical Society.

50. U.S. Office of Naval Records and Library, *Quasi-war between the United States and France*, vol. 5, p. 542.

51. U.S. Office of Naval Records and Library, *Naval Documents related to the Quasi-war between the United States and France*, vol. 6 (Washington, D.C.: U.S. Government Printing Office, 1938), pp. 95, 517.

52. Hyde, *Copper for America*, p. 9.

53. Robert B. Gordon and Patrick M. Malone, *The Texture of Industry* (New York: Oxford University Press, 1994), p. 247.

54. The uncertainty surrounding the exact date of Revere's land purchase, which could have happened at any time between January 1800 and March 1801, is described in Jayne E. Triber, *A True Republican* (Amherst: University of Massachusetts Press, 1998), p. 268, note 4.

55. Elbridge Henry Goss, *The Life of Colonel Paul Revere* (Boston: Plimpton Press, 1902), p. 556; Moreno, "Patriotism and Profit: The Copper Mills at Canton," p. 102. Waterpower issues are discussed in more detail in Chapter Eight.

56. September 1800 memorandum in "Memoranda Book 1797–1801," reel 14, vol. 51.14, RFP.

57. Revere to Eben [Lux], January 13, 1801, and Paul Revere to William Bennoch, May 12, 1802, both in "Letterbook 1783–1800," reel 14, vol. 53.1, RFP.

58. Paul Revere to Benjamin Stoddert, January 17, 1801, "Letterbook 1783–1800," reel 14, vol. 53.1, RFP.

59. Paul Revere to Joshua Humphreys, January 22, 1801, "Letterbook 1783–1800," reel 14, vol. 53.1, RFP.

60. Moreno, "Patriotism and Profit: The Copper Mills at Canton," p. 104.

61. Paul Revere to Harrison G. Otis, January 17, 1801, and Paul Revere to Joshua Humphreys, January 22, 1801, in "Letterbook 1783–1800," reel 14, vol. 53.1, RFP.

62. U.S. Office of Naval Records and Library, *Naval Documents related to the Quasi-war between the United States and France*, vol. 7 (Washington, D.C.: U.S. Government Printing Office, 1938), p. 128.

63. White, *The Jeffersonians*, pp. 75, 265, 267, 142–143; Nettels, *Emergence of a National Economy*, pp. 318–319.

64. Paul Revere to Benjamin Stoddert, March 5, 1801, "Letterbook 1801–1806," reel 14, vol. 53.2, RFP.

65. March 10, 1801 letter from Benjamin Stoddert, "Loose Manuscripts 1746–1801," reel 1, RFP.

66. Paul Revere to William Bartlett, date prior to April 3, 1801, "Letterbook 1801–1806," reel 14, vol. 53.2, RFP. The date and some of the text of the letter are missing, but it immediately precedes an April 3, 1801 letter.

67. Paul Revere letter (addressee missing, but other sources indicate that it was Robert Smith), May 11, 1801, "Letterbook 1801–1806," reel 14, vol. 53.2, RFP.

68. U.S. Office of Naval Records and Library, *Quasi-war between the United States and France*, vol. 7, p. 247.

69. "1801 Furnace," in "Account Book, Boston 1793–1810," reel 6, vol. 10, RFP.

These records are presumably comprehensive, but as always, this cannot be claimed with certainty, given Revere's incomplete and often confusing recordkeeping style.

70. This income figure includes the $10,000 loan, which he received in two installments: $6,000 on October 30, 1801, and $4,000 on February 1, 1802.

71. Revere to Levi Lincoln, April 21, 1801, "Letterbook 1801–1806," reel 14, vol. 53.2, RFP. Note that Levi Lincoln was Revere's good friend and the brother of two of Revere's sons-in-law, as well as a close personal contact who was well placed in the government. Revere wrote to him occasionally over the next few years whenever a contract encountered difficulty. These matters had absolutely nothing to do with Lincoln's job responsibilities as attorney general and it is unclear whether he provided Revere with any assistance. We might speculate that, if nothing else, Lincoln might have advocated for Revere behind the scenes, and it seems likely that he offered some service since Revere continued soliciting his aid.

72. Paul Revere to Robert Smith, October 26, [1801], "Letterbook 1801–1806," reel 14, vol. 53.2, RFP.

73. Paul Revere to Robert Smith, October 26, [1801], and Paul Revere to Robert Smith, May 24, 1802, "Letterbook 1801–1806," reel 14, vol. 53.2, RFP.

74. Paul Revere to Robert Smith, November 27, 1803, "Letterbook 1801–1806," reel 14, vol. 53.2, RFP.

75. Palmer, *Stoddert's War*, p. 121.

CHAPTER SEVEN: The Onset of Industrial Capitalism: Managerial and Labor Adaptations (1802–1811)

1. Revere to Robert Smith, May 24, 1802, "Letterbook 1801–1806," Revere Family Papers (hereafter RFP), microfilm edition, 15 reels (Boston: Massachusetts Historical Society, 1979), reel 14, vol. 53.2.

2. Revere to Joshua Humphreys, December 19, 1803, "Letterbook 1801–1806," reel 14, vol. 53.2, RFP.

3. Revere to Henry Dearborn, September 13, 1809, "Letterbook 1805–1810," reel 14, vol. 53.5, RFP.

4. Jonathan Prude, *The Coming of Industrial Order* (Cambridge: Cambridge University Press, 1983), p. 17.

5. The number of American post offices increased from 75 in 1790, to 903 in 1800, to 2,300 in 1810. Richard R. John, *Spreading the News: The American Postal System from Franklin to Morse* (Cambridge, Mass.: Harvard University Press, 1995), pp. 26–31, 51–57.

6. George Rogers Taylor, *The Transportation Revolution, 1815–1860* (New York: M. E. Sharpe, Inc., 1977), pp. 15–19; David R. Meyer, *The Roots of American Industrialization* (Baltimore: Johns Hopkins University Press, 2003), pp. 28–30; Curtis P. Nettels, *The Emergence of a National Economy, 1775–1815* (New York: Holt, Rinehart, and Winston, 1962), pp. 251–253; Paul A. Gilje, "The Rise of Capitalism in the Early Republic," *Journal of the Early Republic* 16, no. 2 (Summer 1996): 165.

7. The shape of America's growing market economy was not a foregone conclusion: different interests advocated for a neo-mercantilist market featuring strong government control of economic expansion, for unrestricted markets governed by the laws of supply and demand, or for a more traditional "yeoman" producer-republic. The unregulated market won out in the short term and helped promote industrialization, which fostered

additional market growth in turn. Therefore, industrialization served as both a product and an instigator of economic changes. Walter Licht, *Industrializing America* (Baltimore: Johns Hopkins University Press, 1995), pp. xvi–xvii.

8. Meyer, *Roots of American Industrialization*, pp. 7, 25; Licht, *Industrializing America*, pp. 4, xvii; Winifred Rothenberg, *From Market-Places to a Market Economy* (Chicago: University of Chicago Press, 1992), pp. 143–144, 242–244; Winifred Barr Rothenberg, "The Invention of American Capitalism: The Economy of New England in the Federal Period," in *Engines of Enterprise: An Economic History of New England,* ed. Peter Temin (Cambridge, Mass.: Harvard University Press, 2000), p. 75; Nettels, *Emergence of a National Economy,* p. 289; Michel Beaud, *A History of Capitalism, 1500–1980* (New York: Monthly Review Press, 1983), pp. 72–73.

9. As mentioned in earlier chapters, capitalism existed to a lesser degree in the colonial period as well, as illustrated by the use of circulating money, European banks, corporations, and profit calculations. Gilje, "Rise of Capitalism in the Early Republic," pp. 169–173, 178; John Larson, "A Bridge, A Dam, A River: Liberty and Innovation in the Early Republic," *Journal of the Early Republic* 7, no. 4 (Winter 1987): 354; Jan de Vries, "The Industrial Revolution and the Industrious Revolution," *Journal of Economic History* 54, no. 2 (June 1994): 258; Rothenberg, "Invention of American Capitalism," p. 69; Jonathan Prude, "Capitalism, Industrialization, and the Factory in Post-revolutionary America," *Journal of the Early Republic* 16, no. 2 (Summer 1996): 242.

10. England reeled from a widespread anti-machine movement in 1811 and 1812, when manufacturing laborers gathered in the name of Nad Ludd and destroyed countless looms, spinning jennies, and other machines in more than a thousand textile mills. These "Luddite" laborers did not hate all machines, but merely protested specific implementations of labor-saving equipment that exacerbated the unemployment and poverty common at the time. In America no similar movement arose because this category of machinery filled a void. Meyer, *Roots of American Industrialization*, pp. 11, 37; Robert Gross, *The Minutemen and Their World* (New York: Hill and Wang, 1976), pp. 85–87; Licht, *Industrializing America*, pp. 42, 46–47; Thomas Cochran and William Miller, *The Age of Enterprise: A Social History of Industrial America* (New York: Harper Torchbooks, 1961), p. 5; Prude, "Capitalism, Industrialization, and the Factory," p. 245; Rothenberg, "Invention of American Capitalism," pp. 70, 93.

11. Gilje, "Rise of Capitalism in the Early Republic," p. 169; Licht, *Industrializing America*, pp. 13–14; John F. Kasson, *Civilizing the Machine* (New York: Hill and Wang, 1979), pp. 10, 19, 30, 41.

12. Prude, "Capitalism, Industrialization, and the Factory," pp. 247, 251; Alan Dawley, *Class and Community: The Industrial Revolution in Lynn* (Cambridge, Mass.: Harvard University Press, 1976), pp. 222–224.

13. Lindy Biggs, *The Rational Factory* (Baltimore: Johns Hopkins University Press, 1996), pp. 19–21; Thomas C. Cochran, *Frontiers of Change: Early Industrialism in America* (New York: Oxford University Press, 1981), pp. 58–60, 65–69; Kasson, *Civilizing the Machine*, p. 28; Otto Mayr and Robert C. Post, eds., *Yankee Enterprise: The Rise of the American System of Manufactures* (Washington: Smithsonian Institution Press, 1981), p. vii.

14. Nathan Rosenberg, "*Yankee Enterprise: The Rise of the American System of Manufactures,*" ed. Otto Mayr and Robert C. Post (Washington: Smithsonian Institution Press, 1981), p. 57; Nathan Rosenberg, "Technological Change in the Machine Tool Industry,"

Journal of Economic History 22 (1963): 423–443; Merritt Roe Smith, *Harpers Ferry Armory and the New Technology* (Ithaca: Cornell University Press, 1977).

15. Cochran, *Frontiers of Change*, p. 10.

16. See Merritt Roe Smith's *Becoming Engineers in Early Industrial America*, STS Working Paper no. 13 (Cambridge, Mass.: MIT, 1990), esp. pp. 22–24, and Cochran, *Frontiers of Change*, p. 58.

17. From 1774 to 1815 equipment occupied only 4 to 5 percent of capital stock while land and buildings rose from 17 to 27 percent. Meyer, *Roots of American Industrialization*, pp. 59–61.

18. Bank quotation taken from Sean Wilentz, *Chants Democratic* (New York: Oxford University Press, 1984), p. 29. Three separate attempts to create a New York bank to loan money to needy artisans failed, either because the bank wavered in its desire to make these loans or because the bank collapsed. In the ensuing years the number of banks steadily increased, numbering in the thousands by the 1830s. By that point the larger number of banks did increase the quantity of circulating currency as well as available capital, greatly expanding the economy. Also see Gilje, "Rise of Capitalism in the Early Republic," pp. 162–165; Stuart Bruchey, *The Roots of American Economic Growth, 1607–1861* (New York: Harper, 1965), p. 213; Howard B. Rock, *Artisans of the New Republic* (New York: New York University Press, 1984), pp. 165, 168–169; Matthew Roth, *Platt Brothers and Company: Small Business in American Manufacturing* (Hanover, N.H.: University Press of New England, 1994), p. 2; Cochran, *Frontiers of Change*, pp. 68–69; Laura Rigal, *The American Manufactory: Art, Labor, and the World of Things in the Early Republic* (Princeton, N.J.: Princeton University Press, 2001), pp. 55–88.

19. Jefferson changed his tune after the embargo of 1807 and the War of 1812, both in his writings and in his actions. He installed numerous machines on his property at Shadwell and Monticello, including a nail manufactory that used three hearths and a cutting machine; a grist mill; and a textile manufactory that included a carding machine, spinning machine, and loom with a flying shuttle. He also advocated for decentralized manufacturing using the most efficient possible machinery; labor of women, children, or men unable to work on farms; and propagation of agricultural interests. Kasson, *Civilizing the Machine*, pp. 24–25; Cochran and Miller, *The Age of Enterprise*, p. 11.

20. Revere to Robert Smith, November 6, 1802, and Revere Letter, recipient unknown (probably Samuel Brown), November 6, 1802, "Letterbook 1801–1806," reel 14, vol. 53.2, RFP.

21. Revere to Robert Smith, November 6, 1802, "Letterbook 1801–1806," reel 14, vol. 53.2, RFP.

22. Revere to Thomas Ramsden, August 4, 1804, "Loose Manuscripts 1802–1813," reel 2, RFP.

23. Leonard D. White, *The Jeffersonians, A Study in Administrative History* (New York: Macmillan, 1956), pp. 423–425, 472–473; Douglass C. North, *The Economic Growth of the United States* (Upper Saddle River, N.J.: Prentice Hall, Inc., 1961), pp. 38, 46, 55.

24. Revere to Joseph Carson, March 6, 1809, "Letterbook 1805–1810," reel 14, vol. 53.3, RFP.

25. Cochran, *Frontiers of Change*, p. 36.

26. Merritt Roe Smith, "Eli Whitney and the American System of Manufacturing," in *Technology in America*, ed. Carroll W. Pursell Jr. (Cambridge, Mass.: MIT Press, 1989), p. 51.

27. Revere to Hathaway and Davis, March 1, 1805, "Letterbook 1801–1806," reel 14, vol. 53.2, RFP.

28. Thomas Hazard to Revere, June 9, 1808, "Loose Manuscripts 1802–1813," reel 2, RFP; February 26, 1806 entries in Revere's "Boston Wastebook 1804–1811," reel 5, vol. 5, RFP.

29. Joseph Carson to Revere, June 6, 1809, "Loose Manuscripts 1802–1813," reel 2, RFP.

30. Quotes taken from January 1806 and September 12, 1803 letters from Harmon Hendricks to Paul Revere, cited in Mark Bortman, "Paul Revere and Son and their Jewish Correspondents," *Publications of the American Jewish Historical Society* 43 (1953–1954): 199–229. The Soho copper plant technically became a competitor to Revere, although the small size of both manufactories kept them from harming each other. Hendricks did testify against Revere's tariff petition, as described later in this chapter, but the cordial and mutually beneficial nature of their continuing correspondence makes it hard to imagine this as a malicious act.

31. Edgard Moreno, "Patriotism and Profit: The Copper Mills at Canton," in *Paul Revere—Artisan, Businessman, and Patriot* (Boston: Paul Revere Memorial Association, 1988), p. 112.

32. Revere to Josiah Snelling, October 26, 1810, "Letterbook 1809–1810," reel 14, vol. 53.4, RFP.

33. Folsom and Lubar, *The Philosophy of Manufactures*, pp. xxiii, xxvi.

34. These rates were doubled for the extent of the war, and then readjusted to 20 percent in 1816 and to 25 percent in 1818. Charles K. Hyde, *Copper for America* (Tucson: University of Arizona Press, 1998), p. 12; John R. Nelson, *Liberty and Property* (Baltimore: Johns Hopkins University Press, 1987), p. 153; Howard B. Rock, *The New York City Artisan, 1789–1825* (Albany: State University of New York Press, 1989), pp. 143–147; Rock, *Artisans of the New Republic*, pp. 171–172.

35. Revere to Albert Gallatin and Revere to Josiah Quincy, April 3, 1806, "Loose Manuscripts 1802–1813," reel 2, RFP.

36. Revere to Josiah Quincy, February 12, 1807, "Letterbook 1805–1810," reel 14, vol. 53.5, RFP.

37. Report of the Committee of Commerce and Manufactures, January 21, 1808.

38. Revere to Josiah Quincy, December 12, 1808, "Letterbook 1805–1810," reel 14, vol. 53.5, RFP.

39. Revere to James Prince, December 10, 1810, "Letterbook 1805–1810," reel 14, vol. 53.3, RFP.

40. Revere to Henry Dearborn, September 13, 1809, "Letterbook 1805–1810," reel 14, vol. 53.3, RFP.

41. Gabriel Duvall to Revere, September 4, 1810, "Loose Manuscripts 1802–1813," reel 2, RFP. Also see Otis E. Young Jr., "Origins of the American Copper Industry," *Journal of the Early Republic* 3 (Summer 1983): 127; Revere to James Prince, December 10, 1810, "Letterbook 1805–1810," reel 14, vol. 53.3, RFP; Nettels, *Emergence of a National Economy*, p. 274.

42. Revere to Thomas Ramdsen, August 4, 1804, "Loose Manuscripts 1802–1813," reel 2, RFP.

43. Partnership agreement between Paul and Joseph Warren Revere, June 7, 1804, "Loose Manuscripts 1802–1813," reel 2, RFP.

44. Partnership renewal between Paul and Joseph Warren Revere, June 7, 1807, "Loose Manuscripts 1802–1813," reel 2, RFP.

45. Naomi Lamoreaux, "The Partnership Form of Organization: Its Popularity in Early-Nineteenth Century Boston," in *Entrepreneurs: The Boston Business Community, 1700–1850*, ed. Conrad E. Wright and Katheryn P. Viens (Boston: Massachusetts Historical Society, 1997), pp. 270–271, 275, 293–295; Roth, *Platt Brothers and Company*, pp. 12–14.

46. In 1801 only eight manufacturing corporations had received charters in all of America. Michael B. Folsom and Steven D. Lubar, *The Philosophy of Manufactures* (Chapel Hill: University of North Carolina Press, 1985), p. xxvii; Morton J. Horwitz, *The Transformation of American Law* (Cambridge, Mass.: Harvard University Press, 1977), p. 308; Oscar Handlin and Mary Handlin, *Commonwealth* (Cambridge, Mass.: Belknap Press, 1969), p. 162; Pauline Maier, "The Revolutionary Origins of the American Corporation," *William and Mary Quarterly* 50, no. 1 (January 1993): 51, 53, 82.

47. Nettels, *Emergence of a National Economy*, p. 289; Bruce Laurie, *Artisans into Workers* (New York: Hill and Wang, 1989), pp. 44–45; Lamoreaux, "The Partnership Form of Organization," pp. 273–275.

48. Jack Larkin, *The Reshaping of Everyday Life* (New York: HarperPerennial, 1989), p. 59; Bruce Laurie, "'Spavined Ministers, Lying Toothpullers, and Buggering Priests': Third-Partyism and the Search for Security in the Antebellum North," in *American Artisans: Crafting Social Identity*, ed. Howard B. Rock, Paul A. Gilje, and Robert Asher (Baltimore: Johns Hopkins University Press, 1995), pp. 99, 105–106; Robert B. Gordon and Patrick M. Malone, *The Texture of Industry* (New York: Oxford University Press, 1994), pp. 347–348; Prude, "Capitalism, Industrialization, and the Factory," p. 251.

49. H. J. Habakkuk, *American and British Technology in the Nineteenth Century* (Cambridge: Cambridge University Press, 1962), pp. 4–43; Licht, *Industrializing America*, p. 43; Paul A. David, *Technical Choice Innovation and Economic Growth* (Cambridge: Cambridge University Press, 1975), pp. 21–33.

50. Caleb Gibbs to Revere, July 15, 1803, "Loose Manuscripts 1802–1813," reel 2, RFP. Also see Rock, *Artisans of the New Republic*, p. 242.

51. Laurie, *Artisans into Workers*, pp. 38–41, 45; Licht, *Industrializing America*, p. 49; Biggs, *The Rational Factory*, p. 3; Gordon and Malone, *Texture of Industry*, pp. 385–386; Wilenz, *Chants Democratic*, pp. 30–32, 53–54; Ronald Schultz, *The Republic of Labor: Philadelphia Artisans and the Politics of Class, 1720–1830* (New York: Oxford University Press, 1993), p. 171; Nettels, *Emergence of a National Economy*, pp. 235–236; Beaud, *A History of Capitalism*, pp. 66–67; Merritt Roe Smith, "Industry, Technology, and the 'Labor Question' in 19th Century America: Seeking Synthesis," presidential address for the Society for the October 19, 1990 History of Technology Meeting, reprinted in *Technology and Culture* 32, no. 3 (July 1991): 558–559; de Vries, "The Industrial Revolution and the Industrious Revolution," p. 258.

52. Smith, *Harpers Ferry Armory*, pp. 62–67; Nettels, *Emergence of a National Economy*, p. 265.

53. Laurie, *Artisans into Workers*, p. 36; Licht, *Industrializing America*, pp. 49–51, 69; Wilenz, *Chants Democratic*, pp. 49–50, 56–58.

54. These figures are taken from Moreno, "Patriotism and Profit," p. 106. A comparison with Appendix 8 reveals small discrepancies in the number of employees each year. In some cases Revere's records are ambiguous and subject to interpretation, and in

other cases Moreno and I disagree, for example, as to whether twelve months of service beginning in December should count as one year of employment or two.

55. Revere to Joseph Carson, September 8, 1809, and Revere to Joseph Carson, December 28, 1810, both in "Letterbook 1809–1810," reel 14, vol. 53.4, RFP.

56. Wages listed on page beginning with "15 Nov 1805," in "Canton Ledger 1802–6," reel 10, vol. 29, RFP. Also see Horwitz, *The Transformation of American Law*, pp. 186–187; Cochran, *Frontiers of Change*, pp. 27–28.

57. Thomas Dublin, *Women at Work* (New York: Columbia University Press, 1979); Smith, "Eli Whitney and the American System of Manufacturing," pp. 51–53.

58. Revere to Joshua Humphreys, December 19, 1803, "Letterbook 1801–1806," reel 14, vol. 53.2, RFP.

59. Revere to Mr. Bosworth, October 19, 1805; Revere to Mrs. Bosworth, December 21, 1805, "Letterbook 1801–1806," reel 14, vol. 53.2, RFP.

60. "Canton Ledger 1802–6," reel 10, vol. 29, RFP. This ledger is summarized in Appendix 8.

61. Receipt from Paul Revere to Willaby Dexter, September 20, 1803, "Loose Manuscripts 1802–1813," reel 2, RFP; Revere to the Selectmen of the Town of Canton, February 20, 1808, "Loose Manuscripts 1802–1813," reel 2, RFP.

CHAPTER EIGHT: Becoming Industrial: Technological Innovations and Environmental Implications (1802–1811)

1. Revere's copper sheathing lasted for longer than its expected lifetime, as it remained in use at least until 1810, seven years later. However, by that time the sheathing had lost its effectiveness and the *Constitution's* hull was covered with barnacles, mussels, oysters, and seaweed, ruining its speed and maneuverability. Ian W. Toll, *Six Frigates: The Epic History of the Founding of the U.S. Navy* (New York: W. W. Norton, 2006), p. 319; U.S. Office of Naval Records and Library, *Naval Documents related to the United States Wars with the Barbary Powers*, vol. 2 (Washington, D.C.: U.S. Government Printing Office, 1939), pp. 414, 426, 462; *Columbian Centinel*, June 18, 1803.

2. Thomas Hughes, "The Evolution of Large Technological Systems," in *The Social Construction of Technological Systems*, ed. Wiebe Bijker, Thomas Hughes, and Trevor Pinch (Cambridge, Mass.: MIT Press, 1987). Technological systems theory is recapped in additional detail in Appendix 9. In brief, technological systems theory posits five stages of system growth: invention, development, innovation, transfer, and growth. The invention phase generally involves technical improvements to products or processes, the subsequent development and innovation phases facilitate the commercial implementation of these inventions, and the transfer and growth phases enable the system to expand into new areas. Paul Revere's copper-rolling experience most closely follows technological system theory because he was the first American to "invent" this technology, with little help from Britain's example other than the valuable knowledge that such a process actually did exist. The term *radical invention* describes the key groundbreaking development that inaugurates a new system, and Revere's radical invention involved the establishment of his rolling mill and the associated knowledge regarding annealing and work-hardening. Entrepreneurs improve systems by developing and innovating "conservative" inventions that enable the primary technology to function in a real-world setting, and Revere achieved this when he set up the Canton mill; determined how his

rollers and waterwheels might best utilize the suboptimal flow of the Neponset River; learned to heat copper bars and sheets to the proper temperature before rolling them; and so on. The preceding description simplifies Hughes's complex theory, which itself simplifies actual conditions. For example, Hughes indicates that systems do not always proceed through these stages in a linear manner as implied here.

3. Nathan Rosenberg, *Technology and American Economic Growth* (New York: M. E. Sharpe, Inc., 1972), pp. 1–2, 5–6.

4. Thomas C. Cochran, *Frontiers of Change: Early Industrialism in America* (New York: Oxford University Press, 1981), p. 8.

5. S. R. Epstein, "Craft Guilds, Apprenticeship, and Technological Change in Pre-industrial Europe," *Journal of Economic History* 58, no. 3 (September 1998): 703–704; Brooke Hindle, *Emulation and Invention* (New York: W. W. Norton and Company, 1981), pp. 18–19.

6. Doron S. Ben-Atar, *Trade Secrets: Intellectual Piracy and the Origins of American Industrial Power* (New Haven: Yale University Press, 2004), pp. xiv–xv; Hindle, *Emulation and Invention*, p. 129; Brooke Hindle and Steven Lubar, *Engines of Change: The American Industrial Revolution, 1790–1860* (Washington: Smithsonian Institution Press, 1986), pp. 77–79, 81–83.

7. Revere to Josiah Quincy, December 12, 1808, "Letterbook 1805–1810," Revere Family Papers (hereafter RFP), microfilm edition, 15 reels (Boston: Massachusetts Historical Society, 1979), reel 14, vol. 53.5.

8. Court rulings often endorsed entrepreneurial endeavors and ruled that attempts to improve property took precedence over attempts to perpetuate existing uses. Robert B. Gordon and Patrick M. Malone, *The Texture of Industry* (New York: Oxford University Press, 1994), p. 101; Cochran, *Frontiers of Change*, pp. 9, 11, 15.

9. Bond quotations in Neil York, *Mechanical Metamorphosis: Technological Change in Revolutionary America* (Westport: Greenwood Press, 1985), pp. 172–173. Also see Stuart Bruchey, *The Roots of American Economic Growth, 1607–1861* (New York: Harper, 1965), p. 180; Rosenberg, *Technology and American Economic Growth*, pp. 23–24; David R. Meyer, *The Roots of American Industrialization* (Baltimore: Johns Hopkins University Press, 2003), pp. 64–65; Gordon and Malone, *Texture of Industry*, pp. 38, 386; Hindle, *Emulation and Invention*, p. 128; Cochran, *Frontiers of Change*, pp. 11–12.

10. The importance of skilled labor emigration, particularly during America's colonial and post-Revolutionary years, is told in great detail in Ben-Atar, *Trade Secrets*. Also see John F. Kasson, *Civilizing the Machine* (New York: Hill and Wang, 1979), p. 26.

11. Revere to William Ben[noch], May 12, 1802, "Letterbook 1801–1806," reel 14, vol. 53.2, RFP.

12. An identifiable American machine-tool or machine-producing sector appeared between 1840 and 1880. Nathan Rosenberg, "Technological Change in the Machine Tool Industry," *Journal of Economic History* 22 (1963): 416–417.

13. David Jeremy, *Artists, Entrepreneurs, and Machines* (Brookfield: Ashgate Publishing, 1991), p. 5.

14. Various passport documents are all contained in "Loose Manuscripts 1802–1813," reel 2, RFP.

15. All quotations in the following section pertaining to Joseph Warren's trip are taken from "Joseph Warren Revere Journal and Letterbook," reel 15, vol. 56, RFP.

16. Young, "Origins of the American Copper Industry," p. 127; Henry J. Kauffman,

American Copper and Brass (Camden, N.J.: T. Nelson Publishers, 1968), p. 27; Revere to Hollingsworth, September 1, 1808, and Hollingsworth to Revere, May 10, 1809, "Loose Manuscripts 1802–1813," reel 2, RFP.

17. HM Salomon to Revere, May 2, 1810, "Loose Manuscripts 1802–1813," reel 2, RFP.

18. Revere to H. M. Salomon, May 8, 1810, "Loose Manuscripts 1802–1813," reel 2, RFP.

19. HM Salomon to Revere, February 9, 1810, May 2, 1810, and May 31, 1810, and Revere to H. M. Salomon, May 8, 1810; all in "Loose Manuscripts 1802–1813," reel 2, RFP; Mark Bortman, "Paul Revere and Son and their Jewish Correspondents," *Publications of the American Jewish Historical Society* 43 (1953–1954): 199–229.

20. Naomi R. Lamoreaux, "Rethinking the Transition to Capitalism in the Early American Northeast," *Journal of American History* 90, no. 2 (September 2003): 7–8, 10.

21. Historian of technology David Hounshell observed that *mass production* is a grammatically ambiguous term, potentially implying either "large scale production" or "production for the masses." Since *mass production* did not enter widespread usage until the twentieth century, other terms will be used in this book. However, many of the connotations of mass production apply to Revere's operations. David Hounshell, *From the American System to Mass Production* (Baltimore: Johns Hopkins University Press, 1984), pp. 1–12. Also see Hindle, *Emulation and Invention*, pp. 6–8.

22. Ferguson quote taken from Hounshell, *American System*, p. 15. Also see Cochran, *Frontiers of Change*, p. 55; Alfred Chandler, *The Visible Hand* (Cambridge, Mass.: Harvard University Press, 1977), p. 241.

23. Revere to Josiah Snelling, October 26, 1810, "Letterbook 1805–1810," reel 14, vol. 53.3, RFP.

24. Revere letters to Messrs Heywood, Flagg, Howell, July 5, 1802, and August 30, 1802, "Letterbook 1801–1806," reel 14, vol. 53.2, RFP.

25. Pitt Clarke to Revere, May 29, 1810, "Loose Manuscripts 1802–1813," reel 2, RFP.

26. Quote in Revere to Joseph Carson, September 22, 1810, "Letterbook 1805–1810," reel 14, vol. 53.5, RFP. Revere hired two machinists from North Providence, Abraham and David Wilkinston, to install a lathe for him in June 1804. This might pertain to his cannon-turning device, but probably describes a different piece of equipment used for smaller jobs. Receipt to Abraham and David Wilkinson, June 28, 1804, "Loose Manuscripts 1802–1813," reel 2, RFP.

27. Revere to Levi Lincoln, February 26, 1802, "Letterbook 1801–1806," reel 14, vol. 53.2, RFP.

28. Joseph Carson to Revere, September 25, 1810, "Loose Manuscripts 1802–1813," reel 2, RFP.

29. Revere to Beck and Harvey, October 29, 1803, "Letterbook 1801–1806," reel 14, vol. 53.2, RFP.

30. Revere to Beck and Harvey, November 20, 1803, "Letterbook 1801–1806," reel 14, vol. 53.2, RFP.

31. Revere to Hollingsworth, September 1, 1808, and Hollingsworth to Revere, May 10, 1809, "Loose Manuscripts 1802–1813," reel 2, RFP.

32. Revere to Hathaway and Davis, March 1, 1805, "Letterbook 1801–1806," reel 14, vol. 53.2, RFP.

33. "Bank Book Boston 1806–1812," reel 14, vol. 52, RFP.

34. "Learning by doing" is the process of on-the-job experimentation and practice. It is particularly valuable where theoretical knowledge and practical experience are lacking. Paul A. David, *Technical Choice Innovation and Economic Growth* (Cambridge: Cambridge University Press, 1975), pp. 101–111, 163–165.

35. Revere to Joshua Humphreys, December 19, 1803, "Letterbook 1801–1806," reel 14, vol. 53.2, RFP.

36. Paul Revere to Robert Smith, May 24, 1802, "Letterbook 1801–1806," reel 14, vol. 53.2, RFP.

37. Revere to Robert Smith, July 1, 1803, Robert Smith to Revere, July 22, 1803, and Revere to Robert Smith, October 29, 1803, all in "Letterbook 1801–1806," reel 14, vol. 53.2, RFP, and "Loose Manuscripts 1802–1813," reel 2, RFP.

38. Paul Revere to Robert Smith, May 24, 1802, "Letterbook 1801–1806," reel 14, vol. 53.2, RFP; "Bank Book Boston 1806–1812," reel 14, vol. 52, RFP.

39. John S. Morgan, *Robert Fulton* (New York: Mason/Charter, 1977); Cynthia Owen Philip, *Robert Fulton* (New York: Franklin Watts, 1985); Morton J. Horwitz, *The Transformation of American Law* (Cambridge, Mass.: Harvard University Press, 1977), p. 122.

40. John Livingston to Revere, October 8 1808, "Loose Manuscripts 1802–1813," reel 2, RFP; Revere to Livingston, October 12, 1808, "Letterbook 1805–1810," reel 14, vol. 53.3, RFP.

41. John Livingston to Revere, November 25, 1808, "Loose Manuscripts 1802–1813," reel 2, RFP; Robert Fulton to Revere, February 24, March 8, and July 28, 1814, "Loose Manuscripts 1814–1964, Undated Material," reel 3, RFP.

42. Revere to William Torrey, February 10, 1810, and Revere to Robert Fulton, December 15, 1810, "Loose Manuscripts 1802–1813," reel 2, RFP.

43. Revere to Joseph Carson, March 6, 1809, "Letterbook 1805–1810," reel 14, vol. 53.3, RFP; Joseph Carson to Revere, June 6, 1809, "Loose Manuscripts 1802–1813," reel 2, RFP; Revere to George Cabot, September 11, 1809, "Letterbook 1805–1810," reel 14, vol. 53.3, RFP.

44. Charles K. Hyde, *Copper for America* (Tucson: University of Arizona Press, 1998), p. 10; Maxwell Whiteman, *Copper for America: The Hendricks Family and a National Industry, 1755–1939* (New Brunswick: Rutgers University Press, 1971), p. 47.

45. Revere to Robert Smith, June 10, 1803, and Revere to Jacob Crowninshield, March 30, 1805, both in "Letterbook 1801–1806," reel 14, vol. 53.2, RFP.

46. Revere to Robert Smith, October 29, 1803, "Letterbook 1801–1806," reel 14, vol. 53.2, RFP; Revere to Joseph Carson, October 21, 1809, "Letterbook 1805–1810," reel 14, vol. 53.3, RFP; Whiteman, *Copper for America*, pp. 55–56.

47. Interestingly, Revere seems to have purchased more wood than charcoal, which is peculiar because charcoal burns at a far higher temperature and would probably be the only fuel able to actually melt copper or iron. Revere may have included wood purchases for other uses, such as heating his home, in his company records. Edgard Moreno, "Patriotism and Profit: The Copper Mills at Canton," in *Paul Revere—Artisan, Businessman, and Patriot* (Boston: Paul Revere Memorial Association, 1988), pp. 107, 112.

48. Coxe, *Statement of the Arts and Manufactures of the United States of America for the year 1810*, p. lviii.

49. Hindle and Lubar, *Engines of Change*, p. 10; Gordon and Malone, *Texture of Industry*, p. 87.

50. Dirk J. Struik, *Yankee Science in the Making* (New York: Dover Publications, 1991), p. 175; Hindle, *Emulation and Invention*, pp. 6–7.

51. J. E. Conant & Co., *The Plant (Real Estate) of the Revere Copper Co.* (Lowell: Butterfield Printing Company, 1909). Waterpower discussion in Gordon and Malone, *Texture of Industry*, p. 87.

52. Gordon and Malone, *Texture of Industry*, p. 89.

53. Hendricks wrote this letter in response to Revere's tariff proposal, which would have increased the duty on the finished copper products Hendricks imported. He maintained a productive working relationship with Revere after this petition fell through, and after starting his own rolling plant in New Jersey he collaborated with Revere on a subsequent petition. Whiteman, *Copper for America*, p. 78, quoting an undated 1807 letter from Harmon Hendricks to Gordon S. Mumford.

54. Theodore Steinberg, *Nature Incorporated* (Amherst: University of Massachusetts Press, 1991), pp. 33–35. The Neponset and Charles rivers had been connected by a diversion in 1639 to alleviate flooding along the Charles.

55. Massachusetts passed the first mill act in 1713, and a more comprehensive version appeared in 1795. The mill acts reached their greatest extent after revisions in 1825 and 1827, which allowed virtually unlimited flooding of lands both above and below the dam, and permitted dam owners to avoid all damage payments if they could prove that the flooding provided economic benefits. Beginning in 1830, the courts reversed their support of the mill acts in response to widespread protest from small landowners. Horwitz, *The Transformation of American Law, 1780–1860*, pp. 47–52; Jonathan Prude, *The Coming of Industrial Order* (Cambridge: Cambridge University Press, 1983), p. 11; Louis C. Hunter, *A History of Industrial Power in the United States, 1780–1930*, vol. 1 (Charlottesville: University Press of Virginia, 1979), pp. 140, 142, 147, 148; Gary Kulik, "Dams, Fish, and Farmers: The Defense of Public Rights in Eighteenth-Century Rhode Island," in *The New England Working Class*, ed. Herbert Gutman and Donald Bell (Champaign: University of Illinois Press, 1986), pp. 191–194; Gordon and Malone, *Texture of Industry*, p. 102; Robert B. Gordon, *American Iron, 1607–1900* (Baltimore: Johns Hopkins University Press, 1996), pp. 52–54.

56. Kulik, "Dams, Fish, and Farmers," pp. 193–195.

57. A small group of local smiths and millers pulled down Slater's second mill dam in 1792, prompting lengthy legal action. This coalition also increased the height of their own older mill dam months later in an attempt to raise the level of the river enough to produce "backwater" that would submerge Slater's waterwheels. Kulik, "Dams, Fish, and Farmers," pp. 204–205.

58. Horwitz, *Transformation of American Law*, pp. 49–50; Kulik, "Dams, Fish, and Farmers," pp. 195–197.

59. Steinberg, *Nature Incorporated*, pp. 140–141, and Horwitz, *Transformation of American Law*, pp. 34–38.

60. Horwitz, *Transformation of American Law*, pp. 38–45. New Hampshire's water law followed a similar but slower pattern, with common law dominating until about 1820, followed by a hodge-podge of principles that finally led to the reasonable use doctrine in the late 1850s. Steinberg, *Nature Incorporated*, pp. 140–147. The doctrine of prior appropriation ("he who is first in time is first in right") survived longer in the western United States because large land developers and industries preserved it. Donald Worster, *Rivers of Empire* (New York: Oxford University Press, 1985), pp. 89–91.

61. Horwitz, *Transformation of American Law*, p. 35.

62. Quote taken from Revere's eleven pages of unlabeled legal notes, dated March 10, 1804, "Loose Manuscripts 1802–1813," reel 2, RFP.

63. A list of Revere's payments to Leonard and Kinsley was submitted to the clerk's office of the court of common pleas on September 20, 1803. Leonard and Kinsley expense sheet, September 20, 1803, "Loose Manuscripts 1802–1813," reel 2, RFP.

64. The following analysis is taken from Revere's eleven pages of unlabeled legal notes, dated March 10, 1804, "Loose Manuscripts 1802–1813," reel 2, RFP.

65. Ephraim Williams, *Reports of Cases Argued and Determined in the Supreme Judicial Court of the State of Massachusetts from September 1804 to June 1805* (Northampton, 1805), microfilm roll 2, pp. 91–95.

66. Leonard and Kinsley to Revere, April 20, 1808, "Loose Manuscripts 1802–1813," reel 2, RFP.

67. Deposition from Abner Crane, dated "1808?," "Loose Manuscripts 1802–1813," reel 2, RFP.

68. Revere to the Selectmen of the Town of Canton, February 20, 1808, "Loose Manuscripts 1802–1813," reel 2, RFP. This citation applies to all of the quotations in the following paragraph as well.

69. Gordon Wood, *Radicalism of the American Revolution* (New York: Vintage Books, 1993), pp. 347–351.

Conclusion

1. "Cantondale" is an undated loose document in reel 2 of the Revere Family Papers, although "1810" was written in the corner at a later date.

2. The pastoral ideal is discussed at great length in Leo Marx's *The Machine in the Garden* (New York: Oxford University Press, 1964).

3. Indenture documents, March 1, 1811, Revere Family Papers (hereafter RFP), microfilm edition, 15 reels (Boston: Massachusetts Historical Society, 1979), reel 2. Paul Revere III was Paul Revere's grandson and the oldest son of Paul Revere Jr., who still managed his own silver 1 in Boston. Thomas Eayres Jr. was also Paul Revere's grandson, the son of Revere's daughter Frances, who had died in 1799, and Thomas Eayres Sr., the promising young silversmith who apprenticed with Paul Revere, started his own shop, and eventually succumbed to mental illness.

4. Elbridge Henry Goss, *The Life of Colonel Paul Revere* (Boston: Plimpton Press, 1902), pp. 579, 590–593, 611; Jayne E. Triber, *A True Republican* (Amherst: University of Massachusetts Press, 1998), p. 194; Esther Forbes, *Paul Revere and the World He Lived In* (Boston: Houghton Mifflin, 1969), pp. 416–418, 440–441, 443–444.

5. Revere's relationship with Paul Jr. had always been close, beginning with Paul Jr.'s apprenticeship as a silversmith and continuing when Paul Jr. took over the silver shop and helped in the foundry. Paul Jr. had worked for some years in his own silver shop and bell foundry and occasionally required small amounts of financial help from his father. Triber, *A True Republican*, p. 181.

6. Paul Revere's last will and testament, dated March 14, 1818, reel 3, RFP.

7. *Transactions of the American Medical Association*, 1850, as quoted in Triber, *A True Republican*, p. 182.

8. A reverberatory furnace separates the fuel from the material to be smelted, thereby preventing the infusion of impurities into the final product.

9. Otis E. Young Jr., "Origins of the American Copper Industry," *Journal of the Early Republic* 3 (Summer 1983): 131; Goss, *Life of Colonel Paul Revere*, pp. 573–576; Canton historical society website at http://www.canton.org/history/revere1.htm; Revere House website at http://www.paulreverehouse.org/copper/index.shtml. The Revere Family Papers contain many volumes pertaining to Joseph Warren's managerial and business practices.

10. Alan Dawley, *Class and Community: The Industrial Revolution in Lynn* (Cambridge, Mass.: Harvard University Press, 1976), pp. 220–223; Nathan Rosenberg, "Why in America?" in *Yankee Enterprise: The Rise of the American System of Manufactures*, ed. Otto Mayr and Robert C. Post (Washington: Smithsonian Institution Press, 1981), p. 59; A. E. Musson, "British Origins," in Mayr and Post, *Yankee Enterprise*, p. 43; Walter Licht, *Industrializing America* (Baltimore: Johns Hopkins University Press, 1995), pp. 42, 46–47.

11. Direct government sponsorship of manufacturing never received much notice among historians. Most American studies focus on the growth of the military-industrial complex in the twentieth century, and some have explored the military's support of early developments such as interchangeable parts, railroads, canals, and machine tools. The earliest timeframe for most of this work is the early nineteenth century, which featured critical events such as the drive toward interchangeable manufacturing or the establishment of West Point and a military engineering academic program, both after 1811. The importance of these events and others like them cannot be overemphasized, but Revere's example illustrates that the story actually began earlier and involved other institutions besides the army. One form of direct government aid, the influence of military research and purchases upon technological and industrial growth, has received some attention. Several studies have investigated the way that European military forces became huge consumers of ordnance, a process that gave a huge impetus to mining, metallurgy, and machine production. See Merritt Roe Smith's introduction to *Military Enterprise and Technological Change* (Cambridge, Mass.: MIT Press, 1985) as well as Alex Roland's bibliographic essay in the same volume. Also see William McNeill, *The Pursuit of Power* (Chicago: University of Chicago Press, 1982).

12. Brooke Hindle, *Emulation and Invention* (New York: W. W. Norton and Company, 1981), p. 11.

13. Hindle, *Emulation and Invention*, p. 16, quoting from John Adams's "Discourses on Davila," 1790–1791.

Index

Page numbers in italics indicate illustrations